Photonic Crystals: Systems and Applied Principles

Photonic Crystals: Systems and Applied Principles

Edited by **Jason Penn**

CLANRYE
INTERNATIONAL

New Jersey

Published by Clanrye International,
55 Van Reypen Street,
Jersey City, NJ 07306, USA
www.clanryeinternational.com

Photonic Crystals: Systems and Applied Principles
Edited by Jason Penn

International Standard Book Number: 978-1-63240-408-4 (Hardback)

Printed in the United States of America.

Contents

Permissions

List of Contributors

Preface

Renowned researchers from across the globe have contributed significant research inferences in this extensive book. Photonic crystals are described as the periodic optical nanostructures that are designed to affect the motion of photons in a similar way that periodicity of a semiconductor crystal affects the motion of electrons. This book introduces the concepts of innovative systems and materials to the readers. It provides comprehensive information regarding characterization approach of photonic crystal waveguides, photonic crystal lasers as well as plasmonics. Photonic crystal materials have the potential of enabling all-optical computer circuits and can also be utilized to design ultralow-power light sources. There is an in-depth account on lasers from microscopic cavities in photonic crystals acting as reflectors for intensifying the collisions between photons and atoms. This leads to lazing, but these lasers are optically-pumped, meaning that these lasers are themselves driven by other lasers. Furthermore, the physical principles behind the phenomenon of slow light in photonic crystal waveguides, as well as their practical limitations, have also been elucidated. The nature of slow light propagation, its bandwidth restraint, coupling of modes and specific kind terminating photonic crystals with metal surfaces enabling propagation in surface plasmon-polariton waves have also been described. The aim of this book is to provide an analysis of these mentioned issues to provide readers with a better understanding about photonic crystals.

This book is the end result of constructive efforts and intensive research done by experts in this field. The aim of this book is to enlighten the readers with recent information in this area of research. The information provided in this profound book would serve as a valuable reference to students and researchers in this field.

At the end, I would like to thank all the authors for devoting their precious time and providing their valuable contribution to this book. I would also like to express my gratitude to my fellow colleagues who encouraged me throughout the process.

Editor

Part 1

Photonic Crystal Laser and Photonic Crystal Characterization

1

The Optical Transmission of One-Dimensional Photonic Crystals Containing Double-Negative Materials

Petcu Andreea Cristina
*National Research and Development Institute for Gas Turbines Bucharest**
Romania

1. Introduction

A crystal is a periodic arrangement of atoms or molecules. The pattern with which the atoms or molecules are repeated in space is the crystal lattice. The crystal presents a periodic potential to an electron propagating through it, and both the constituents of the crystal and the geometry of the lattice dictate the conduction properties of the crystal.

Importantly, however, the lattice can also prohibit the propagation of certain waves. There may be gaps in the energy band structure of the crystal, meaning that electrons are forbidden to propagate with certain energies in certain directions. If the lattice potential is strong enough, the gap can extend to cover all possible propagation directions, resulting in a complete band gap. For example, a semiconductor has a complete band gap between the valence and conduction energy bands.

The optical analogue is the photonic crystal, in which the atoms or molecules are replaced by macroscopic media with differing dielectric constants, and the periodic potential is replaced by a periodic dielectric function (or, equivalently, a periodic index of refraction). If the dielectric constants of the materials in the crystal are sufficiently different, and the absorption of light by the materials is minimal, then the refractions and reflections of light from all of the various interfaces can produce many of the same phenomena for photons (light modes) that the atomic potential produces for electrons. One solution to the problem of optical control and manipulation is thus a photonic crystal, a low-loss periodic dielectric medium. (Joannnopoulos et al, 2008).

There has been growing interest in the development of easily fabricated photonic band gap materials operating at the optical frequencies. The reason for the interest in photonic band gap materials arises from the possible applications of those materials in several scientific and technical areas such as filters, waveguides, optical switches, cavities, design of more efficient lasers, etc. (Li et al., 2008; Wang et al., 2008).

The simplest possible photonic crystal consists of alternating layers of material with different dielectric constants: a one-dimensional photonic crystal or a multilayer film. This arrangement is not a new idea. Lord Rayleigh (1887) published one of the first analyses of

---
* The publishing fee for Andreea Petcu was paid by the National Research and Development Institute for Gas Turbines Bucharest

the optical properties of multilayer films. This type of photonic crystal can act as a mirror for light with a frequency within a specified range, and it can localize light modes if there are any defects in its structure. These concepts are commonly used in dielectric mirrors and optical filters. (Joannnopoulos et al., 2008)

Recently, photonic crystals containing metamaterials have received special attention for their peculiar properties (Deng & Liu, 2008). One kind of metamaterials is double-negative materials (DNG) whose electric permittivity ε and magnetic permeability μ are simultaneously negative (Veselago, 1968), which can be used to overcome optical diffraction limit, realize super-prism focusing and make a perfect lens (Pendry, 2000). Another kind of metamaterials is single-negative materials (SNG), which include the mu-negative media (MNG) (the permeability is negative but the permittivity is positive) and the epsilon-negative media (ENG) (the permittivity is negative but the permeability is positive). These metamaterials possess zero-effective-phase gap and can be used to realize easily multiple-channeled optical filters (Zhang et al., 2007).

In this chapter are analzed one-dimensional photonic crystals composed of two layers: A=dielectric material (TiO$_2$) or A=epsilon-negative material (ENG) and B=double negative material (DNG), from the point of view of their optical transmission. In the case in which A is a dielectric material are used the following materials properties: the magnetic permeability $\mu_A=1$ and the electric permittivity $\varepsilon_A=7.0225$. To describe the epsilon-negative material (ENG) it is used a transmission-line model (Eleftheriads et al., 2002): the magnetic permeability $\mu_A=3$ and the electric permittivity $\varepsilon_A=1-100/\omega^2$. For the double-negative material (DNG) are used the following material properties: the magnetic permeability $\mu_B=1-100/\omega^2$ and the electric permittivity $\varepsilon_B=1.21-100/\omega^2$. Ω is the angular frequency measured in GHz.

An algorithm based on the transfer matrix method (TMM) was created in MATLAB and used to determine the optical transmission of the cosidered photonic structures. In the simulations the angular frequency ω takes values from 0 to 9 GHz. In this chapter is analized the influence of various defects upon the optical transmission of the photonic crystal: the type of the material used in the defect layer, thickness and position of the defect layer upon the optical transmission.

2. Calculus method

The transfer matrix method (TMM) is widely used for the description of the properties of stacked layers and is extensively presented in (Born & Wolf, 1999). The transfer matrix method (TMM) provides an analytical means for calculation of wave propagation in multilayer media. This method permits exact and efficient evaluation of electromagnetic fields in layered media through multiplication of 2 × 2 matrices. The TMM is usually used as an efficient tool to analyse uniform and non-uniform gratings, distributed feedback lasers and even one-dimensional photonic crystals. The solution of the coupled mode (coupled modes TE and TM) equations is represented by a 2 x 2 transfer matrix which relates the forward and backward propagating field amplitudes. The grating structure is divided into a number of uniform grating sections which each have an analytic transfer matrix. The transfer matrix for the entire structure is obtained by multiplying the individual transfer matrices together.

Such an algorithm was implemeted in MATLAB and used to determine the optical transmission of the cosidered photonic structures. In this approach were considered isotropic layers, nonmagnetic and a normal incidence of the incident light.

Let us consider for investigation a stack of m layers perpendicular on the OZ axis as it can be seen in figure 1.

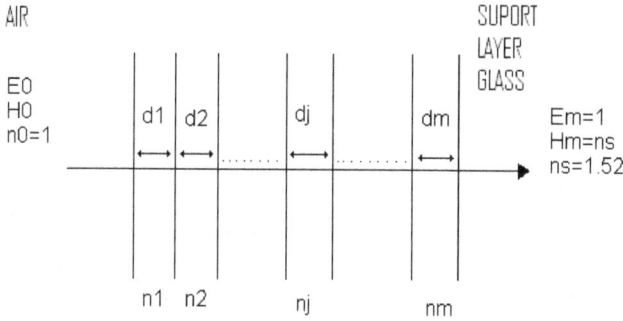

Fig. 1. Schematic representation of a multilayered structure

Using the notations given in figure 1, it was considered known the refractive index of the medium from where the beam of light emerges $n_0=1$ (air), the refractive index of the medium in which the beam of light exits $n_s=1.52$ (glass), the intensities of the electric and magnetic fields E_m and H_m in the support layer of glass.

To determine the electric and magnetic fields in the air, E_0 and H_0, the following system was solved:

$$\begin{bmatrix} E_0 \\ H_0 \end{bmatrix} = M_1 M_2 ... M_m \begin{bmatrix} E_m \\ H_m \end{bmatrix} \tag{1}$$

where

$$M_j = \begin{bmatrix} \cos\phi_j & \sqrt{\dfrac{\mu_j}{\varepsilon_j}} \cdot \sin\phi_j \\ -\dfrac{1}{\sqrt{\dfrac{\mu_j}{\varepsilon_j}}} \cdot \sin\phi_j & \cos\phi_j \end{bmatrix}, j = 1..m \tag{2}$$

and $\phi_j = \omega \cdot \dfrac{1}{v} \cdot \sqrt{\varepsilon_j \mu_j} \cdot d_j$ the phase variation of the wave passing the layer j. ε_j, μ_j and d_j are the electric permittivity, the magnetic permeability, respective the thickness of the layer. v is the phase speed and ω is the angular frequency. The relationship between the wavelength λ and the angular frequency ω is $\lambda = \dfrac{v}{f} = \dfrac{2\pi v}{\omega}$.

Be multiplying the $M_1, M_2, ..., M_m$ matrices we obtain a final matrix of the following shape:

$$M = \begin{bmatrix} M_{11} & M_{12} \\ M_{21} & M_{22} \end{bmatrix} \tag{3}$$

Using the matrix obtained above we calculate the optical transmission with the following relationship:

$$T = \left(\frac{2}{\left| M_{11} + M_{22} + i\left(M_{12} - M_{21}\right) \right|} \right)^2 \qquad (4)$$

The MATLAB code is based on these equations for optical transmission calculus.

3. Structures design

The studied structures were generated starting from the one-dimensional Thue-Morse sequence (Allouche & Shallit, 2003). The one-dimensional Thue-Morse sequence of N order, TM_N , is a binary sequence of two symbols 'A' and 'B'. TM_{N+1} is generated from TM_N in which we substitute 'A' with 'AB' and 'B' with 'BA'. Thus TM_0 = {A}, TM_1 = {AB}, TM_2 = {ABBA}, TM_3 = {ABBABAAB} etc.

In the simulations were used one-dimensional photonic crystals of the following types: $(AB)_{16}$, $(ABBA)_8$ and $(ABBABAAB)_4$, where A and B are two isotropic media with the refractive indexes n_A and n_B.

The following particular cases of one-dimensional photonic crystal composed of two types of layers were considered:

a. The first case: A=dielectric material (TiO$_2$) and B=double negative material with the permeability μ_A=1, respectively μ_B=1-100/ω^2 and the permittivity ε_A=7.0225, respectively ε_B=1.21-100/ω^2.
b. The second case: A=epsilon negative material and B=double negative material with the permeability μ_A=3, respectively μ_B=1-100/ω^2 and the permittivity ε_A=1-100/ω^2, respectively ε_B=1.21-100/ω^2.

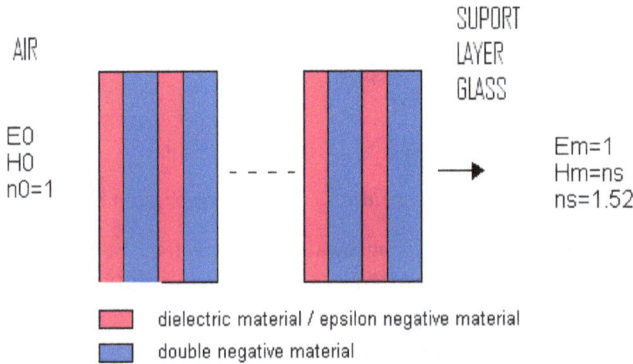

Fig. 2. The one-dimensional analyzed photonic crystal's structure

It was also considered introducing a defect layer in the structure of the analyzed photonic crystals. In the simulations were used both a A type defect layer and a B type defect layer. The position and the thickness of the defect layer will vary so to observe its influence upon the optical transmission of the photonic crystal. The new structure of the crystal is presented in figure 3.

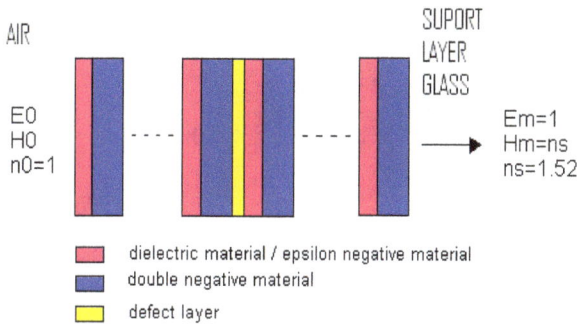

Fig. 3. The structure of the photonic crystal with a defect layer

4. MATLAB simulation results

4.1 Optical transmission of 1D PC without defects

In these experiments the thickness of the layers were kept comstant - dA =24mm and dB =6mm - and the frequency varied $\omega \in (0,9)GHz$. As it is specified earlier there will be two cases: case 1 in which A is a dielectric material and case 2 in which A is an epsilon negative material. In both cases B is a double negative material (Ramakrishna, 2005).

4.1.1 Case 1: A is a dielectric material

Figure 4 presents the optical transmission of the three types of the studied photonic crystals versus of the frequency of the light beam so that we can observe easily the band-gaps.

One easily observes that the $(AB)_{16}$ type photonic crystals have two band gaps, the $(ABBA)_8$ type crystals have three band-gaps and the $(ABBABAAB)_4$ type crystals have four band-gaps. The values are given in table 1.

Crystal's type	Band-gaps (GHz)	The mid-gap frequency of the gap (GHz)	The width of the band-gaps (GHz)
$(AB)_{16}$	$\omega \in (1.2, 2.4)$	1.8	1.2
	$\omega \in (6.5, 7.5)$	7	1
$(ABBA)_8$	$\omega = 0.8$	0.8	0
	$\omega \in (1.2, 1.5)$	1.35	0.3
	$\omega \in (7, 7.5)$	7.25	0.5
$(ABBABAAB)_4$	$\omega = 0.58$	0.58	0
	$\omega \in (1.25, 1.55)$	1.4	0.3
	$\omega \in (1.9, 5)$	3.45	3.1
	$\omega \in (7, 7.5)$	7.25	0.5

Table 1. The photonic band-gaps for $(AB)_{16}$, $(ABBA)_8$ and $(ABBABAAB)_4$ one-dimensional photonic crystal

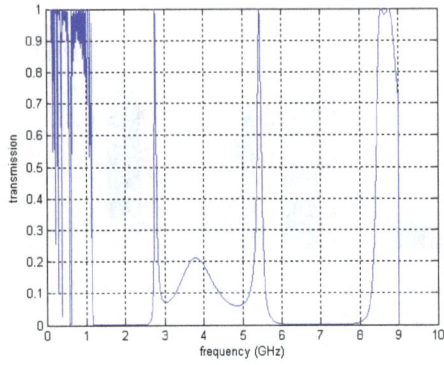

a)

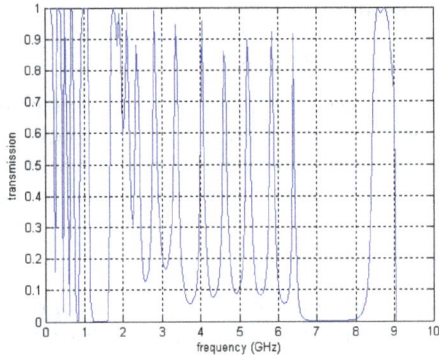

b)

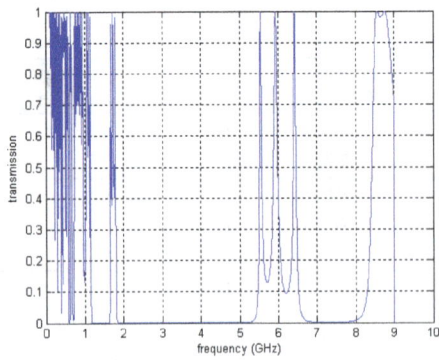

c)

Fig. 4. The optical transmission function of the frequency of the incident radiation for 1D PC of different types: a) $(AB)_{16}$; b) $(ABBA)_8$; c) $(ABBABAAB)_4$

It is also observed that the (AB)$_{16}$ type crystals have wider band-gaps. The (ABBA)$_8$ type crystals have the smallest band-gaps. The maximum band-gap is obtained for a (ABBABAAB)$_4$ type structure.

4.1.2 Case 2: A is an epsilon negative material

Figure 5 presents the optical transmission of the three types of the studied photonic crystals versus the frequency of the light beam so that to observe easily the band-gaps and the localized states.

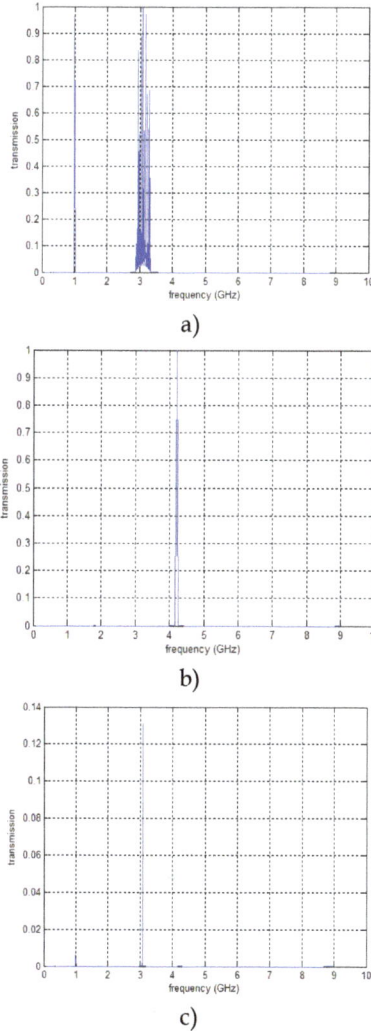

a)

b)

c)

Fig. 5. the optical transmission function of the frequency of the incident radiation for 1D PC of different types: a) (AB)$_{16}$; b) (ABBA)$_8$; c) (ABBABAAB)$_4$

From the graphics above it is observed that for the $(AB)_{16}$ type photonic crystal it is obtained a localized state with an optical transmission of 96% at the frequency of 0.99 GHz and a group of localized states with high optical transmission between the frequencies of 2.5 GHz and 3.5 GHz. One of these localized states has an optical transmission of 100%. This localized state is obtained at 3.07 GHz. For the $(ABBA)_8$ type structure it is obtained a localized state at 4.2 GHz with an optical transmission of 100%. The results obtained for the $(ABBABAAB)_4$ type crystal are not important because the optical transmissions of the localized states are very small (under 15%).

4.2 The optical transmission of 1D PC with defects

In this case it was introduced, in the structure of the one-dimensional photonic crystal, first an A type defect layer and then a B type defect layer. As before it is considered first that A is a dielectric material and then A is an epsilon negative material. B remains a double negative material in both cases.

In these experiments the thicknesses of the layers were kept constant - dA =24mm and dB =6mm - and the frequency was varied $\omega \in (0,9) GHz$.

4.2.1 Case 1: A a dielectric material

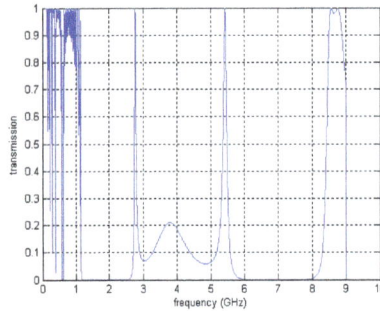

a)

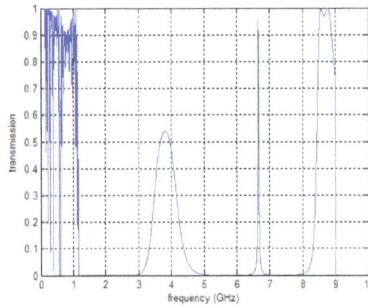

b)

c)

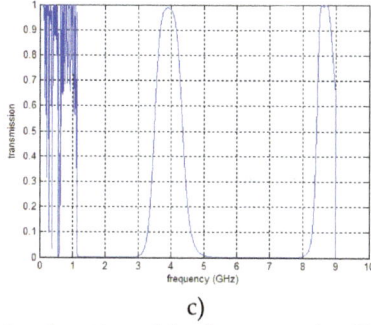

Fig. 6. The optical transmission function of the frequency for 1D PC of (AB)$_{16}$ type:
a) without defect; b) with A type defect; c) with B type defect.

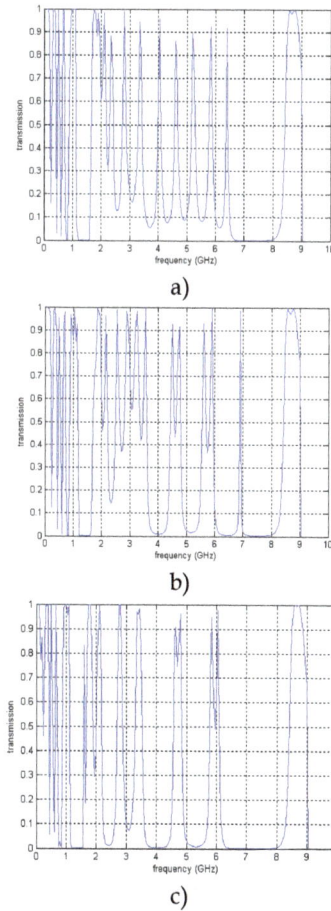

a)

b)

c)

Fig. 7. The optical transmission function of the frequency for 1D PC of (ABBA)$_8$ type:
a) without defect; b) with A type defect; c) with B type defect.

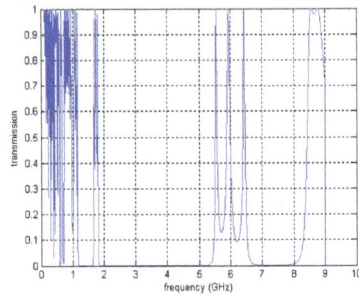

a)

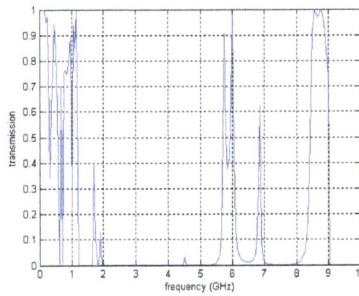

b)

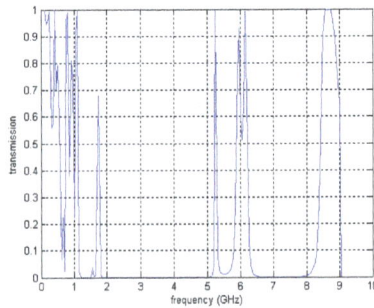

c)

Fig. 8. The optical transmission function of the frequency for 1D PC of $(ABBABAAB)_4$ type: a) without defect; b) with A type defect; c) with B type defect.

From the figures above it can be seen that only for the $(AB)_{16}$ type photonic crystal significant results were obtained.

4.2.2 Case 2: A an epsilon negative material

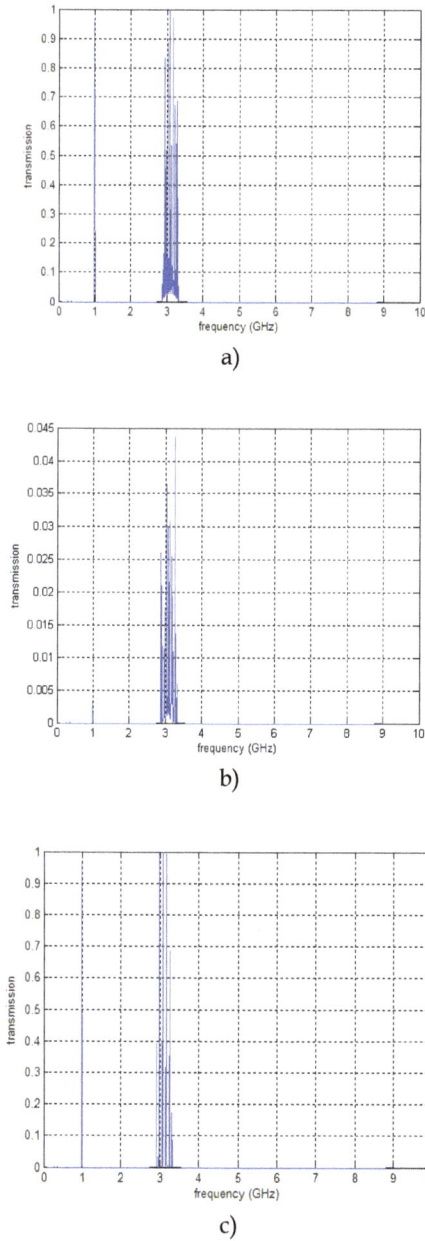

a)

b)

c)

Fig. 9. The optical transmission function of the frequency for 1D PC of (AB)$_{16}$ type: a) without defect; b) with A type defect; c) with B type defect.

Inserting a A type defect layer lowers the optical transmission in the localized state obtained at 0.99 GHz while the insertion of a B type defect layer doesn't have a big influence on the optical transmission of the (AB)$_{16}$ type photonic crystal.

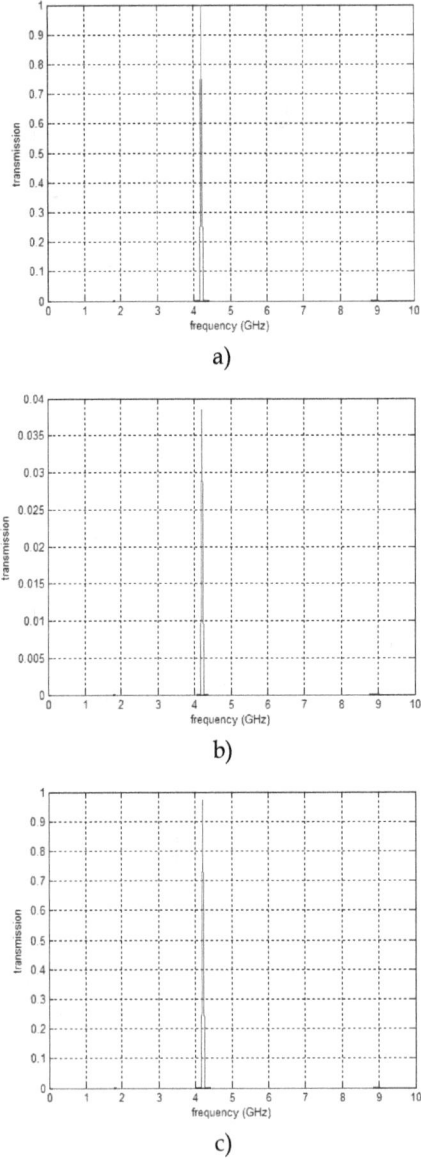

a)

b)

c)

Fig. 10. The optical transmission function of the frequency for 1D PC of (ABBA)$_8$ type: a) without defect; b) with A type defect; c) with B type defect.

From the figure above it is observed that like in the case of a $(AB)_{16}$ type crystal in the case of a $(ABBA)_8$ type crystal the insertion of a A type defect layer lowers the optical transmission (from 100 % to 4%) in the localized state obtained at 4.2 GHz while the insertion of a B type defect layer has little influence upon the optical transmission of the photonic crystal.

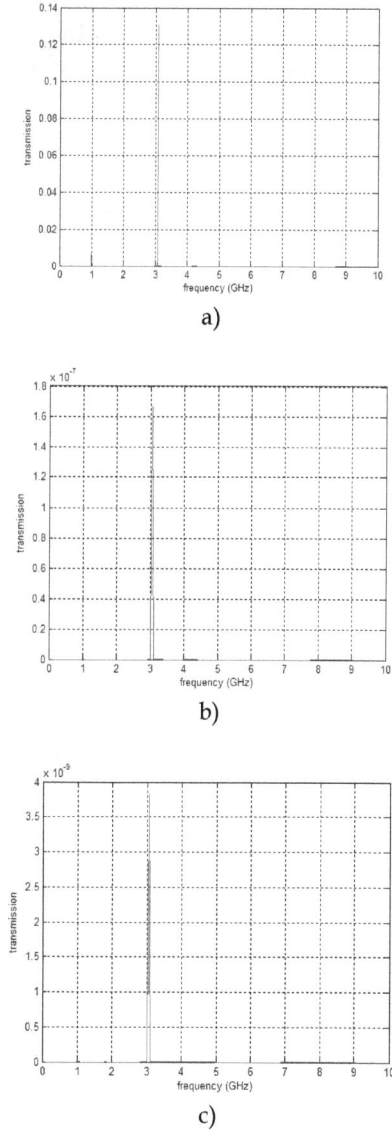

a)

b)

c)

Fig. 11. The optical transmission function of the frequency for 1D PC of $(ABBABAAB)_4$ type: a) without defect; b) with A type defect; c) with B type defect.

Analyzing the graphics above it was observed that by inserting a defect layer of A type or B type the optical transmission of the (ABBABAAB)$_4$ type photonic crystal in the localized state becomes even smaller (under 0.001 %).

Considering the results obtained till now the following types of photonic crystals will be studied farther more:

a.   The (AB)$_{16}$ type crystal with a A type defect layer (A a dielectric material)
b.   The (AB)$_{16}$ type crystal with a B type defect layer (A a dielectric material)
c.   The (AB)$_{16}$ type crystal with a B type defect layer (A a epsilon negative material)

4.3 The optical transmission of 1D PC of (AB)$_{16}$ type with a defect layer

The influence of the thickness and the position in which is inserted the defect layer in the structure of the photonic crystal will be analyzed. Only the three cases enumerated above will be considered.

4.3.1 The (AB)$_{16}$ type crystal with a A type defect layer (A is a dielectric material)

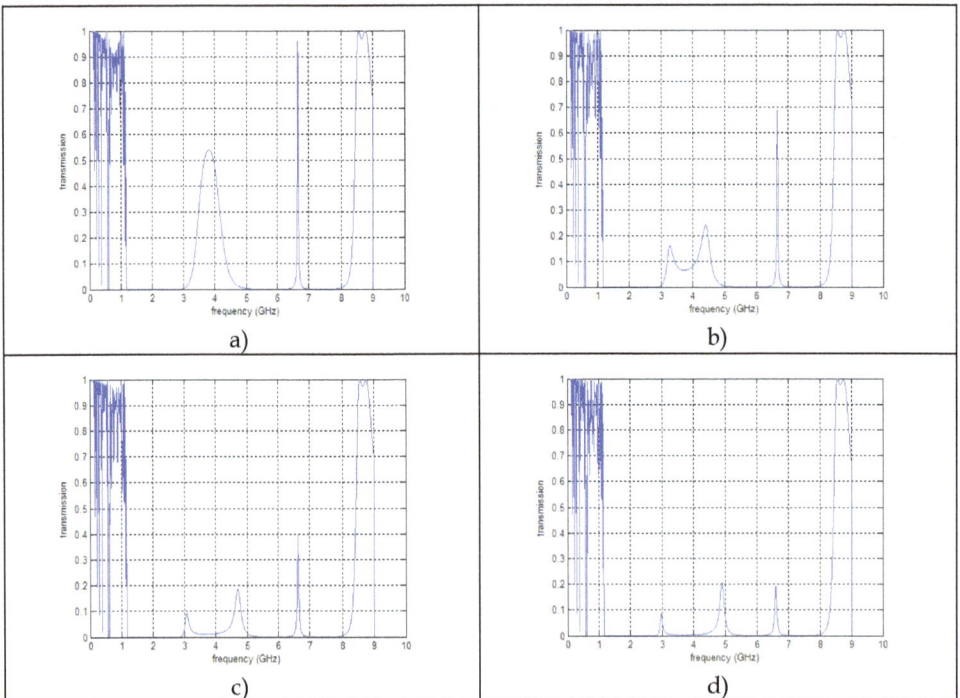

Fig. 12. The variation of the position of the defect layer: a) (AB)$_8$A(AB)$_8$; b) (AB)$_9$A(AB)$_7$; c) (AB)$_{10}$A(AB)$_6$; d) (AB)$_{11}$A(AB)$_5$

From the simulations it is observed that the best results are reached if the defect layer is inserted after 8 (AB) groups. A localized state with high optical transmission (95%)is obtained at a frequency of 6.65 GHz.

Keeping the defect layer on position 8 (after 8 (AB) groups) we will now vary its thickness to see its influence upon the optical transmission or the frequency of the localized state.

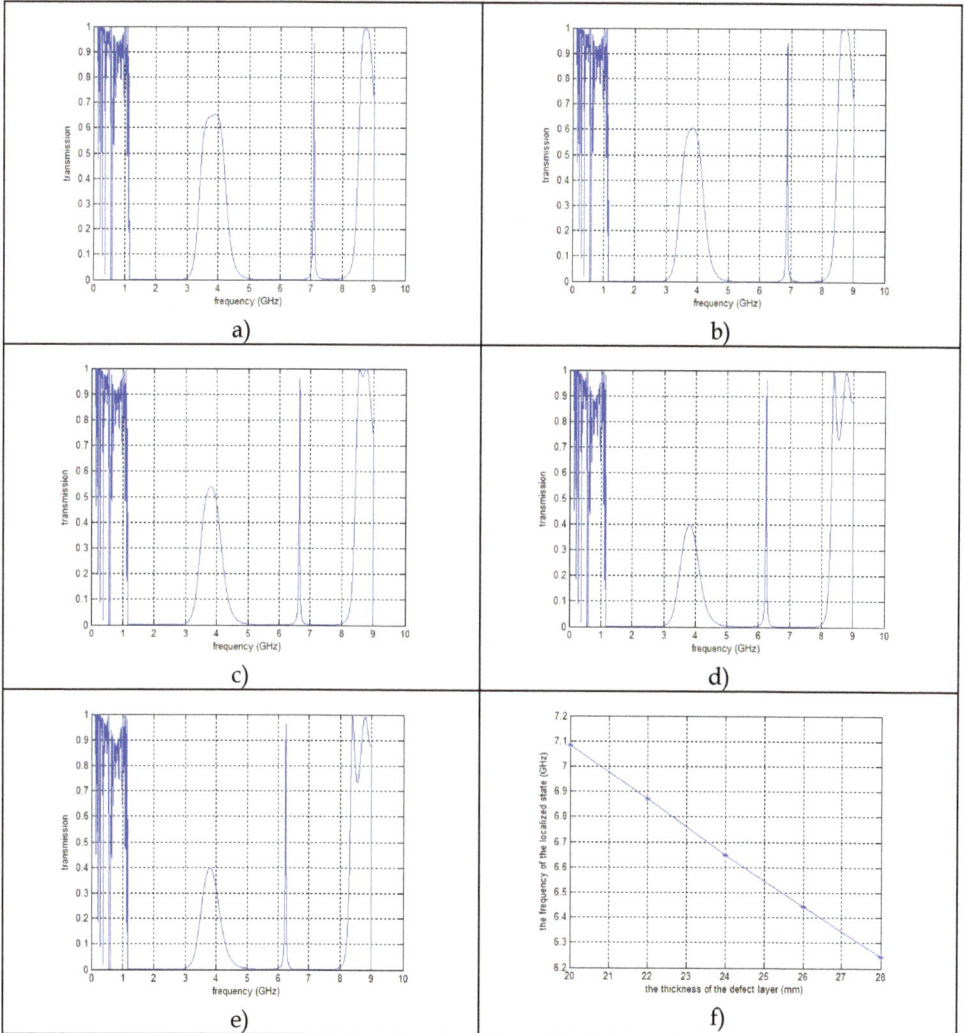

Fig. 13. The variation of the thickness of the A type defect layer a) dA=20 mm; b) dA=22 mm; c) dA=24 mm; d) dA=26 mm; e) dA=28 mm; f) the frequency of the localized state vs. the thickness of the defect layer

Analyzing the graphics above it is reached the conclusion that by increasing the thickness of the defect layer the localized state moves to right and by decreasing the thickness of the defect layer the localized state moves to left. The optical transmission of the localized state is not significantly influenced.

Father on the thickness of all the A type layers in the structure of the crystal is varied to see its influence upon the optical transmission of the photonic crystal.

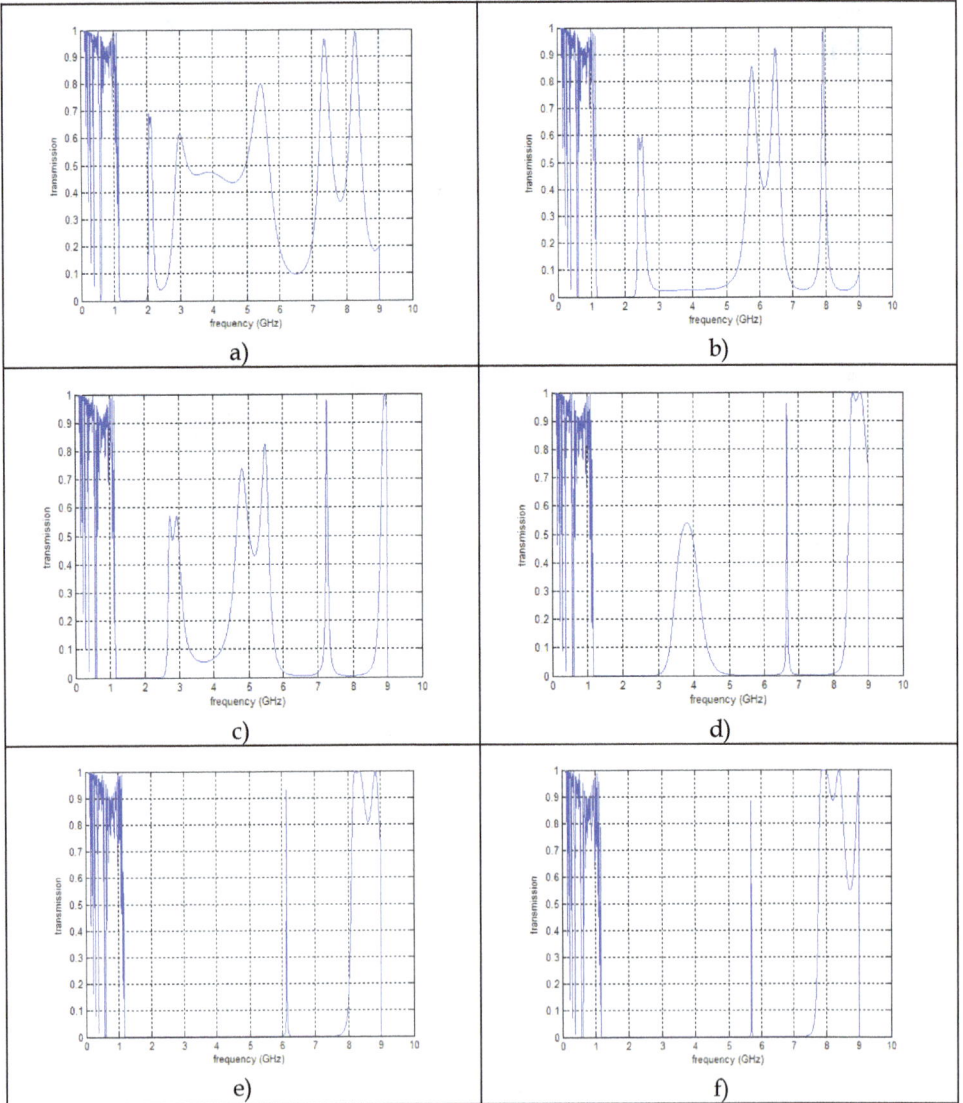

a)

b)

c)

d)

e)

f)

g)

h)

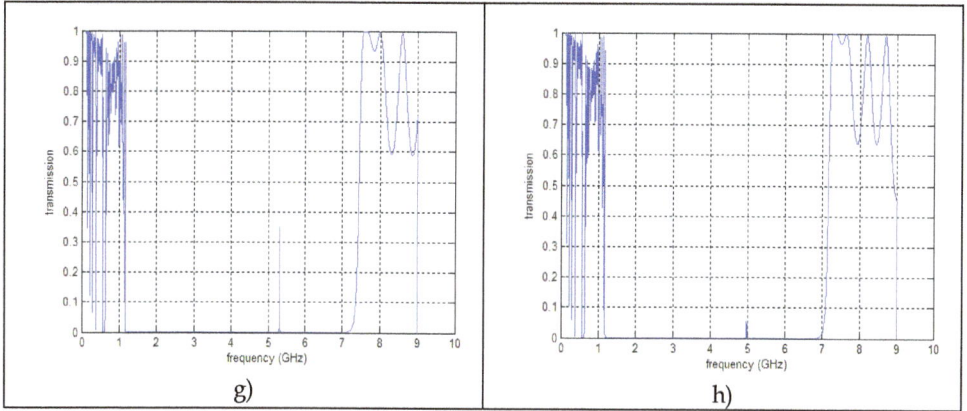

Fig. 14. The variation of the thickness of the A type layer: a) dA=20 mm; b) dA=22 mm; c) dA=23 mm; d) dA=24 mm; e) dA=25 mm; f) dA=26 mm; g) dA=27 mm; h) dA=28 mm

It is observed that by varying the thickness of all A type layers we influence both the optical transmission and the frequency of the localized state that was initially obtained at 6.65 GHz. By increasing the thickness of the layers the optical transmission of the localized state decreases and the localized state moves to left – the localized state is obtained at smaller values of the frequency (as it can be seen in figure 15 a) and b)).

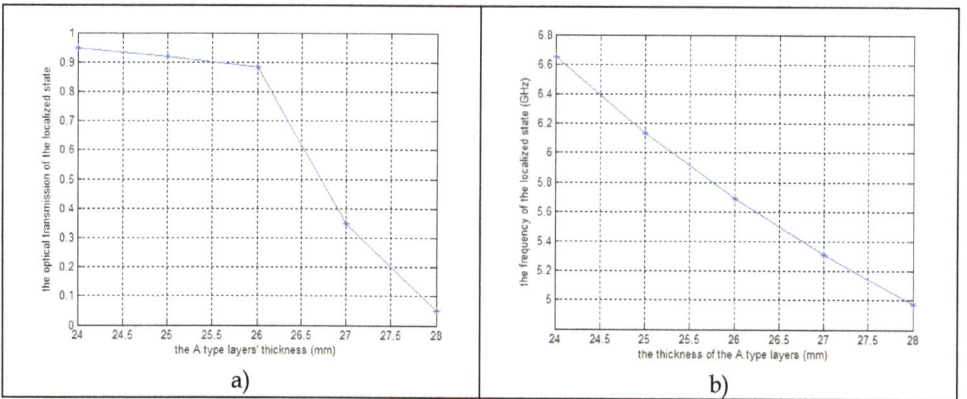

a)

b)

Fig. 15. a) the optical transmission of the localized state vs. the thickness of the A type layers in the structure of the $(AB)_8A(AB)_8$ type photonic crystal; b) the frequency at which is obtained the localized state vs. the thickness of the A type layers in the structure of the $(AB)_8A(AB)_8$ type photonic crystal.

4.3.2 The $(AB)_{16}$ type crystal with a B type defect layer (A is a dielectric material)

In figure 16 are presented some of the results obtained by changing the position of the defect layer. It is observed that the best position to insert the B type defect layer is after 8 (AB) groups.

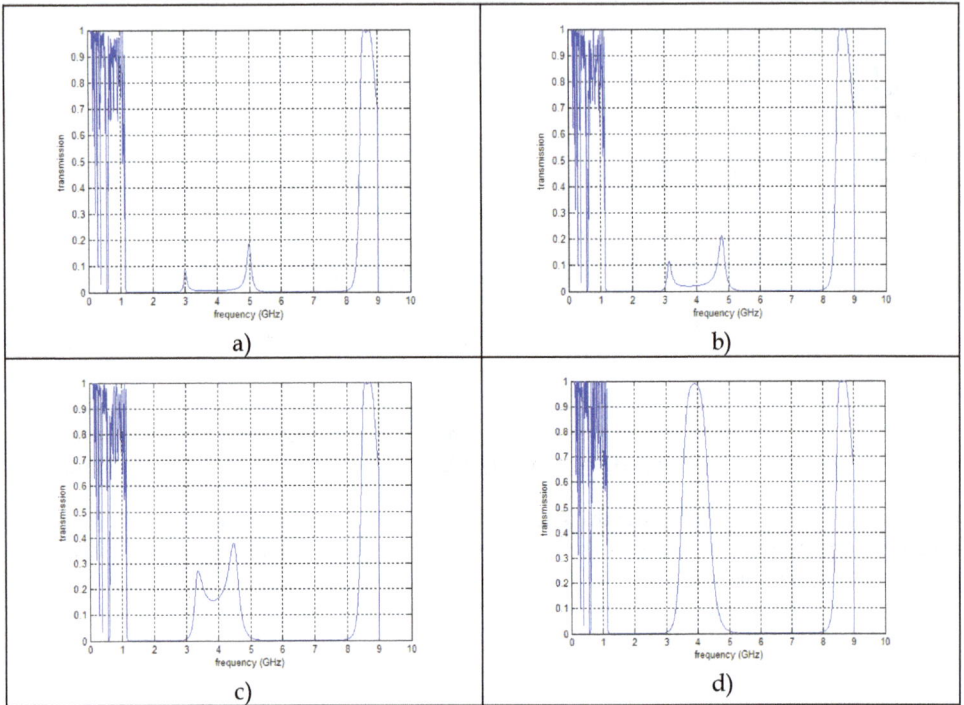

Fig. 16. The variation of the position of the defect layer: a) $(AB)_5B(AB)_{11}$; b) $(AB)_6B(AB)_{10}$; c) $(AB)_7B(AB)_9$; d) $(AB)_8B(AB)_8$

Keeping the defect layer on position 8 (after 8 (AB) groups) its thickness will be varied to see its influence upon the optical transmission of the $(AB)_8B(AB)_8$ type photonic crystal.

It is observed that by increasing the thickness of the B type defect layer a localized state is obtained. The optical transmission and the frequency at which is obtained this state depend on the thickness of the defect layer as it can be seen in the graphics bellow. By increasing the thickness of the defect layer the localized state moves to right – the state is obtained at higher values of the frequency (figure 18 b)). The values of the localized state's optical transmission have mainly an ascending trajectory (figure 18 a)).

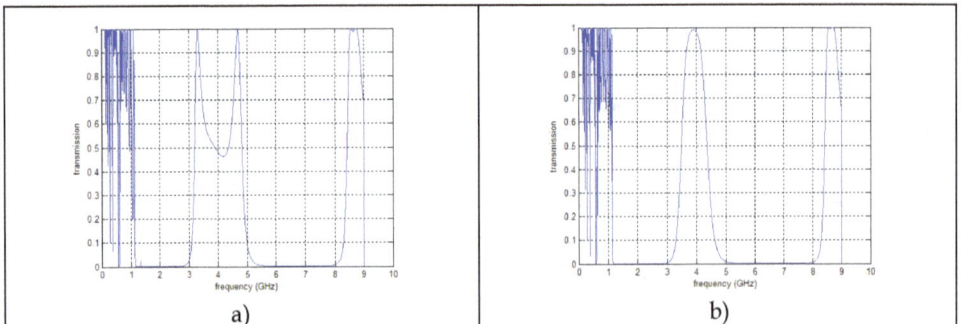

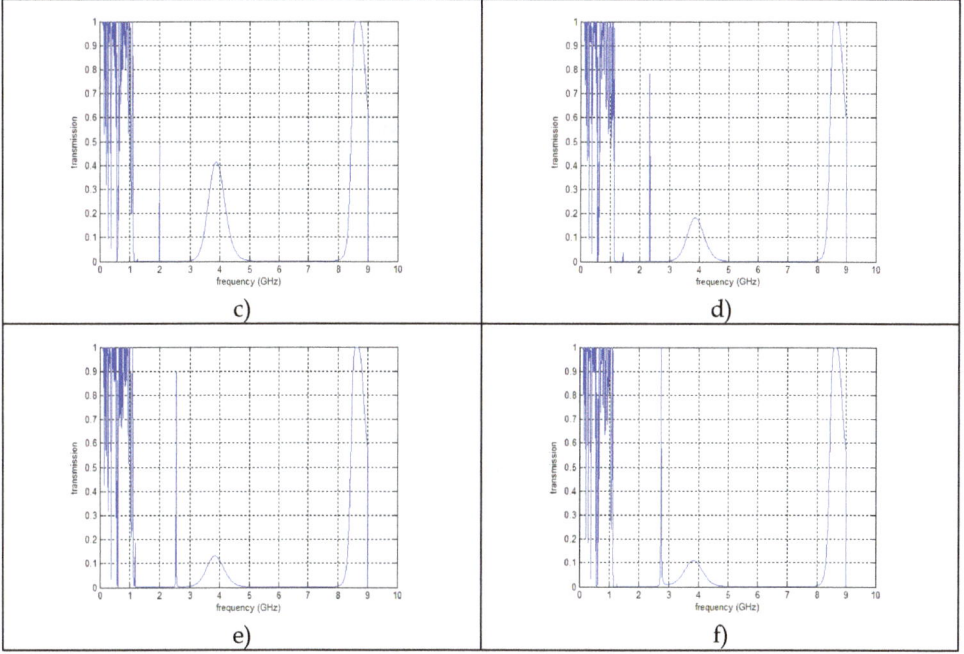

Fig. 17. The variation of the thickness of the B type defect layer: a) dB=3 mm; b) dB=6 mm; c)dB=9 mm; d) dB=12 mm; e) dB=14 mm; f) dB=16 mm;

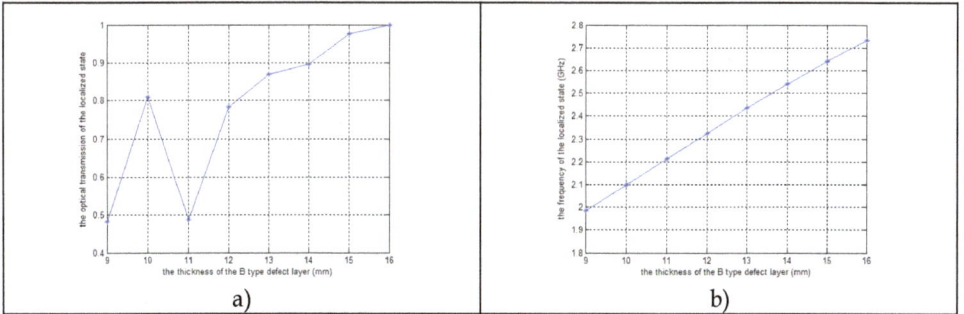

Fig. 18. a) the optical transmission of the localized state vs. the thickness of the B type defect layer in the structure of the $(AB)_8A(AB)_8$ type photonic crystal; b) the frequency at which is obtained the localized state vs. the thickness of the B type defect layer in the structure of the $(AB)_8A(AB)_8$ type photonic crystal.

Father on the thickness of all the B type layers in the structure of the crystal is varied to see its influence upon the optical transmission of the photonic crystal.

By increasing the thickness of all the B type layers in the structure of the $(AB)_8A(AB)_8$ type photonic crystal there are no longer obtained localized states but a series of band-gaps.

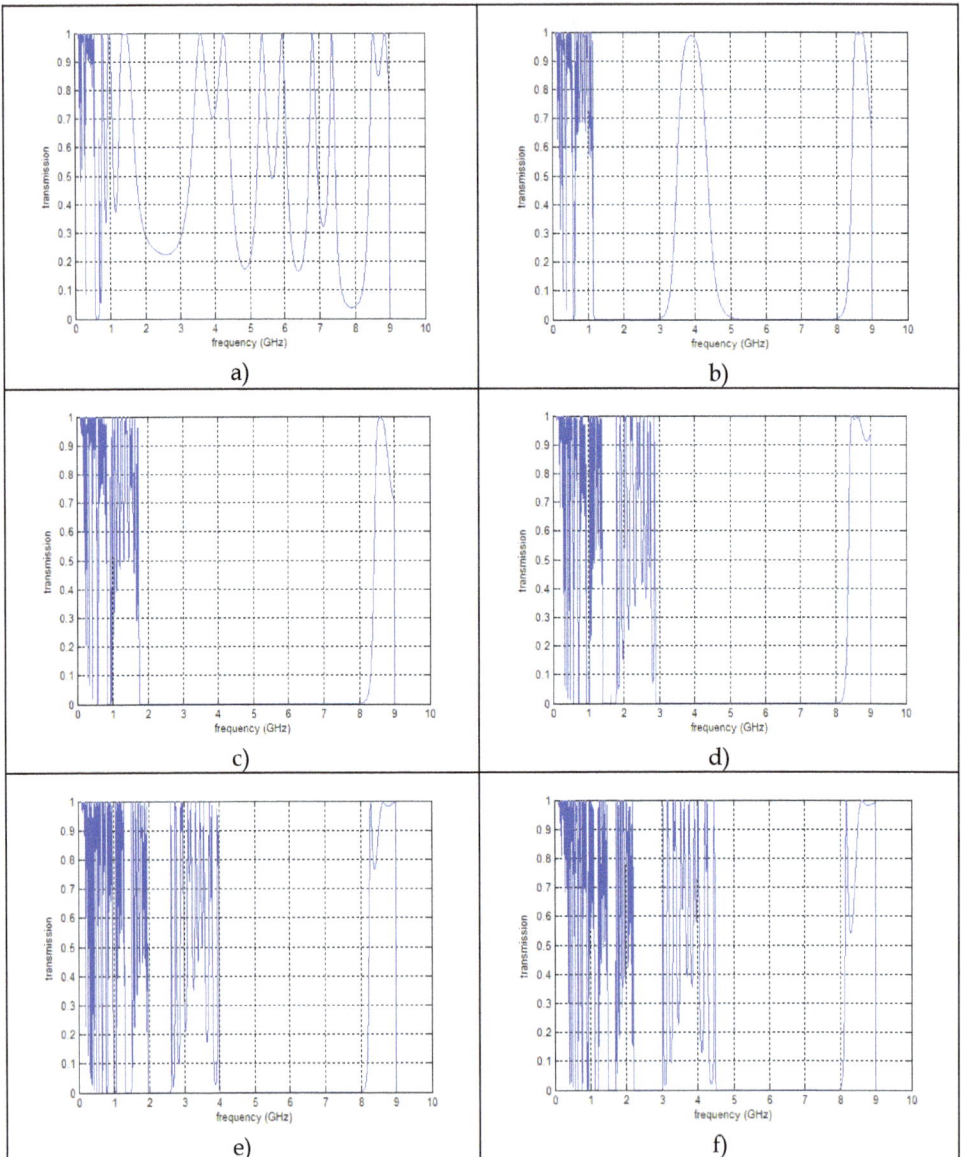

Fig. 19. The variation of the thickness of the B type layer: a) dB=3 mm; b) dB=6 mm;
c) dB=9 mm; d) dB=15 mm; e) dB=21 mm; f) dB=24 mm;

Layer thickness (mm)	Band-gaps (GHz)	The width of the band-gaps (GHz)
6	$\omega \in (1.2, 2.7)$	1.5
	$\omega \in (5.5, 7.6)$	2.1
9	$\omega \in (0.85, 0.93)$	0.08
	$\omega \in (1.8, 7.5)$	5.7
15	$\omega \in (0.7, 0.72)$	0.02
	$\omega \in (0.935, 0.995)$	0.06
	$\omega \in (1.41, 1.76)$	0.35
	$\omega \in (2.9, 8.1)$	5.2
21	$\omega \in (0.78, 0.8)$	0.02
	$\omega \in (0.98, 1.02)$	0.04
	$\omega \in (1.3, 1.45)$	0.15
	$\omega \in (1.95, 2.6)$	0.65
	$\omega \in (4, 8)$	4
24	$\omega \in (0.745, 0.755)$	0.01
	$\omega \in (0.89, 0.92)$	0.03
	$\omega \in (1.15, 1.2)$	0.05
	$\omega \in (1.5, 1.7)$	0.2
	$\omega \in (2.2, 3)$	0.8
	$\omega \in (4.5, 8)$	3.5

Table 2.

4.3.3 The $(AB)_{16}$ type crystal with a B type defect layer (A an epsilon negative material)

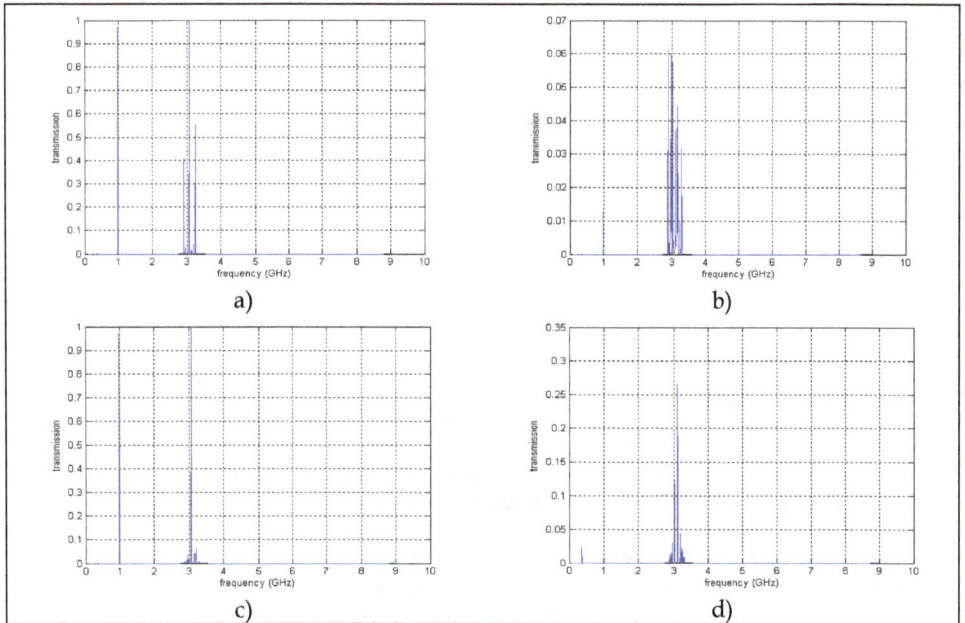

Fig. 20. The variation of the position of the defect :a) $(AB)_4B(AB)_{12}$; b) $(AB)_5B(AB)_{11}$;
c) $(AB)_6B(AB)_{10}$; d) $(AB)_7B(AB)_9$

In figure 20 it is showed the influence of the position of the defect layer upon the transmission of the 1D PC. If the defect layer is inserted after an even number of (AB) groups the transmission in the localized states is near 100%. If the defect layer is inserted after an ode number of (AB) groups the transmission is low (under 30%).

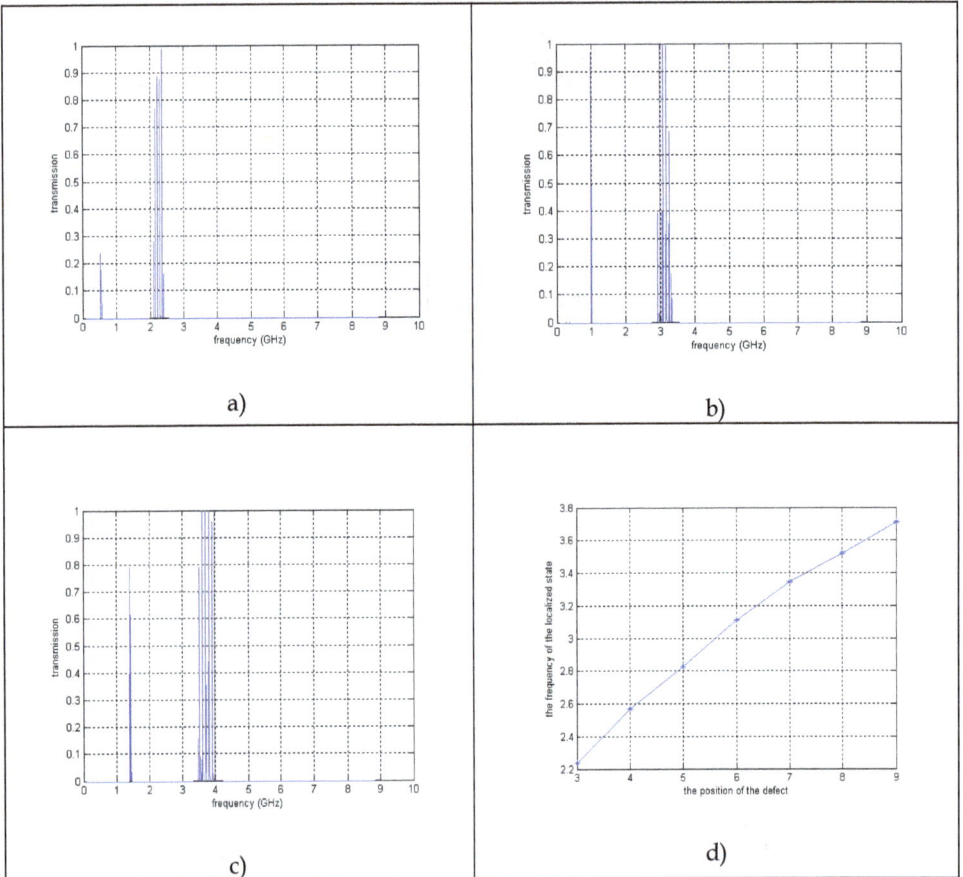

Fig. 21. The variation of the thickness of the B type layer a) dB=3 mm; b) dB=6 mm; c) dB=9 mm; d) the frequency of the localized state vs. the position of the defect

Now it is analyzed the influence of the thickness of the B type defect layer upon the optical transmission of the 1D PC. The defect layer is inserted after 8 (AB) groups. The variation of the thickness of the defect layer doesn't influence the optical transmission of the crystal but if it is varied the thickness of all B type layers in the crystal structure are obtained some interesting results as it can be seen in the following graphics.

From the graphics above it is observed that the thickness of the defect layer influences the value of the frequency at which we obtain the localized states (figure 21 d).

5. Conclusions

The insertion of a double negative material defect layer in the structure of the one-dimensional photonic crystal structure generates localized states. The optical transmission in these localized states and the value of the frequency at which the localized states are obtained depend on the thickness and position in which we have inserted the defect layer.

Analyzing the simulations results it is reached the conclusion that the best photonic crystal configurations are the following three:

a. The $(AB)_{16}$ type crystal with a A type defect layer (A a dielectric material)
b. The $(AB)_{16}$ type crystal with a B type defect layer (A a dielectric material)
c. The $(AB)_{16}$ type crystal with a B type defect layer (A a epsilon negative material)

From the results obtained it can be said that the studied structures can be used as optical filters.

6. References

Allouche, J. P. & Shallit, J. (2003). *Automatic Sequences: Theory, Applications, Generalizations*, pp. 152-153, Cambridge University Press, Cambridge, England, 2003

Born M. & Wolf, E. (1999). *Principles of Optics*, 7th ed., Sect. 1.6, pp. 54–74, Cambridge U. Press, Cambridge, England,1999

Eleftheriades G. V.; Iyer A. K. & Kremer P. C. (2002). *Planar negative refractive index media using periodically L-C loaded transmission lines*. IEEE Transactions on Microwave Theory and Techniques,Vol. 50, No. 12, pp. 2702–2712, 2002

Deng X. & Liu N. (2008). *Resonant tunneling properties of photonic crystals consisting of single-negative materials*. Chinese Science Bulletin, Vol. 53, No. 4, pp. 529-533, Feb. 2008

Joannnopoulos, J. D.; Johnson, S. G.; Winn, J. N. & Meade, R. D. (2008). *Photonic Crystals. Molding the Flow of Light*, Second Edition, Princeton University Press, 2008

Li, X.; Xie, K. & Jiang, H. (2008). *Transmission properties of one-dimensional photonic crystals containing double-negative and single-negative materials*, Chinese Optics Letters, Vol. 6, No. 2, 2008

Pendry J. B. (2000). *Negative refraction makes a perfect lens*. Physics Review Letters, Vol. 85, No. 18, pp. 3966–3969, 2000

Ramakrishna, S.A. (2005). *Physics of negative refractive index materials*, Reports on Progress in Physics 68, pp. 449-521, 2005

Veselago V. G. (1968). *The electrodynamics of substances with simultaneously negative values of ε and μ*. Soviet Physics Uspekhi, Vol.10, No. 4, pp. 509–514, 1968

Wang, Z.Y.; Chen, X.M.; He, X.Q. & Han, S.L. (2008). *Photonic crystal narrow filters with negative refractive index structural defects*, Progress in Electromagnetics Research, PIER 80, pp. 421-430, 2008

Zhang H. Y.; Zhang Y. P.; Wang P. et al. (2007). *Frequency response in photonic heterostructures consisting of single-negative materials.* Journal of Applied Physics, Vol. 101, No. 1, pp. 013111, 2007

Angular-Resolved Optical Characteristics and Threshold Gain Analysis of GaN-Based 2-D Photonics Crystal Surface Emitting Lasers

Shih-Wei Chen[1,2], Tien-Chang Lu[1], Ting-Chun Liu[1],
Peng-Hsiang Weng[1], Hao-Chung Kuo[1] and Shing-Chung Wang[1]
[1]Department of Photonic and Institute of Electro-Optical Engineering
National Chiao Tung University, Hsinchu
[2]Green Energy & Environment Research Labs
Industrial Technology Research Institute, Hsinchu
Taiwan, R.O.C.

1. Introduction

Photonic crystal (PhC) surface emitting lasers (PCSELs) utilizing Bragg diffraction mechanism have considerable amounts of publication during the past few years[1,2,3,4]. Such PhC lasers have many excellent advantages to attract the attention especially in controlling the specific lasing modes such as longitudinal and transverse modes, lasing phenomenon over the large area, and narrow divergence beam. Therefore, we fabricated the GaN-based PCSELs devices with AlN/GaN distributed Bragg reflectors (DBR) and analyzed the PhC laser characteristics caused by the surrounding PhC nanostructure. However, there were many theoretical methods calculating the photonic band diagrams and the distribution of electric or magnetic field of the PhC nanostructure in the past few years, such as 2-D plane wave expansion method (PWEM)[2,5], finite difference time domain (FDTD)[6,7], transfer matrix method, and multiple scattering method (MSM), etc. Many different advantages and limitations occur while using these methods. Therefore, in our case, we applied the MSM and PWEM to calculate the PhC threshold gain and photonic band diagram by using our PCSEL device structure.

In this chapter, the fabrication process of PhC lasers will be introduced in section 2. They can be divided into two parts, the epitaxial growth and the device fabrication. Section 3 will show the the foudamental mode characteristics of PhC laser, such as laser threshold pumping power, far-field pattern, MSM theoretical calculation methods, and divergence angles. Section 4, in the Bragg diffraction mechanism, each PhC band-edge mode is calculated and exhibits other type of wave coupling mechanism. Section 5, the photinc band diagrams of foundamental and high order lasing modes can be observed by the angular-resolved μ-PL (AR μ-PL) system. Comparing with the theoretical calculation resulted by PWEM and the experiment results of photonic band diagrams measured by AR μ-PL, they can be well matched and show the novel PhC characteristics. Besides, the fundamental and high order PhC lasing modes would be calculated in this section.

2. Fabrication processes

Here, the fabrication processes are composed of two parts. One is the epitaxial growth on sapphire substrates by metal organic chemical vapour deposition (MOCVD), including a 29-pair distributed Bragg reflectors (DBR), a p-GaN layer, multi-quantum wells, a n-GaN, and a un-doped GaN layer, etc. Another one is to fabricate the PhC nanostructure on the epitaxial wafers by the E-beam lithography system and inductive coupled plasma reactive ion etching (ICP-RIE) system. Finally, the GaN-based photonic crystal surface emitting laser (PCSEL) devices with AlN/GaN DBR are performed.

2.1 Growth of nitride-based reflectors and micro-cavity

The detail growth process and experiment parameters of the micro-cavity and nitride-based DBR on sapphire substrates by metal organic chemical vapor deposition (MOCVD) are described as follows:

First, the substrate was thermally cleaned in the hydrogen ambient for 5 min at 1100 °C. And then, a 30 nm-thick GaN nucleation layer was grown at 500°C. The growth temperature was raised up to 1100 °C for the growth of a 2 μm-thick GaN buffer layer. The subsequent epitaxial structure consisted of a 29-pair of quarter-wave AlN/GaN DBR grown at 1100 °C, a 7-lamda cavity (λ = 410 nm) which includes a 860 nm-thick Si-doped n-GaN layer, 10 pairs $In_{0.2}Ga_{0.8}N$/GaN (2.5 nm/12.5 nm) MQWs, a 24 nm-thick AlGaN layer as the electron blocking layer, a 110 nm-thick Mg-doped p-GaN layer, and a 2 nm-thick p^+ InGaN layer as the contact layer. The AlN/GaN super-lattices (SL) inserted in the stacks of 29-pair AlN/GaN layers are fabricated because they can release the strain during the growth of AlN/GaN DBR and further improve interface and raise reflectivity of the DBR. Besides, the AlN/GaN DBR can play the role of the low refractive index layer to confine the optical field in the active region in the whole structure. And then, the AlGaN electron blocking layer was served to reduce the electron overflow to the p-GaN layer. The reflectivity spectrum of the AlN/GaN DBR is shown in Fig. 1. It shows the highest reflectivity of the DBR is about 99% at 416 nm. The stop-band of the DBR is as wide as about 25 nm. Fig. 2. is (a) the OM and (b) cross-sectional TEM images of the as-grown micro-cavity sample.

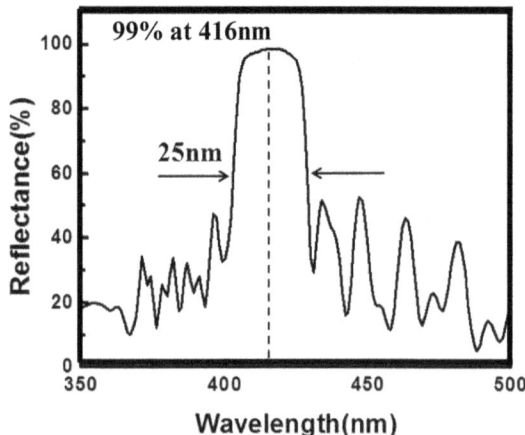

Fig. 1. The reflectivity spectrum of the AlN/GaN

Angular-Resolved Optical Characteristics and Threshold Gain Analysis of GaN-Based 2-D Photonics
Crystal Surface Emitting Lasers

29

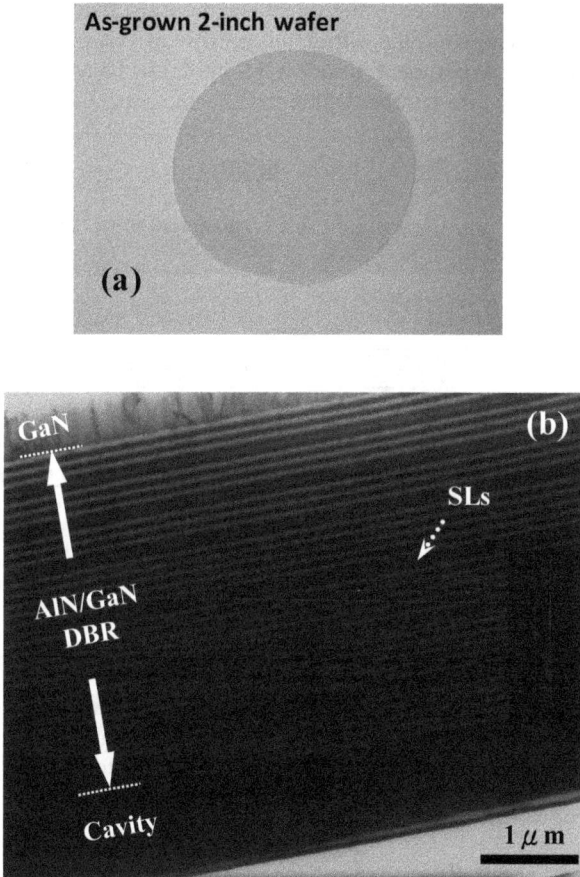

Fig. 2. (a) OM and (b) cross-sectional TEM images of the as-grown micro-cavity sample.

2.2 The fabrication process of photonic crystal surface emitting lasers (PCSELs)

The PhC nanostructure was fabricated on the epitaxial wafers by the following process steps as shown in Fig. 3. In the beginning, the hard mask SiN_x 200 nm was deposited on as-grown samples by PECVD. Then, PMMA layer (150 nm) was spun by spinner and exposed by using E-beam writer to form a soft mask. The pattern on the soft mask was transferred to SiN_x film to form the hard mask by using ICP-RIE (Oxford Plasmalab system 100), and then, the PMMA layer was removed by dipping ACE. The pattern on hard mask was transferred to GaN by using ICP-RIE (SAMCO RIE-101PH) to form the PhC layer. In order to remove the hard mask, the sample is dipped in BOE. Finally, the PCSEL devices have been fabricated as shown in Fig. 4. Fig. 5. shows the plane-view (a) and the cross section (b) of SEM images of our PCSELs. Although the hole profiles of PhC nanostructure etched through the MQWs region are not perfect due to the lateral plasma etching by ICP-RIE shown in Fig. 5(b), the PhC nanostrusture near the sample surface which has smooth etching profile show the largest coupling effects of the light field in the MQW region of about 100nm thickness. Therefore the diffraction profiles of PhC nanostructure still can be observed in the following experiment. Besides, the minimun hole diameter and maximum depth of the PhC nanostructure are about 40nm and 1 μ m, respectively.

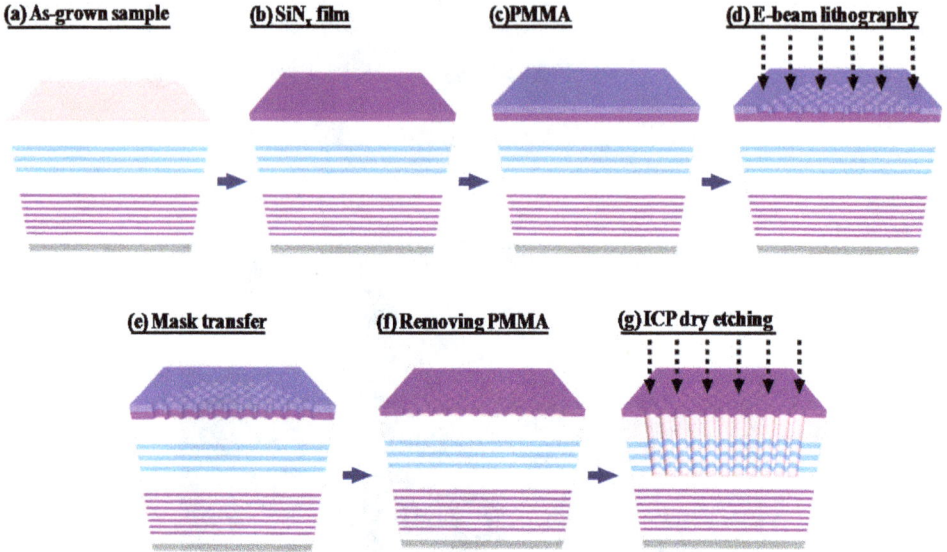

Fig. 3. PCSEL fabrication flowcharts: (a) as-grown sample structure, (b) deposit SiNx film by PECVD, (c) spin on PMMA, (d)E-beam lithography, (e) PhC patter transfer to SiN$_x$ layer, (f) remove PMMA by Acetone, and (g) PhC patterns transfer to GaN layer.

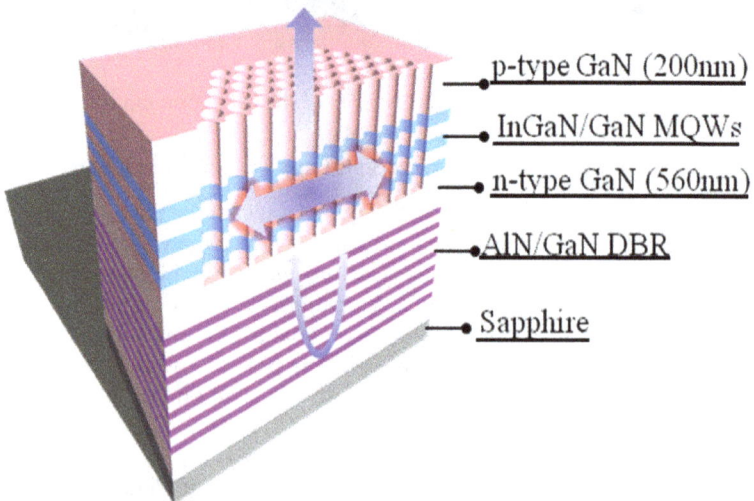

Fig. 4. The GaN-based PCSEL devices with AlN/GaN DBRs

(a)

(b)

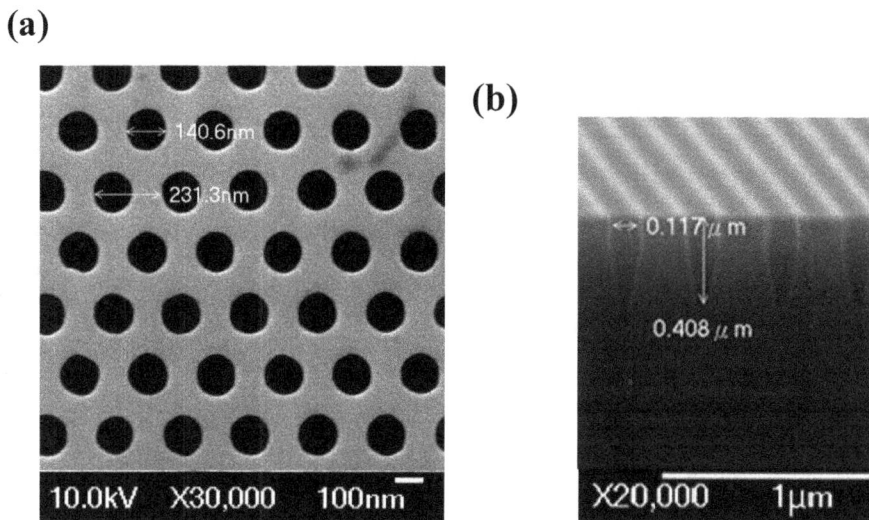

Fig. 5. SEM images of PCSELs: (a) plane view. (b) cross-section view.

3. Optical measurement system and the foundatment mode of PhC laser

Section 3.1, the angular-resolved μ-PL (AR μ-PL) system will be introduced, including the pumping lasers, light paths, and so on. Then, using the AR μ-PL system, the characteristics of foundament mode PhC laser would be shown in Section 3.2 and 3.3, such as threshold characteristics and far field patterns, etc. Furthermore, by adopting the multiply scattering method (MSM), the threshold gains of foundamental modes PhC lasers can be calculated in Section 3.4.

3.1 Angular-resolved μ-PL (AR μ-PL)

This section would intorduce the angular-resolved μ-PL (AR μ-PL) system which is designed for multiple applications. As shown in Fig. 6, it can observe two optical pump sources, including a frequency tripled Nd:YVO4 355 nm pulsed laser with a pulse width of ~0.5ns at a repetition rate of 1KHz and 325 nm He-Cd continuous wavelength (CW) laser; two optical pump incidence paths, two collecting PL method and two way to collect sample surface image are as well observed. The samples are pumped by the laser beam with an incident angle from 0 degree to 60 degrees normally from the sample. The laser spot size is about 50 μm in diameter covering the whole PhCs pattern area. The PL spectrum of the samples can be collected by a 15 X objective len and coupled into a spectrometer with a charge-coupled device (Jobin-Yvon iHR320 Spectrometer) or a fiber with a 600 μm core. The resolution is about 0.07 nm for the spectrometer. Fig. 6. shows the setup of the AR μ-PL system. The GaN-based PCSELs were placed in a cryogenics controlled chamber to perform PL experiment at low temperature. The temperature of the chamber can be controlled from room temperature (300 K) down to 77 K via the liquid nitrogen.

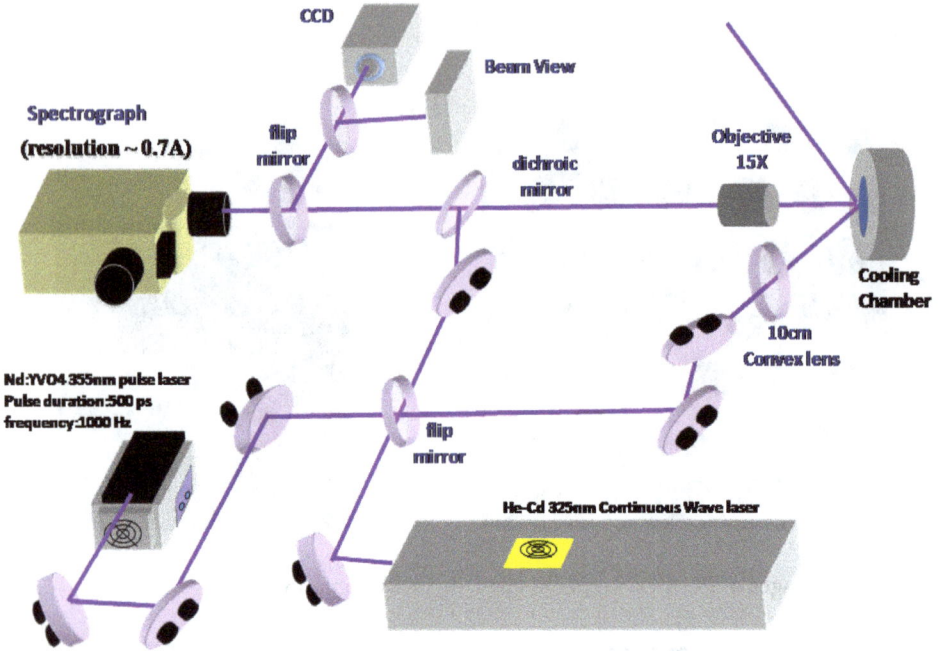

Fig. 6. The angular-resolved μ-PL (AR μ-PL) system

3.2 Threshold characteristics of fundamental mode of PhC lasers

In the optical pumped experiments of PCSEL devices, the lasing action was clearly observed in several devices with different lasing wavelength ranging from 395 nm to 425 nm. Fig. 7 shows the output emission intensity versed the pumping energy density with the PhC lattice constant of about 254nm. In the figure, the clear threshold pumping energy shows at the threshold pumping energy density of 2.8 mJ/cm², and a peak power density of 5.6 MW/cm². When the laser pumping energy exceeds the threshold energy, the laser output intensity increases abruptly and linearly with the pumping energy. Fig. 8 shows the excitation energy dependent emission spectrums of 0.8 E_{th}, 1 E_{th}, 1.2 E_{th}, and 1.3 E_{th}. These spectrums clearly show the transition behavior from spontaneous emission to stimulated emission. Furthermore, above the threshold, only one dominant peak wavelength of 419.7 nm with a linewidth of 0.19 nm can be observed.

3.3 Far field patterns (FFP) of PhC fundamental mode lasers

The lasing area of the GaN-based 2-D PCSEL, obtained by a CCD camera, is relatively large and covers near the whole area of PhC pattern with only one dominant lasing wavelength as shown in Fig. 9. It's interesting to note that the threshold power density of GaN-based 2-D PCSEL is in the same or even better order than the threshold of the GaN-based VCSEL we have demonstrated recently[8]. Unlike the small emission spots observed in the GaN-based VCSELs, the large-area emission in 2-D PCSEL has great potential in applications and requires high power output operation.

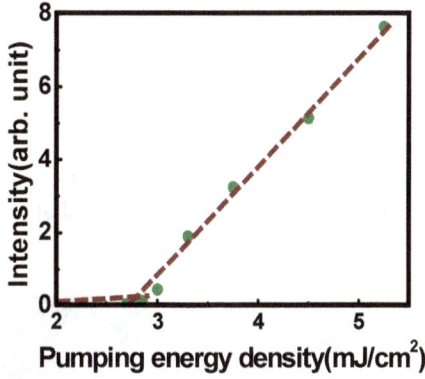

Fig. 7. Laser intensity as a function of pumping energy density

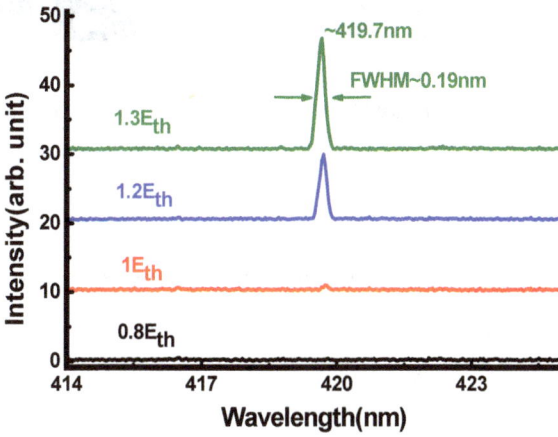

Fig. 8. The lasing spectrums under different pumping energy densities

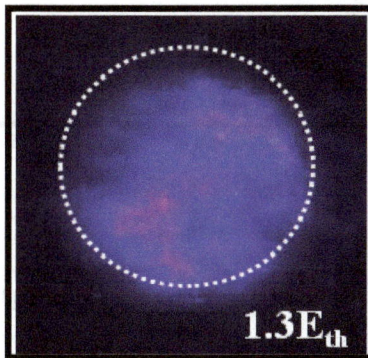

Fig. 9. The lasing CCD image is at 1.3 Eth and the dash circle is the PhC nanostructure region of about 50μm

The far-field patterns (FFP) of the laser were detected by an angular-resolved optical pumped system as shown in Fig. 10. In this figure, the lasing far field profiles with different distances from the sample surface were measured. When we increased the measurement distance from the sample surface, the lasing spot sprits of four points with two axes, Γ-M and Γ-K directions, indicated that the lasing has strong direction and energy concentration properties in real space. Then, we re-plotted the lasing spot sizes as a function of the measurement distance as shown in Fig. 11. In the figure, it shows the divergence angle of PCSEL determined by the distance of two lasing spot axes of about 5.6 degrees. It is smaller than edge emitting laser (~10^0~20^0) and VCSEL(8^0).

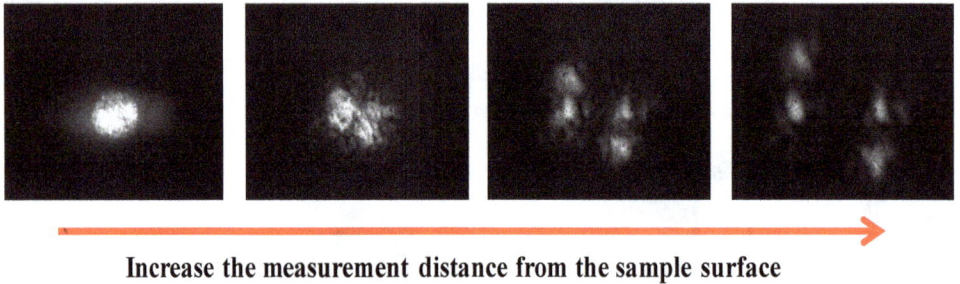

Increase the measurement distance from the sample surface

Fig. 10. The far field pattern with different distance from the sample surface collected by objective lens

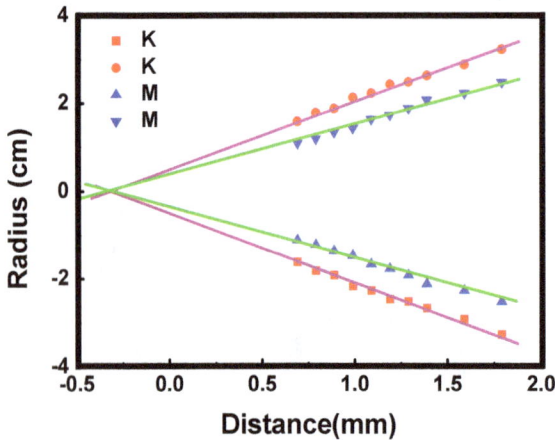

Fig. 11. The divergence angle between the two axes.

3.4 Threshold gain analysis by multiple scattering method (MSM)

This section would introduce the multiple scattering method (MSM) shown below:

The simulation structure is composed of finite two-dimensional PhCs nanostructure with triangular-lattice patterns and parallel cylinders placed in a uniform GaN-based material.

The complex dielectric constant is the light amplification in GaN-based material shown as
follows:

$$\varepsilon_{backgound}(\omega) = \varepsilon_{GaN}(\omega) - i\frac{2c\sqrt{\varepsilon_{GaN}}}{\omega}k_a''$$ (1)

where ε_{GaN} represents the dielectric constant varied with frequency of light and k_a''
represents the amplitude gain coefficient of the material. A point source transmitted
monochromatic waves are placed at the original point. The total system matrix can be
obtained as below[9]:

$$\Gamma_n^i A_n^i - \sum_{j=1,j\neq i}^{N} \sum_{l=-\infty}^{\infty} G_{l,n}^{i,j} A_l^j = T_n^i$$
(2)

The A_n^i and T_n^i are matrixes representing expansion coefficients of scattering waves and
incident waves, respectively. Here, according to main dipole oscillation in the GaN active
region, only the transverse electric (TE) mode polarization (polarization direction
perpendicular to the cylinder axis) is considered[10]. Eq. (2) could be simplified to an eigen
value problem: MA=T. If the value of vector A / T is divergent, the laser oscillation
condition would be achieved. Therefore, det(M)=0 is the complex determinant equation
which is used to search for a pair of variables threshold amplitude gain k_{am}'' and
normalized frequency from $k = \omega/c$ in Eq. (1).

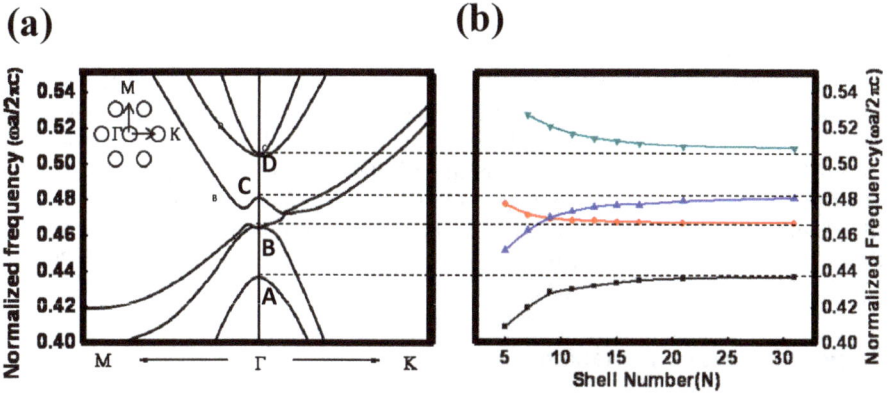

Fig. 12. (a) Photonic band diagram of a PhC triangular lattice with TE mode polarization
calculated by PWEM near the first Γ band edges showing four different modes; (b)
Normalized frequencies of lasing modes calculated by MSM for different PhC shells
(N values).

According to PWEM, the first Γ band edge of photonic band diagram with the PhC
triangular lattice and TE mode polarization are calculated as shown in Fig. 12(a). We can
find four different band edges causing four resonant modes (A – D) since modes B and D
are doubly degenerate. In Fig. 12(b), the normalized frequencies of lasing modes is
calculated for different PhC shells (N values) by MSM., where the parameter N is

represented as the number of cylinder layers in the Γ-M direction. The dashed lines of Fig. 12 represent different resonant modes of A, B, C and D at the Γ band edge. It can be observed that the resonant mode frequencies calculated by MSM will approach to band edge frequencies calculated by PWEM when the shell number increases. Therefore, we could obtain more accurate results when the layer number goes beyond 20. Because of the shapes of photonic band diagrams, the blue-shifted or red-shifted trends of normalized frequencies are increased with the shell numbers in Fig. 12(b).

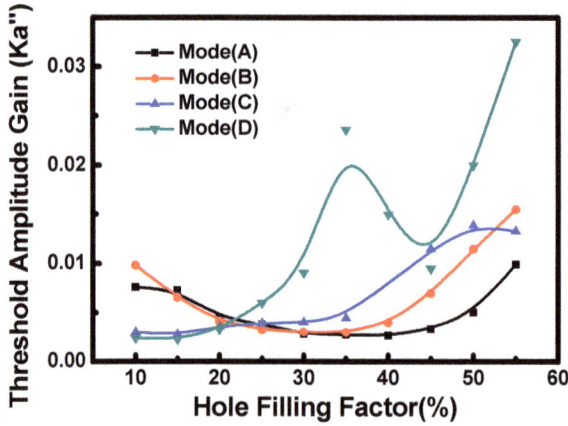

Fig. 13. Threshold amplitude gain of four modes as a function of the hole filling factor. The inset shows the lasing mode at Γ point in the PhC plane using Bragg diffraction scheme[10].

Fig. 13 shows the threshold amplitude gain of modes A-D as a function of the hole filling factor calculated by MSM. The confinement factor and effective refractive index are 0.865 and 2.482 for guided modes in the calculation, respectively. Hence, real parts of ε_{GaN} and ε_{Hole} are 7.487 and 3.065 for the GaN material and PhC air holes[11,12]. In the figure, the mode A and B have the lowest threshold gain for hole filling factors of about 35% and 30%; besides, mode C and D have the lowest threshold gain for hole filling factors of about 10% and 15%. This result shows that the proper hole filling factor can control the PhC mode selection.

4. Bragg diffraction mechanism

According to Bragg diffraction theory, the first order Bragg diffraction with 2-D PhC triangular lattice will be introduced in Section 4.1. The high order diffraction mechanism will be shown in Section 4.2 together with K2 and M3 PhC modes.

4.1 First order Bragg diffraction in 2-D PhC triangular lattice[6,13]

Fig. 14(a) shows a photonic band diagram with PhC triangular lattice. Among the points (A), (B), (C), (D), (E), and (F) in band diagram, each of them presents different lasing modes, including Γ1, K2, M1, Γ2, K2, and M2, which can control the light propagated in different lasing wavelength and band-edge region. A schematic diagram of the PhC nanostructure in reciprocal space transferred from real space are shown in Fig. 14(b). The parameter of a is the PhC lattice constant. The *K1* and *K2* are the Bragg vectors with the same magnitude, $|K|=2\pi/a_0$. Considering the TE modes in the 2-D PhC nanostructure, the diffracted light wave from the PhC structure must satisfy the Bragg's law and energy conservation:

$$k_d = k_i + q_1 K_1 + q_2 K_2, \qquad q_{1,2} = 0, \pm 1, \pm 2, \dots \qquad (3)$$

$$\omega_d = \omega_i \qquad (4)$$

where k_d is a xy-plane wave vector of diffracted light wave; k_i is a xy-plane wave vector of incident light wave; $q_{1,2}$ is order of coupling; ω_d is the frequency of diffracted light wave, and ω_i is the frequency of incident light wave. Eq. (3) represents the momentum conservation, and Eq. (4) represents the energy conservation. When both equations are satisfied, the lasing behavior would be observed.

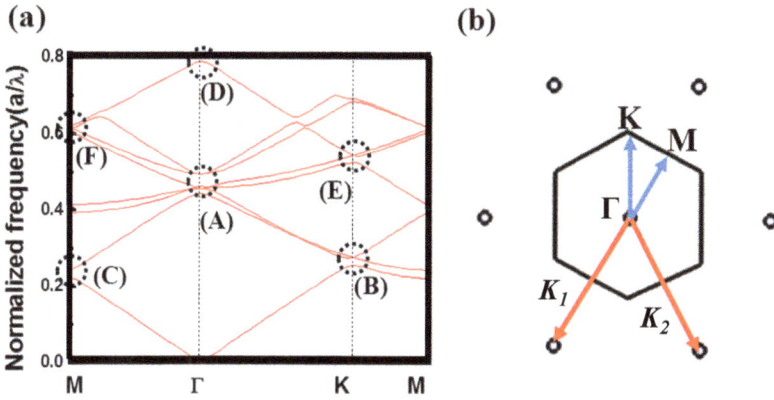

Fig. 14. (a) The band diagram of PhC with triangular lattice; (b) The schematic diagram of PhC with triangular lattice in reciprocal space.

In the calculation, the PhC band-edge lasing behavior would occur at specific points on the Brillouin-zone boundary, including Γ, M, and K which would split and cross. At these PhC lasing band-edge modes, waves propagating in different directions would be coupled and increase the density of state (DOS). Each of these band-edge modes exhibits different types of wave coupling routes. For example, only the coupling at point (C) involves two waves, propagating in the forward and backward directions as shown in Fig. 15(c). In different structures, all of them show similar coupling mechanism but different lasing behaviors. However, they can be divided into six equivalent Γ-M directions. It means that the cavity can exist independently in three different directions to form three independent lasers. Point (B) has an unique coupling characteristic as shown in Fig. 15(b). It forms the triangular shape resonance cavity propagating in three different directions while comparing with the conventional DFB lasers. On the other hand, the point (B) can also be six Γ-K directions in the structure shown two different lasing cavities in different Γ-K directions coexisted independently. In Fig. 15(a) point (A), the coupling waves in in-plane contain six directions of 0°, 60°, 120°, -60°, -120°, and 180°. According to the first order Bragg diffraction theory, the coupled light can emit perpendicular from the sample surface as shown in Fig. 16. Therefore, the PhC devices can function as surface emitting lasers.

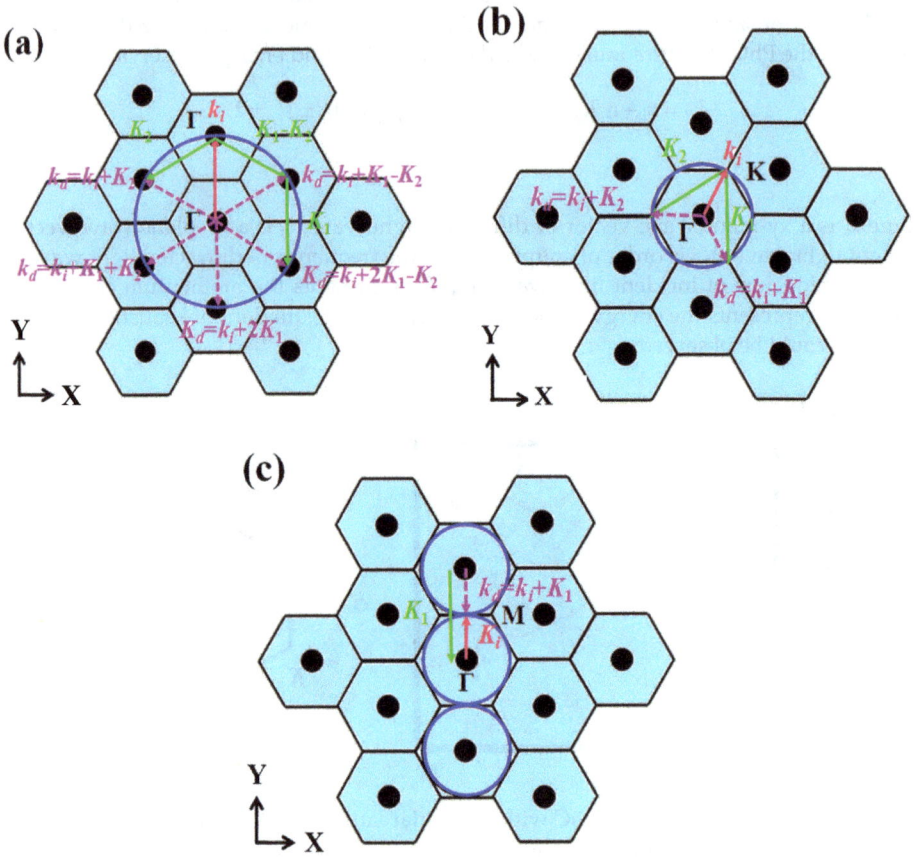

Fig. 15. Wave vector diagram at points (A), (B), (C) in Fig. 13(a); k_i and k_d indicate the incident and diffracted light wave.

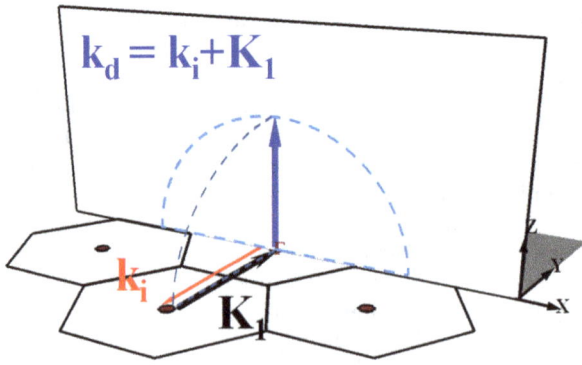

Fig. 16. The wave vector diagram at point (A) in vertical direction.

4.2 High order Bragg diffraction in 2-D PhC with triangular lattice

At point (E) which satisfies the Bragg's law, Fig. 17(a) and (b) show the in-plane and vertical diffraction of the light wave diffracted in three Γ-K directions to three K′ points. In the wave-vector diagram of one K′ point, the light wave is diffracted to an angle tilted 30° normally from the sample surface as shown in Fig. 17(b). Therefore, the lasing behavior of K2 mode would emit at this specific angle of about 30°.

At point (F), Fig. 18(a) and (b) represented the in-plane and vertical diffraction that the light wave is diffracted in two different Γ-M directions and reaches to three M′ points. Fig. 18(b) shows the wave-vector diagram of one M′ point where the light wave is diffracted into three independent angles tilted of about 19.47°, 35.26°, and 61.87° normally from the sample surface.

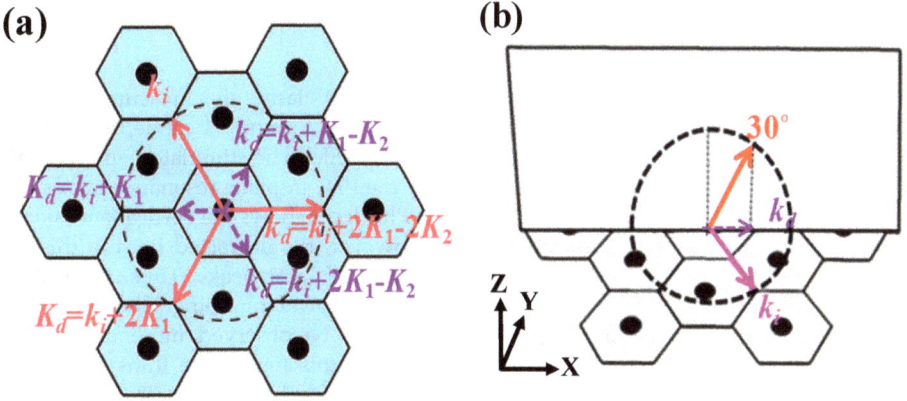

Fig. 17. Wave vector diagram of (a) in-plane and (b) vertical direction at point (E) (or K2 mode); k_i and k_d indicate incident and diffracted light wave.

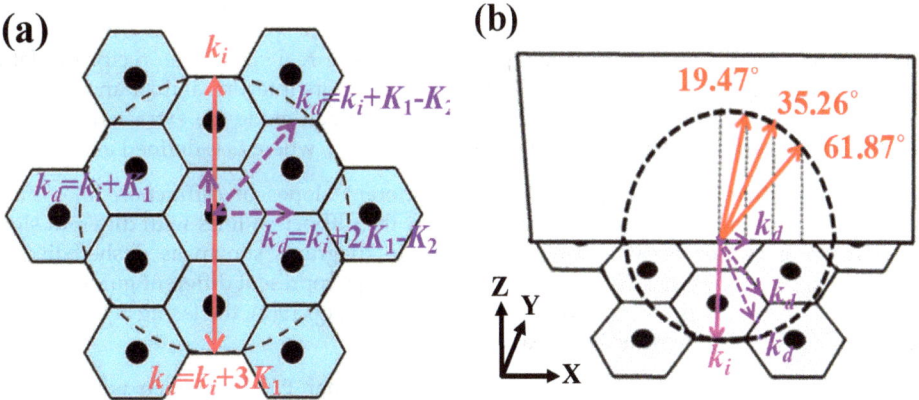

Fig. 18. Wave vector diagram of (a) in-plane and (b) vertical direction at point (F) (or M3 mode); k_i and k_d indicate incident and diffracted light wave.

5. Angular-resolved optical characteristics at different band-edge modes

Section 5.1 shows the transformation method from angular-resolved measurement data to the AR μ-PL diagrams. In Section 5.2, the AR μ-PL diagrams and the divergence angles of Γ1, K2, and M3 modes are introduced.

5.1 Data normalization

After measurements by the angular-resolved measurement system, we transformed the AR μ-PL spectrums to obtain the guided modes dispersion relation (reduced frequency $u=\Lambda/\lambda_0$ as y-axis versus in-plane wave vector, $k_{//}$, as x-axis) by the relation $k_{//}= k_0*sin\theta$. In addition, each wavelength, $I_{PL}(\delta)$, is normalized relatively to its integrated intensity[14]. The normalized AR μ-PL diagram reveals the clear dispersion relation of guided modes and detaily figures out the relative excitation and out-coupling efficiency.

5.2 AR μ-PL diagram

Pumped by the YVO4 pulse laser and the He-Cd CW laser, the measured dispersion diagrams at Γ1 mode are observed as shown in Fig. 19. Around Γ1 mode, the dash lines represent the simulated photonic band diagram by PWEM. The stimulated emission of the lasing phenomenon from the devices provided by the PhC in-plane resonance routes can be observed by a YVO4 pulse laser in Fig. 19(a). The PhC laser shows the vertical emission near the normal direction from the sample surface. However, the diffracted lines in this figure cannot be observed clearly because of high intensity of laser peaks. Thus, the diffracted emissions are measured by a He-Cd CW laser with a lower pumping intensity from the PCSEL devices. Therefore, the diffracted pattern can be observed more clearly in the measured dispersion diagram shown in Fig. 19(b). In this figure, the transverse upward curving lines derived from the Fabry-Perot effect provided by the device structure and modulated by the interference of the DBR layers. The electric field propagating in the PhC structure could be described as a Bloch mode: $E(r) = \Sigma G \; E_G \times exp\; [i(k_{//} + G)\bullet r]$ to explain the observed diffraction patterns caused by a PhC nanostructure, where E_G is the electric field component corresponding to harmonic reciprocal lattice vector G, and $k_{//}$ is the in-plane wave vector of the Bloch mode. The reciprocal lattice in K space is a 2-D PhC triangular lattice rotated by 30° with respect to the direct lattice in real space. The reciprocal lattice vectors can be written as: $G = q_1 K1+ q_2 K2$, where q_1 and q_2 are integers, and $K1$ and $K2$ are the two reciprocal lattice basis vectors. Harmonics of the Bloch mode are extracted if their in-plane wave vectors are within the light cone: $|k_{//} + G| < k_0$, where k_0 is defined as $2\pi/a$.

In Fig. 19(b), there are several groups with different slopes of diffraction lines in the dispersion diagram. Different dispersion modes of the diffraction lines with different slopes can be well matched to calculated photonic band diagrams shown as dashed lines by PWEM. The parallel diffraction lines with the same slope represent different guide modes in the in-plane direction. By comparing the Fig. 19(a) with Fig. 19(b), the lasing actually occurs at the third guided mode near the Γ1 band edge.

In Fig. 20, the measured AR μ-PL diagrams of another PCSEL device with different PhC structure near the K2 modes along the Γ-K direction are measured. By using YVO4 pulse laser pumping, Fig. 20(a) reveals the lasing peaks in the AR μ-PL diagram. Besides, the AR μ-PL diagram is shown in Fig. 20(b) pumped by a CW He-Cd laser. In the figure, the

Angular-Resolved Optical Characteristics and Threshold Gain Analysis of GaN-Based 2-D Photonics
Crystal Surface Emitting Lasers

41

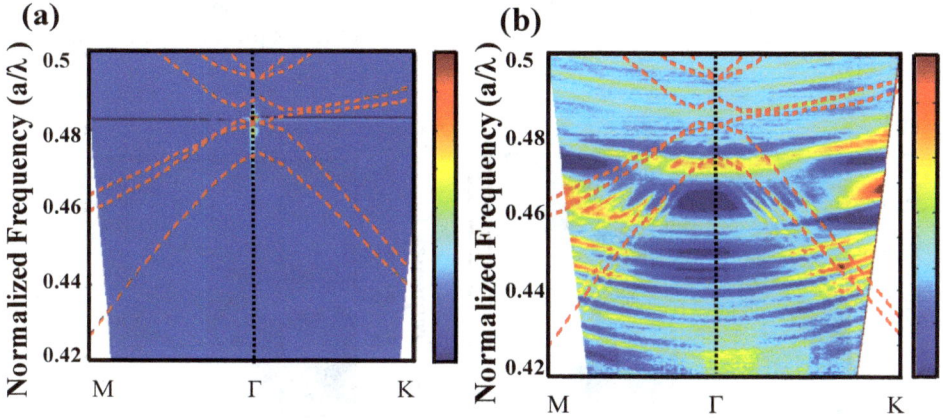

Fig. 19. The measured AR μ-PL diagram near the Γ1 mode ((a) pumped by YVO4 pulse laser; (b) pumped by He-Cd laser), the dash lines represent the calculated photonic band diagram.

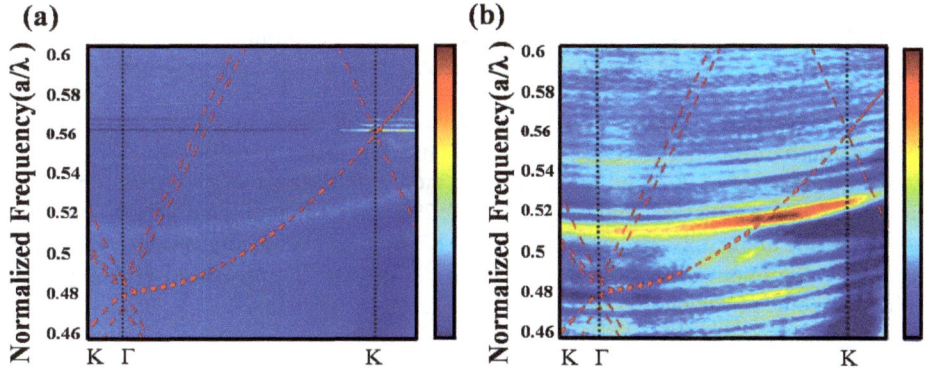

Fig. 20. The measured AR μ-PL diagram near the K2 mode ((a) pumped by YVO4 pulse laser; (b) pumped by He-Cd laser). The dash lines represent the calculated photonic band diagrams.

diffracted lines can be observed and well matched to the calculated 2-D TE-like photonic band diagram, by using parameters of r/a = 0.285, a = 210 nm, n_b = 2.560, n_a = 2.343, and n_{eff} = 2.498 for calculation shown as the dash lines in Fig. 20. In addition, the experiment results show the lasing beam emission angle of about 29 degree off from the normal along the Γ-K direction, which is exactly matched to the estimated value of about 30 degree derived in the previous section. Furthermore, we measured another PCSEL devices exhibited characteristics of M3 band edge mode along the Γ-M direction. The measured dispersion diagrams pumped by a YVO4 pulse laser and a He-Cd CW laser are shown in Fig. 21(a) and (b), respectively. The lasing peaks can be clearly seen in Fig. 21(a). The diffracted patterns can be observed in Fig. 21(b) and well matched by using parameters of r/a = 0.204, a = 230 nm, n_b = 2.617, n_a = 1.767, and n_{eff} = 2.498. Shown as the dash lines in Fig. 21, the emission angle of lasing beam was about 59.5 degree off from the normal along the Γ-K direction, which was also quite matched to one of the estimated values of about 61.87

degree derived in the previous section. The reason that only one emission angle was obtained could be due to the fact that we only measured the AR μ-PL diagram along one Γ-M direction.

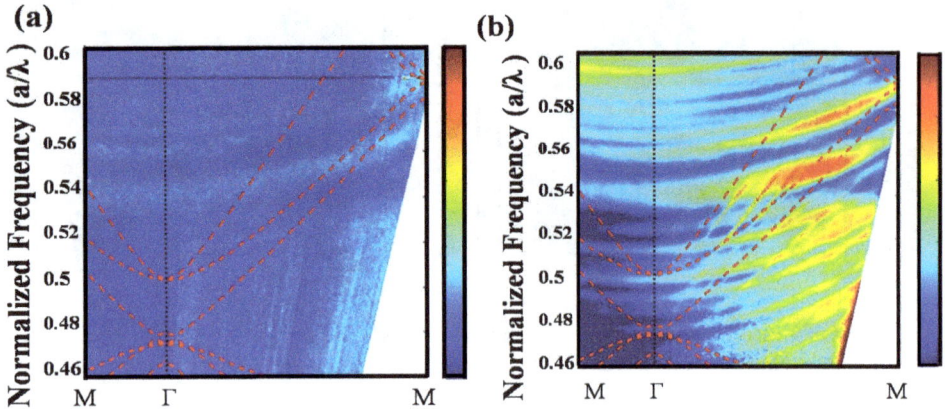

Fig. 21. The measured AR μ-PL diagram near the M3 mode ((a) pumped by YVO4 pulse laser; (b) pumped by He-Cd laser). The dash lines represent the calculated photonic band diagrams.

Each of PhC band-edge modes exhibited specific emission angle by different type of wave coupling mechanism in Fig. 19(a), Fig. 20(a), and Fig. 21(a). Finally, Fig. 22 shows the divergence angles of Γ1, K2, and M3 band-edge modes on the normal plane from the sample surface. The lasing emission angles are about 0°, 29°, and 59.5°, and the divergence angles of laser beams are about 1.2°, 2.5°, and 2.2° for (Γ1, K2, and M3) band edge modes, respectively. Due to the alignment difficulties in the AR μ-PL system, the measured emission angles might have some offset values of about 1° to 2°.

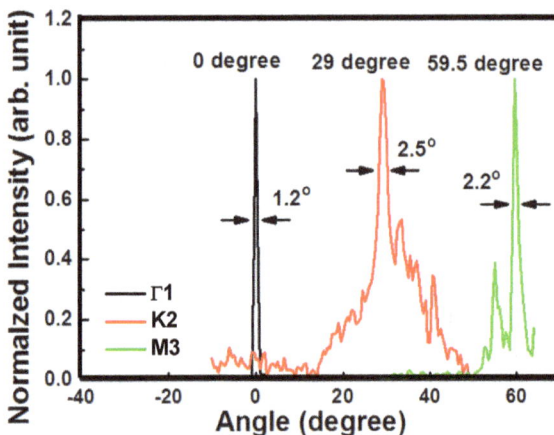

Fig. 22. The emission angles and divergence angles of Γ1, K2, and M3 band-edge modes on the normal plane from the sample surface.

6. Conclusion

In the chapter, the optical pumped of GaN-based 2-D photonic crystal surface emitting lasers (PCSELs) with AlN/GaN distributed Bragg reflectors (DBR) are fabricated and measured. The laser has a 29-pair bottom DBR which plays the role of a low refractive index layer to enhance the coupling between photonic crystal (PhC) nanostructure and electrical field in the whole cavity. Therefore, the lasing action can be achieved in the optical pumping system. Each of these laser devices emits only one dominant wavelength between 395 nm and 425 nm.That normalized frequency of PhC lasing wavelength can be well matched to these three band-edge frequencies (Γ1, K2, M3) indicated that the lasing action can only occur at specific band-edges. In the angular-resolved μ-PL (AR μ-PL) system, the diffracted lines in the AR μ-PL diagrams of PCSEL devices can be further matched to the calculated 2-D TE-like photonic band diagram calculated by PWEM. These three band-edge frequencies (Γ1, K2, M3) have different emission angles in the normal direction of about 0°, 29°, and 59.5° and are further confirmed by the Bragg theory. The divergence angles of the (Γ1, K2, M3) modes are about 1.2°, 2.5°, and 2.2°. Moreover, according to multiple scattering method (MSM), the resonant mode frequencies will approach to band edge frequencies compared with plane wave expansion method (PWEM). In addition, the threshold gain of four resonant modes varies with the filling factor. This result shows that the proper hole filling factor can control the PhC mode selection. Finally, all of these calculation and experiment results indicate that GaN-based PCSELs could be a highly potential optoelectronic device for lasers in the next generation.

7. Acknowledgment

The authors would like to gratefully acknowledge A. E. Siegman at Stanford for his fruitful suggestion. The study was supported by the MOE ATU program, Nano Facility Center and, in part, by the National Science Council in Taiwan

8. References

[1] Imada, M.; Noda,S.; Chutinan, A.; Tokuda, T.; Murata, M. & Sasaki, G. Coherent two-dimensional lasing action in surface-emitting laser with triangular-lattice photonic crystal structure. *Applied Physics Letters*, Vol. 75, (1999), pp. 316-318, ISSN 0003-6951.

[2] Noda, S.; Yokoyama, M.; Imada, M.; Chutinan, A. & Mochizuki, M. Polarization Mode Control of Two-Dimensional Photonic Crystal Laser by Unit Cell Structure Design. Science, Vol. 293, (2001), pp. 1123-1125, ISSN 0036-8075.

[3] Ryu, H. Y.; Kwon, S. H.; Lee, Y. J. & Kim, J. S. Very-low-threshold photonic band-edge lasers from free-standing triangular photonic crystal slabs. *Applied Physics Letters*, Vol. 80, (2002), pp. 3476-3478, ISSN 0003-6951.

[4] Turnbull, G. A.; Andrew, P.; Barns, W. L. & Samuel, I. D. W. Operating characteristics of a semiconducting polymer laser pumped by a microchip laser. *Applied Physics Letters*, Vol. 82, (2003), pp. 313-315, ISSN 0003-6951.

[5] Sakai, K.; Miyai, E.; Sakaguchi, T.; Ohnishi, D.; Okano, T. & Noda, S. Lasing band-edge identification for a surface-emitting photonic crystal laser. *IEEE Journal on Selected Areas in Communications*, Vol. 23, (2005), pp. 1335-1340, ISSN 0733-8716.

[6] Imada, M.; Chutinan, A.; Noda S. & Mochizuki M. Multidirectionally distributed feedback photonic crystal lasers. *Physical Review B*, Vol. 65, (2002), pp. 195306, ISSN 1098-0121.

[7] Yokoyama M. & Noda S. Finite-difference time-domain simulation of two-dimensional photonic crystal surface-emitting laser. *Optics Express*, Vol. 13, (2005), pp. 2869-2880, ISSN 1094-4087.

[8] Wang, S. C.; Lu, T. C.; Kao, C. C.; Chu, J. T.; Huang, G. S.; Kuo, H. C.; Chen, S. W.; Kao, T. T.; Chen, J. R. & Lin, L. F. Optically Pumped GaN-based Vertical Cavity Surface Emitting Lasers: Technology and Characteristics. *Japanese Journal of Applied Physics*, Vol. 46, (2007), pp. 5397-5407, ISSN 0021-4922.

[9] Nojima, S. Theoretical analysis of feedback mechanisms of two-dimensional finite-sized photonic-crystal lasers. *Journal of Applied Physics*, Vol. 98, (2005), pp. 043102, ISSN 0021-8979.

[10] Lu, T. C.; Chen, S. W.; Lin, L. F.; Kao, T. T.; Kao, C. C.; Yu, P.; Kuo, H. C.; Wang, S. C. & Fan, S. H. GaN-based two-dimensional surface-emitting photonic crystal lasers with AlN/GaN distributed Bragg reflector. *Applied Physics Letters*, Vol. 92, (2008), pp. 011129, ISSN 0003-6951.

[11] Chen, S. W.; Lu, T. C.; Hou, Y. J.; Liu, T. C.; Kuo H. C. & Wang, S. C. Lasing characteristics at different band edges in GaN photonic crystal surface emitting lasers. *Applied Physics Letters*, Vol. 96, (2010), pp. 071108, ISSN 0003-6951.

[12] Chen Y. Y. & Ye, Z. Propagation inhibition and wave localization in a two-dimensional random liquid medium. *Physical Review E*, Vol. 65, (2002), pp. 056612, ISSN 1539-3755.

[13] Notomi, M.; Suzuki, H. & Tamamura, T. Directional lasing oscillation of two-dimensional organic photonic crystal lasers at several photonic band gaps. *Applied Physics Letters*, Vol. 78, (2001), pp. 1325-1327, ISSN 0003-6951.

[14] Soller, B. J.; Stuart, H. R. & Hall, D. G. Energy transfer at optical frequencies to silicon-on-insulator structures. *Optics Letters*, Vol. 26, (2001), pp. 1421-1423, ISSN 0146-9592.

3

980nm Photonic Microcavity Vertical Cavity Surface Emitting Laser

Yongqiang Ning and Guangyu Liu
Changchun Institute of Optics, Fine Mechanics and Physics
Chinese Academy of Sciences Changchun
China

1. Introduction

Vertical-Cavity Surface-Emitting Laser (VCSEL) is a type of semiconductor laser with laser beam perpendicular to the surface of the semiconductor substrate, as shown in Fig.1(a) [1]. VCSEL has many advantages, such as non-divergence output beam, fabrication and test on wafer, easy two-dimensional integration, and single longitudinal mode work. VCSEL is composed of an active region sandwiched between top and bottom highly reflective DBR mirror [2,3]. Generally high power VCSEL could be realized through large emission window, but suffers multi-mode operation due to the inhomogeneous current distribution across the active region. On the other hand single-mode operation is required in many applications including optical communications. Single-mode can transport longer distance and meet the requirements of high-speed data transmission [4,5]. Several approaches such as confined aperture less than 3µm, proton implantation, oxide and proton implantation mixed structure have been reported to achieve single-mode VCSEL. Due to the small aperture of emission window, these VCSELs are lasing at low output power. Besides the requirements of high output power and single mode operation, the wavelength range of VCSEL is broadened by applying InAs quantum dots or InGaAsN quantum well of the wavelength range of 1300nm and nitride quantum well of the blue light range for the applications of fiber communication and display.

In the past few years photonic crystal materials became of a great interest due to their powerful properties allowing for previously unknown flexibility in shaping the light. On the contrary to conventional edge emitting laser, the cavity length of VCSEL is of the size of optical wavelength. This brings VCSEL actually into microcavity field, where spontaneous emission is believed not to be an intrinsic atomic property anymore. Spontaneous emission can be enhanced or inhibited by tailoring the electromagnetic environment that the atom can radiate into. In a conventional edge emitting laser made of large cavity, most of the spontaneous emission is lost to free space as radiation modes and only a small fraction couples to the resonant mode of the cavity formed by the mirrors. Therefore, significant stimulated emission output can only be obtained when the input power crosses a threshold to overcome the free-space loss. In a wavelength-sized microcavity, the photon-mode density develops singularities, just as in the case of carrier confinement in quantum well. In this case, a single spectrally distinct mode can receive most or all of the spontaneous emission, indicating threshold-free stimulation. The rate of spontaneous emission is enhanced in such a microcavity, due to the change in the mode density. Photons whose energies lie within the band gap of photonic crystal cannot propagate through the structure. A point defect in the photonic crystal structure will generate localized state inside the band gap and form a microcavity. All the photons corresponding to the wavelength of the defect can propagate in the crystal. An example of such microcavity is DBR with high-reflectivity mirrors in the direction of the guided modes.

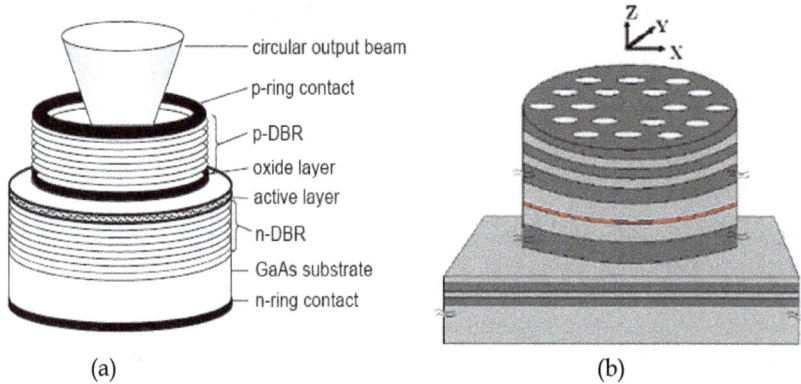

Fig. 1. VCSEL structure ((a): conventional VCSEL, (b): PhC-VCSEL)

The localization of electromagnetic models in single or multiple defects enabled to build photonic-crystal fibers, photonic planar waveguides, filters, splitters etc. Among these novel photonic crystal structures, photonic crystal-based VCSEL (PhC-VCSEL), as shown in Fig.1(b), is becoming an alternative approach and attracting more and more attention. These devices have strong potential due to their unique properties, which make them a perfect choice for many applications. These properties include stable single-mode operation [6], high-speed modulation [7] and polarization control [8]. However, to guarantee the efficient use of photonic crystals one needs careful consideration of the photonic crystal structure, which actually form a microcavity to modulate the spontaneous emission characteristics of VCSEL. Typical PhC-VCSELs consist of a classical VCSEL cavity surrounded by Distributed Bragg Reflectors (DBRs) of high reflectivity. The photonic crystal has a form of cylindrical holes located in various parts of the device. In the simplest case—and therefore the most popular one—the holes are etched in the top DBR. However, there are other possibilities like drilling the whole structure or placing the holes solely in the cavity, which can improve some properties of PhC-VCSEL but although constitutes a technological challenge. Photonic crystal structure with defects at the center was incorporated into the top layer to form microcavity, which provide lateral light confinement and also the modulation to the photon mode. However, large optical loss due to deeply-etched air holes still remains as a problem. The large optical loss is undesirable because it increases not only threshold current but also operating current level. High operating current can limit maximum single-mode output power via heating problem and lead to higher electrical power consumption.

Traditional VCSELs suffer a major drawback of the instability of the polarization, which generally attributed to the symmetric device structure. The polarization of a VCSEL tends to randomly follow one of the crystal axes and fluctuates with current. For applications such as 10-Gbit/s-class high-speed modulation1 and free-space interconnect using polarization-dependent optical components, a pinned polarization gives better performance. The competition between the modes with orthogonal polarizations can lead to polarization switching and mode hopping [9,10]. Such behavior is unacceptable for many practical applications such as intra-cavity frequency doubling, where other elements are polarization-dependent. Several approaches for polarization control have been reported based on the introduction of anisotropy to either gain or losses. These approaches include asymmetric shape resonator, metal-semiconductor gratings, or sub-wavelength grating by directly etching the top surface. In order to make use of the PhC structure for polarization control in VCSELs, PhC with elliptic air holes has been reported with polarization mode suppression ratio (PMSR) of over 20 dB in [11]. Triangular lattice PhC has been implemented with air holes elongated either along CK or CM directions. Disadvantages of etching photonic crystal holes include increased resistance and optical losses leading to higher threshold currents and voltage.

In this paper two-dimensional photonic crystal structure of hexagonal lattice of air holes on the top DBR reflector was introduced in VCSEL to suppress higher order mode operation. Defect structure of photonic crystal was created by filling one air hole (H_1 microcavity) or seven air holes (H_2 microcavity) to investigate the mode characteristics of VCSEL. With the proper selection of hole depths, diameters, and arrangement, this index confinement can be exploited to create single mode photonic crystal defect VCSELs that have the potential for low threshold currents and high output powers. The specific parameters of hexagonal lattice were optimized to achieve high Q factor of the microcavity.

2. Model and calculation

2.1 Photonic crystal micro-cavity VCSEL model

The active region of 980nm VCSEL was composed of three 8nm thick $In_{0.2}Ga_{0.8}As$ quantum well layers with 10nm thick GaAs barrier layer. $Al_{0.98}Ga_{0.02}As$ layer is incorporated between the P-type DBR and the active region to form lateral oxidation and provide both current and optical confinements. The reflectors were DBR mirrors with the reflectivity higher than 99%. In this work a periodic arrangement of air holes on the top DBR reflector was designed to form two-dimensional photonic crystal structure. Two kinds of lattice defect were produced to evaluate the Q factor of the microcavity. Schematic diagram of the structure was shown in Fig. 2.

Generally there were two types of two-dimensional periodic arrangement of photonic crystals: hexagonal lattice and square lattice. Under the similar lattice parameters of hole depth, diameter and distance, hexagonal lattice was suggested to obtain photonic band gap easily than the square lattice does. Once the photonic band gap was created, the band gap of hexagonal lattice was wider than that of square lattice. Therefore hexagonal lattice was often used in the design of PhC-VCSEL. When one or several holes were removed from the lattice, the periodicity of the lattice structure was destroyed. The simplest way is to remove one air hole from the center of the lattice. This created the H1 cavity, shown in Figure 3(a). The second photonic crystal defect structure, H2 microcavity, was to remove seven air holes from the center, as shown in Figure 3(b). In our simulation the air hole was etched through the top DBR and stop above lateral oxidation layer. The period of air holes was chosen to be 5.5µm for the easiness of fabrication.

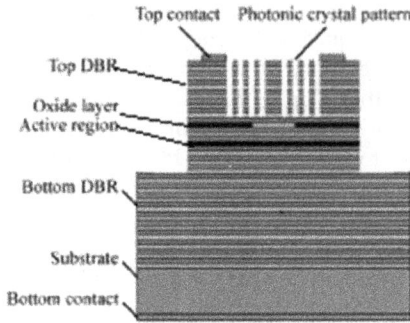

Fig. 2. Photonic crystal microcavity VCSEL

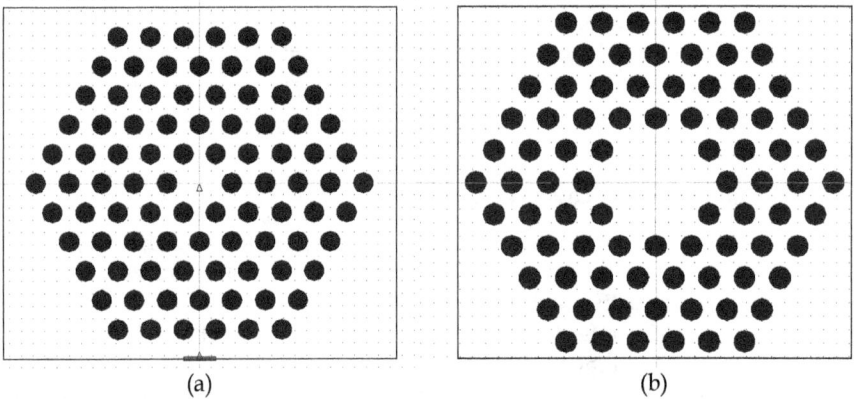

| (a) | (b) |

Fig. 3. Photonic crystal micro-cavities (a: H1 microcavity and b: H2 micro-cavity)

2.2 Analysis of single-mode condition

Photonic crystal defect structure with several holes missing at the center was similar to photonic crystal fiber where the solid center was surrounded by periodic arrangement of air holes, as shown in Figure 4. The characteristics of microcavity was only determined by the arrangement of air holes and the configuration of defect. There is no active material in the PhC structure. Therefore, the theory of photonic crystal fiber was used to investigate the normalized frequency of PhC defect structure in this work.

In the theory of photonic crystal fiber, the normalized frequency was expressed as following:

$$V_{eff} = (2\pi a / \lambda)(n^2{}_0 - n^2{}_{eff})^{1/2} \qquad (1)$$

Where a is the lattice period, λ is the wavelength, n_0 is the refractive index of the cavity center, n_{eff} is the external refractive index of the photonic crystal cladding.

Fig. 4. Photonic crystal fiber

According to photonic crystal theory, the following requirement of normalized frequency should be met to achieve single-mode operation.

$$V_{eff} < 2.405 \qquad (2)$$

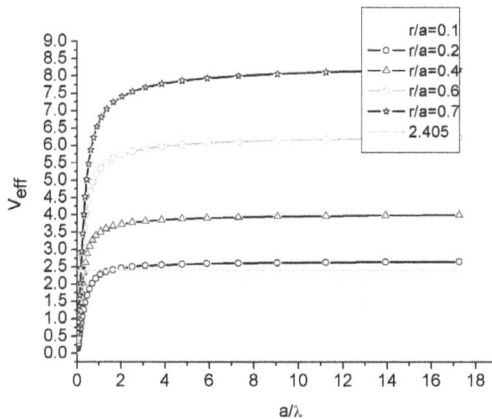

Fig. 5. Normalized frequencies V_{eff} of H1microcavity at different filling ratio

According to the analysis above, the normalized frequency of H1 photonic crystal microcavity was calculated. The filling ratio r/a of 0.1, 0.2, 0.4, 0.6 and 0.7 was suggested for the calculation of normalized frequency, as presented in Fig.5. It was shown that H1 photonic crystal micrcavity meets the requirement of single-mode operation when the filling ratio was less than 0.1. Obviously, smaller filling ratio was beneficial to single-mode operation. But too small filling ratio would cause additional difficulty in the fabrication process of photonic crystal structure. In the above calculation the hole depth was set to be infinite. However, the thickness of VCSEL chip is reasonably around 150µm like conventional edge emitting diode laser chip. It is very difficult, if required small filling ratio, to etch through the entire chip. And the mechanical strength of the device and the electrical properties would be deteriorated significantly. So the reliable hole depth was limited, which was not the case of identical photonic crystal fiber. Therefore the calculation above based on the theory of photonic crystal fiber should be modified as follows:

$$V_{eff} = (2\pi a / \lambda)\left[n^2_0 - (n_0 - \gamma\Delta n)^2\right]^{1/2}$$

(3)

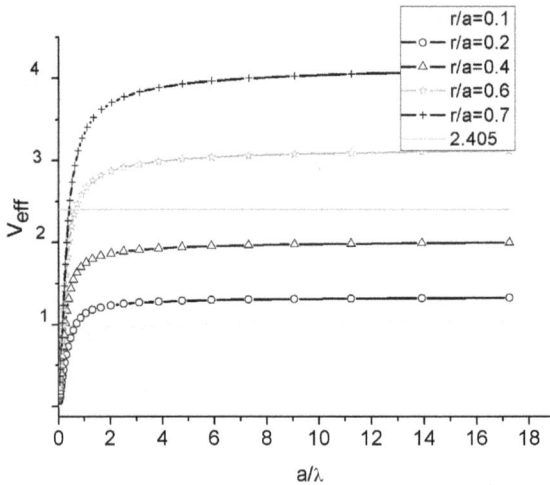

Fig. 6. Normalized frequencies V_{eff} of H1microcavity at different filling ratio based on modified calculation

The normalized frequencies V_{eff} of H1microcavity were calculated based on the modified model of equation (3) for different filling ratio, as shown in Fig.6. The corresponding etching depth factor γ was 0.3. Comparing Fig.5 and Fig.6 carefully, it was observed that single mode operation was realized for a filling ratio of 0.4. Though this filling ratio corresponds originally to multi-mode operation when the hole depth was set to be infinite as shown in Fig.5. This enables the fabrication of H1 microcavity much more easily while single mode operation was still maintained.

Similar to the above analysis, H2 microcavity with seven holes in the center missing was calculated, as shown Fig.4. The normalized frequencies was as following

$$V_{eff} = (2\pi\sqrt{3}a / \lambda)\left[n^2_0 - (n_0 - \gamma\Delta n)^2\right]^{1/2} \tag{4}$$

For single mode operation, the normalized frequency of H2 microcavity was smaller than that of H1 microcavity. At a filling ratio of 0.1, the output is single mode. This result might caused by relatively weak confinement of H2 microcavity compared with H1 microcavity.

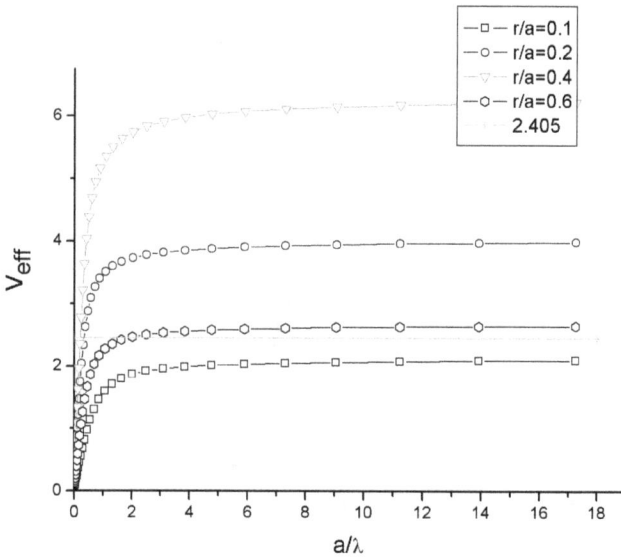

Fig. 7. Normalized frequencies V_{eff} of H2 micro-cavity at different filling ratio

2.3 Quality factor Q of microcavity

The cavity mode volume was reduced greatly in a photonic crystal microcavity, which improved the coupling of light field with the cavity mode. High quality factor of microcavity could be realized due to the light confinement provided by the photonic band gap in the lateral direction.

Cavity quality factor Q was an important parameter for evaluating photonic crystal VCSEL. Quality factor implies the ability of a microcavity to store energy. Obviously a photonic crystal microcavity with high Q was the purpose of an ideal design. Q was defined as:

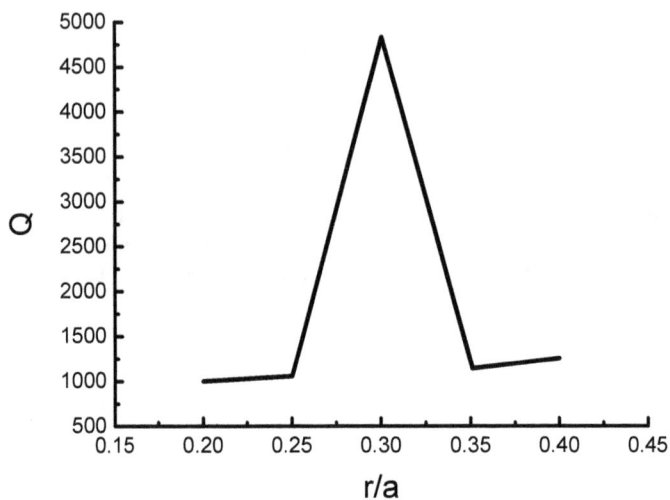

(a)

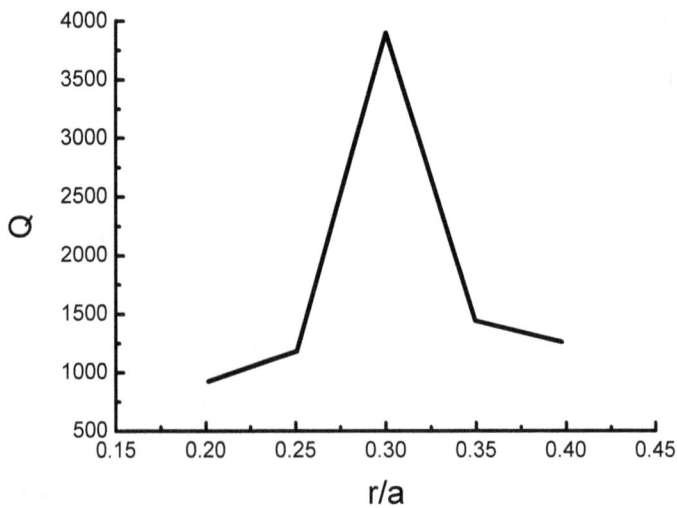

(b)

Fig. 8. Quality factor of microcavity at different filling ratio (a: H1 microcavity, and b: H2 microcavity)

$$Q = 2\pi \frac{qh\nu}{\left(-\dfrac{dh}{dt}\right)h} \tag{5}$$

Where q is the total number of photon in microcavity, h is Planck constant, ν is the resonant frequency. Now the quality factors Q of H1 and H2 microcavity were calculated for different filling ratio, as shown in Fig.8.

It was shown that the Q value reach a maximum of 4832(H1) and 3931(H2) when the filling ratio was 0.3.

3. Conclusion

Photonic crystal micro-cavity VCSEL with hexagonal lattice of air holes was discussed in regarding the quality factor and the requirement of single mode operation. The normalized frequencies of two types of microcavities (H1 and H2) were calculated based on modified theory of photonic crystal fiber. A filling ratio of 0.4 for H_1 microcavity was considered to be a good choice for single mode operation when the etching depth factor was 0.3. For H_2 microcavity, the filling ratio less than 0.1 were necessary for single mode operation. The difference between the filling ratios for H_1 and H_2 microcavities might suggest weak confinement of H_2 microcavity. Quality factors Q of two microcavities were calculated to be 4832(H_1) and 3931(H_2) respectively.

4. References

H Soda, K Iga, C Kitahara,*et al.*GaInAsP/InP Surface Emitting Injection Lasers[J], *Jpn J Appl Phys*,1979,18(12):2329-2330.

R.S.Geels, S.W.Corzine, L.A.Coldren. InGaAs Vertical-Cavity Surface-Emitting Lasers, *IEEE J. Quantum Electron.*, 1991,27:1359-1362.

Kenichi Iga, Vertical-Cavity Surface-Emitting Laser:Its Conception and Evolution[J],*Japanese Journal of Applied Physics*,2008, 47(1):1-10.

Li-Gao Zei, Stephan Ebers, Joerg-Reindhardt Kropp, *el a1.* Noise Performance of Multimode VCSELs[J]. *Journal of Lightwave Technology*.2001,19(6):884-887.

Maria Susana Torre, Cristina Masoller, K. Alan Shore, Synchronization of unidirectionally coupled multi-transverse-mode vertical-cavity surface-emitting lasers[J], *Journal of the Optical Society of America B*.2004,21(10):1772-1880.

D. S. Song, S. H. Kim, H. G. Park, c. K. Kim, and Y. H. Lee, "Single-fundamental-mode photonic-crystal verticalcavity surface-emitting lasers," Appl. Phys. Lett. 80, 3901 – 3903 (2002).

T. S. Kim, A. J. Danner, D. M. Grasso, E. W. Young, and K. D. Choquette, "Single fundamental mode photonic crystal vertical cavity surface emitting laser with 9 GHz bandwidth," Electron. Lett. 40, 1340 – 1341 (2004).

D. S. Song, Y. J. Lee, H. W. Choi, and Y. H. Lee, "Polarization-controlled, single-transverse-mode, photoniccrystal, vertical-cavity, surface-emitting lasers," Appl. Phys. Lett. 82, 3182 – 3184 (2003).

K.D. Choquette, R.P. Schneider, K.L. Lear, R.E. Leibenguth, IEEE J. Sel. Top. Quantum Electron. 1 (1995) 661.

K. Panajotov, B. Ryvkin, J. Danckaert, M. Peeters, H. Thienpont, I. Veretennicoff, IEEE Photon. Technol. Lett. 10 (1998) 6.

D.S. Song, Y.J. Lee, H.W. Choi, Y.H. Lee, Appl. Phys. Lett. 82 (2003) 3182.

4

Dynamic All Optical Slow Light Tunability by Using Nonlinear One Dimensional Coupled Cavity Waveguides

Alireza Bananej[1], S. Morteza Zahedi[1],
S. M. Hamidi[2], Amir Hassanpour[3] and S. Amiri[4]

[1]*Laser and Optics Research School, NSTRI*
[2]*Laser and Plasma research Institute, Shahid Beheshti University, Evin, Tehran*
[3]*Department of physics, K. N. Toosi University of Technology, Tehran*
[4]*Institute for Research in Fundamental Sciences, Tehran*
I. R. Iran

1. Introduction

Light propagation at slow speed, delaying light, has a tremendous role in a wide range of advanced technology from peta-bit optical networks to even laser fusion technology [1,2]. Recently, it was realized that delaying of an optical signal is useful for a number of applications such as optical buffering, signal processing, optical sensing and enhancement of optical nonlinearity in materials [3]. Also, as a consequence of important role of fiber lasers in laser fusion technology, coherent beam combining of several fiber lasers for achieving high power output is a crucial problem in future fusion technology. This task can be overcome by using proper engineered photonic components for achieving specific group delay time [2].

Generally, the speed of light is related to the refractive index of the propagation medium. From the basic definition of group velocity, $v_g = \dfrac{d\omega}{dk}$, It can be shown that group velocity of light in a medium can be derived as [4]:

$$v_g = \frac{d\omega}{dk} = \frac{c}{n + \omega(\dfrac{dn}{d\omega})}$$

Where, ω is the light frequency, c is the speed of light in vacuum, n is the refractive index of the medium and k is the propagation constant. Evidently, as a reason of limited optical materials and hence, limited refractive indices, it is not possible to slow down the speed of light sufficiently or adjust it to a specific desired value. However, it is obvious that by controlling the rate of refractive index changing, $\dfrac{dn}{d\omega}$, the speed of light can be controlled and in a special case when $\dfrac{dn}{d\omega} \gg 1$, slow down the speed of light can be obtained sufficiently.

During recent years different approaches have been done by scientists for generating of slow light. Electromagnetically induced transparency (EIT) is the first technique for achieving group delay and slow light in vapor mediums [5-9]. The group delay of incident pulses in an EIT system was studied first by Kasapi *etal* in lead vapor. Also, in 1999 Hau *etal* showed 7.05 μs delay in an EIT system which consists of condensed cloud of sodium atoms [10].

As a reason of complicated situation for slow light generation in EIT systems, it is important for real applications that these phenomena could be occurred in room temperature and in compact and solid materials.

Therefore, Spectral holes due to coherence population oscillation (CPO) in room temperature and in solids, can be considered as one of the interesting methods for slow light generation [11]. Slow light generation based on CPO has been studied in various material systems. So, ultraslow light, 57.5m/s, in ruby crystal has been studied by Boyd's group at Rochester [12].

On the other hand, during recent years a new kind of optical waveguides have been considerable attracted both theoretical and experimental attention due to their intense applications not only in data transferring but also in optical data processing [13]. Coupled cavity optical waveguides (CCWs) can be considered as the latest proposal mechanism for optical waveguiding which have been introduced by Yariv *etal* at Caltech [14]. This new kind of waveguides is based on the periodic dielectric structures as photonic crystals, which is separated by the high quality factor cavities that are coupled to each of the nearest neighbor in multiple spatial dimensions [15]. Due to existence of the cavities along the structure and as a result of the overlapping of the evanescent fields, light can be propagated through the CCWs [16]. Actually, it was investigated theoretically and experimentally that CCWs exhibit more advantages over conventional optical waveguides. One of the most important features of CCW is that due to strong optical confinement in defect medium, and high slope of transmission at resonance wavelengths, group velocity at the edge of each resonance modes can be reduced considerably [17]. Slow light generation in CCWs offers several practical advantageous like, design freedom, direct integration with other optoelectronics devices and ability to slow down light in a desired region of wavelengths at room temperature [18].

On the other hand, it should be noticed that the CCWs structures for stopping light suffer a fundamental trade off between the transmission and the optical delay bandwidth [19,20]. Therefore, in the field of slow light technology, delay-bandwidth product is an important parameter which should be considered greatly.

Furthermore, from system points of view, in future photonic circuits, adjustability of the optical properties of components is a great and important bottleneck which many scientists have been proposed special methods. In CCWs, the optical properties such as group velocity, dispersion and its higher order can be modulated through different mechanism such as electro optic effect, free-carrier injection and thermo optic effect [21]. As a consequence of necessity for all optical networks in future, dynamic all optical processing, controlling light by light, can be considered as one of the crucial bottlenecks for future all optical systems [22].

In the following chapter we will investigate the optical properties of the resonance modes such as, group delay, group velocity and bandwidth-delay product (BDP), in nonlinear one dimensional CCWs (1D-NCCWs) which defect mediums consist of intensity dependent-refractive index material. It can be seen by tuning the input optical intensity, the interested parameters such as, group velocity, delay time and specially delay-bandwidth product can be tuned to any desired values.

2. Theory

1D-NCCW is formed by placing optical resonator in a linear array, to guide light through whole of the structure by photon hopping between adjacent resonators. As a result of overlapping the evanescent field in the defect medium, cavity zone, electric field enhancement in this region can be obtained. Tight binding (TB) approximation (like using it in solid state physics) is used to describe the mechanism of waveguiding of the structure.

The basic structure for 1D-NCCW is as:

$$n_0 \left| (HL)^N HD(HL)^N HD(HL)^N H \right| Glass$$

Where, H and L denotes for high index (TiO$_2$) and low index (SiO$_2$) materials as n_H=2.33 and n_L=1.45, respectively for construction the basic resonators. Also, N is number of the repetition for the basic structure. It can be shown that by increasing N, light confinement in defect layer will be increased. Fig.1. Shows the basic structure of 1D-NCCW.

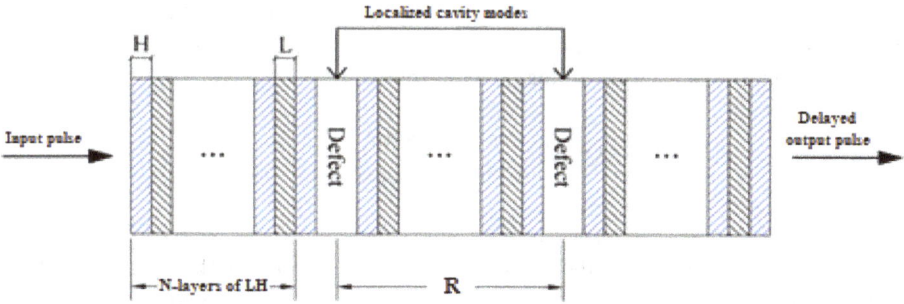

Fig. 1. Schematic illustration of 1D-NCCW with two defects.

As an example, defect layer, D is consist of CdSe which is an intensity dependent refractive index material.

The input pulse comes from left hand and during passing through whole of the 1D-NCCW, experiences delay time. This phenomenon is as a reason of reduction the speed of light propagation and confinement in the defect layer. It can be seen the enhancement of the electric field component of the incident light in the defect layer [21].

Nonlinear refraction is commonly defined either in terms of the optical field intensity I as [23]:

$$n_d = n_{d0} + \gamma I \tag{1}$$

Or in terms of average of the square of the optical electric field <E²> as:

$$n_d = n_{d0} + n_2 \langle E^2 \rangle \tag{2}$$

Where n_{d0} is the ordinary linear refractive index and γ is the nonlinear refractive index coefficient and n_2 is the nonlinear refractive index. The conversion between n_2 and γ can be written as [23]:

$$n_2 [cm^3 / erg] = (cn_0 / 40\pi)\ \gamma [m^2 / W] = 238.7\ n_0\ \gamma [cm^2 / W] \tag{3}$$

For CdSe γ=-147cm²/W and n_{d0}=2.56. By choosing $\lambda_0 = 1.55\mu m$ as the practical optical communication wavelength, and choosing quarter-wavelength optical thickness for high index and low index materials ($n_H d_H = n_L d_L = \dfrac{\lambda_0}{4}$, where $\lambda_0 = 1.55\mu m$), TiO2 and SiO2 respectively (for constructing the basic resonators), it can be derived the central frequency for the basic resonator as $\Omega = 1.21 \times 10^{15} Hz$.

According to TB approximation, in the presence of defect layer between each resonators with initial half-wavelength optical thickness ($n_{d0} d_d = \dfrac{\lambda_0}{2}$), the central frequency of each resonator, will be split to two eigen frequency due to coupling of the individual cavity modes [15]. In the case which I_{in}=0, the transmission properties of the 1D-NCCW has been investigated by transfer matrix method (TMM) which is widely used for calculating the optical properties of alternative stack layers. TMM is based on solving the Maxwell's equations in each individual layer and considering the continuity conditions for the electric and magnetic components of the incident electromagnetic field yields to obtain the characteristics matrix of each individual layer as following:

$$M_q = \begin{pmatrix} \cos\delta_q & i\sin\delta_q / n_q \\ i n_q \sin\delta_q & \cos\delta_q \end{pmatrix} \tag{4}$$

where n_q and δ_q denotes the refractive index and phase thickness ($\delta_q = 2\pi n_q d_q / \lambda_0$) of the qth layer respectively and λ_0 is the wavelength of the incoming light. By multiplication of the characteristics matrix of each layer, transfer matrix of the multilayered structure can be obtained. Therefore, by using the TMM method the electric and magnetic components of the input and output signal through the whole of structure can be obtained. Hence, optical properties such as transmission, phase characteristics and dispersion and its higher order such as group velocity and third order dispersion of the structure can be derived. Fig.2 shows the transmission spectrum of the structure in the case which $I_{input} = 0$.

For investigating the optical properties of the 1D-NCCW in the presence of input optical intensity signal, Eq.1 is used for determination of refractive index of the nonlinear defect layer. As an approximation method the transmission spectral characteristics of the 1D-NCCW can be obtained in the presence of nonlinear phenomena in the defect lyers. Painou *etal* and Johnson *etal*, have confirmed the convergence and correctness of this approximate approach [24]. Fig.3 shows the effect of increasing the input intensity on the position of the twin mini transmission frequencies band (resonance frequencies). It can be seen by increasing the input optical intensity, (form 0 to 6.4mW/cm²); the resonance modes shift toward right side in the frequency domain (blue shift in wavelength domain).

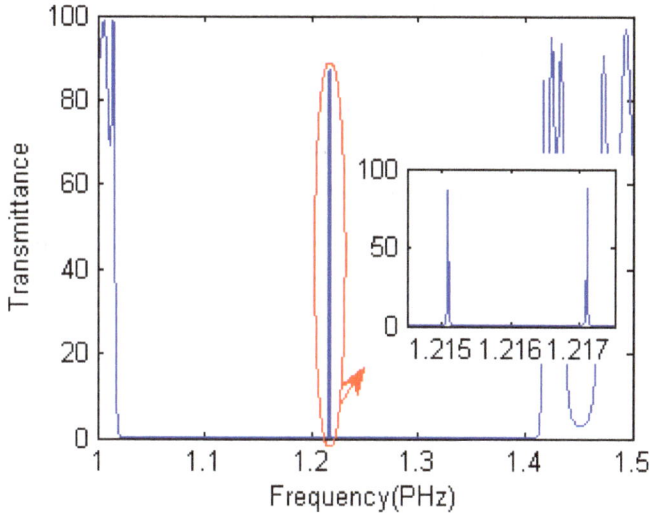

Fig. 2. Illustration of twin resonance bands in transmission spectrum of 1D-NCCW when $I_{input} = 0$.

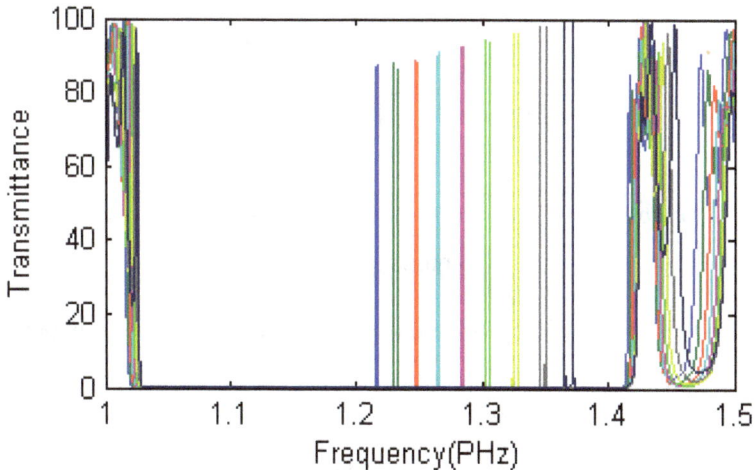

Fig. 3. Right side frequency shifting of the transmitted twin resonance modes when the input intensity changes from 0 to 6.4mW/cm² .

Under TB approximation, it can be shown that the group velocity at the resonance bands can be written as [25]:

$$V_g = -\kappa R\Omega \sin(kR) \tag{5}$$

where κ indicates coupling factor that is the value related to the overlapping of the electric field between the localized modes. $R = 4.8 \times 10^{-6}\,m$, is the separation between each of defect mediums and k is the wave vector of the light traveling in 1D-NCCW.

The coupling factor can be written as [25]:

$$\kappa = \frac{(\Delta\omega_{res})_{FWHM}}{2\Omega} \tag{6}$$

which indicates the inverse proportionality of the coupling factor to the quality factor (Q) of cavity modes ($\kappa \approx \dfrac{1}{2Q}$). Fig.4 indicates the variation of the Q for the resonance modes in the presence of input power.

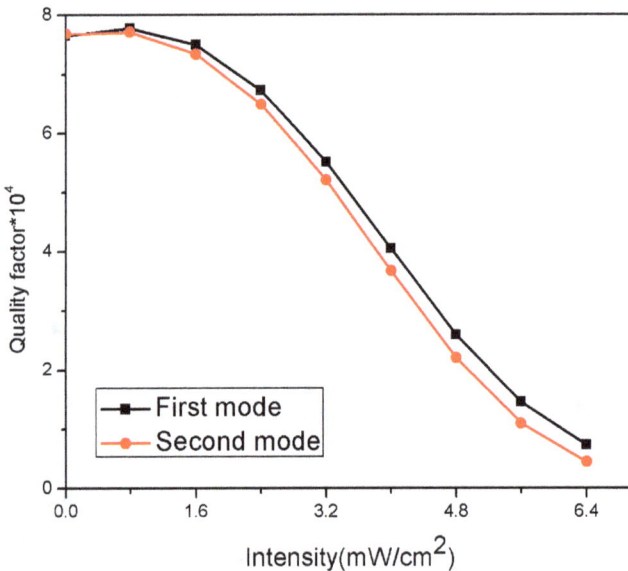

Fig. 4. variation the Q factor of the twin resonance modes versus optical input power.

As a result of increasing the input optical intensity, Q factor of each resonance modes of changes and hence, according to Eqs.5,6 coupling coefficient and group velocity at resonance modes alternate. Fig.5 shows the slow down factor, $S=V_g/C$, as a function of the input optical intensity, where C is the speed of light in vacuum.

It can be seen when the input optical intensity changes from 0 to 6.4mW/cm², the slow down factor can be tuned from 0.43×10^{-4} to 5.6×10^{-4} for the first resonance mode and 0.43×10^{-4} to 9.2×10^{-4} for the second mode. Group delay of resonance modes propagation through the 1D-NCCW, can be derived as following:

$$\tau_{group} = \frac{V_{group}}{L} \tag{7}$$

Where L is the length of the 1D-NCCW and equal to 1.42×10^{-5} m.

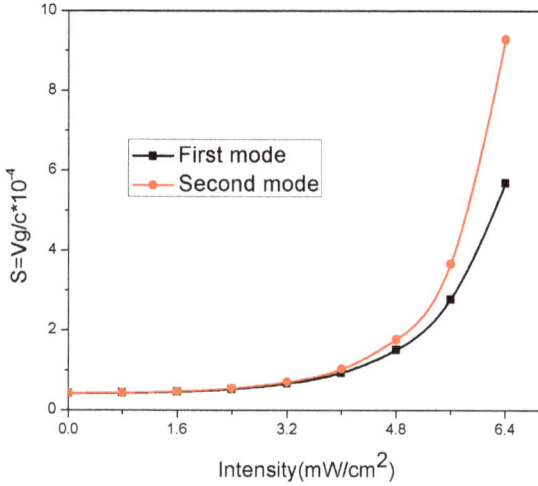

Fig. 5. Slow down factor increasing as a function of input intensity.

As a consequence of Eqs.5,7, the input intensity dependent of group delay of the resonance modes during propagation through the 1D-NCCW is obvious. Fig.6 shows group delay for the first and second resonance modes when the input intensity changes from 0 to 6,4 mW/cm². As an example, for the second mode, it can be seen by adjusting the input intensity between 0 to 6.4mW/cm², the group delay can be tuned from 1.1ns to 0.05ns.

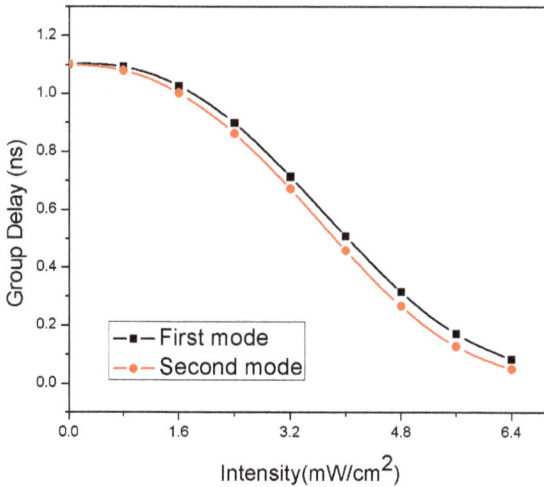

Fig. 6. Group delay versus of input intensity.

As mentioned before, photonic structures suffer from fundamental trade off between transmission bandwidth (FWHM of the transmitted resonance modes) and the optical delay, as delay bandwidth product (DBP).

The DBP variation in 1D-NCCW when the optical input intensity changes, is shown in Fig.7 As an example, for the second mode, it can be seen the DBP can be tuned from 17.4 to 15.6 by increasing the input optical intensity up to 6.4mW/cm².

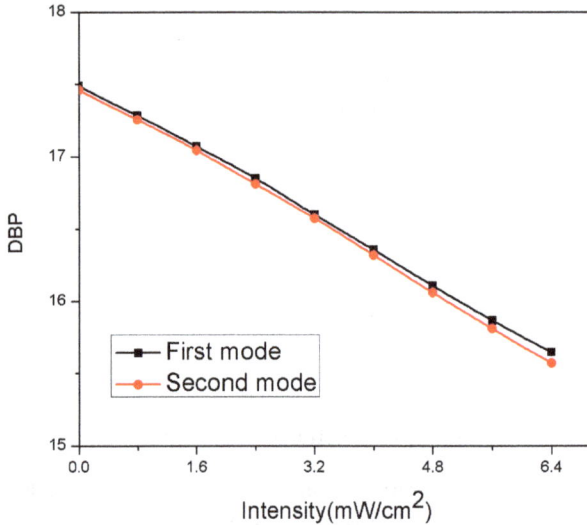

Fig. 7. DBP at the resonance modes versus input intensity.

3. Conclusion

The results of our theoretical study show that 1D-NCCWs can be considered as one of the best candidates for future all optical processing and high energy laser applications.

It can be seen that by introducing nonlinear defect medium in the one dimensional photonic crystals, three important tasks can be achieved. Firstly, in the presence of defect medium, a new mechanism for transferring optical data as coupled cavity waveguide can be achieved. It can be seen as a reason of intensity dependent of the refractive index in the defect medium, the resonance frequency (mini transmission band) can be tuned in the photonic bandgap zone of the basic one dimensional photonic crystal.

Secondly, as a reason of high slope of transmission at resonance bands, slow light in the order of 10^{-4} and group delay in nanosecond regime can be generated. It can be seen, as a reason of increasing the optical input intensity, not only the magnitude of the speed of light propagation and hence its group delay can be tuned, but also the propagation wavelength can be adjusted in wide range of photonic bandgap zone.

Thirdly, from practical system point's of view, as a consequence of importance of DBP parameter, and necessity to tune it for different users with various technical requirements, the DBP parameter can be dynamically tuned by adjusting the optical input intensity.

4. References

[1]   E. Perra and J.R. Lowell, Toward Applications of slow light technology, Opt. Photonics news, 18 (2007) 40-45.

[2]   L. Daniault, M. Hanna, L. Lombard, Y. Zaouter, E. Mottay, D. Goular, P. Bourdon, F. Druon and P. Georges, Coherent beam combining of two femtosecond fiber chirped-pulse amplifiers, Optics Letters,36 (2011) 621-623.

[3]   Slow light: Science and applications, J. B. Khurgin and R. S. Tucker (Ed.), CRC Press, 2009.

[4]   Y. Zhao, H.W. Zhao, X.Y. Zhang, B. Yuan and S. Zhang, New mechanism of slow light and their applications, Optics & Laser Technology, 41(2009) 517-525.

[5]   M. Fleischhauer, A. Imamoglu, J.P. Marangos, Electromagnetically induced transparency: Optics in coherent media, Reviews of Modern Physics, 77 (2005) 633.

[6]   A. Kasapi, M. Jain, G.Y. Yin, S.E. Harris, Electromagnetically Induced Transparency: Propagation Dynamics, Physical Review Letters, 74 (1995) 2447.

[7]   M. Xiao, Y.-q. Li, S.-z. Jin, J. Gea-Banacloche, Measurement of Dispersive Properties of Electromagnetically Induced Transparency in Rubidium Atoms, Physical Review Letters, 74 (1995) 666.

[8]   O. Schmidt, R. Wynands, Z. Hussein, D. Meschede, Steep dispersion and group velocity below c/3000 in coherent population trapping, Physical Review A, 53 (1996) R27.

[9]   J.P. Marangos, Topical review Electromagnetically induced transparency, journal of modern optics, 45 (1998) 471±503.

[10] L.V. Hau, S.E. Harris, Z. Dutton, C.H. Behroozi, Light speed reduction to 17 metres per second in an ultracold atomic gas, Nature, 397 (1999) 594-598.

[11] M. bigelow, Ph.D Thesis, Ultra-slow and superluminal light Propagation in Solids at Room Temperature, The Institute of Optics The College School of Engineering and Applied Sciences University of Rochester, New York, 2004.

[12] M.S. Bigelow, N.N. Lepeshkin and R.W. Boyd, Observation of ultra slow light propagation in a ruby crystal at room temperature, Physical review letters, 90 (2003) 113903.

[13] A. Yariv, Y. Xu, R.K. Lee, A. Scherer, Coupled-resonator optical waveguide: a proposal and analysis, Optics Letter, 24 (1999) 711-713.

[14] J.K.S. Poon, J. Scheuer, Y. Xu, A. Yariv, Designing coupled-resonator optical waveguide delay lines, J. Opt. Soc. Am. B, 21 (2004) 1665-1673.

[15] S.M. Hamidi, A. Bananej, M.M. Tehranchi, Adjustable dispersion compensating in one dimensional coupled resonator planar optical waveguides, Optics Communications, 281 (2008) 4917-4925

[16] G. Ma, J. Shen, K. Rajiv, S. Tang, Z. Zhang, Z. Hua, Optimization of two-photon absorption enhancement in one-dimensional photonic crystals with defect states, Applied Physics B: Lasers and Optics, 80 (2005) 359-363.

[17] T. Krauss, Slow light in photonic crystal waveguides, Journal of Physics D: Applied Physics, 40 (2007) 2666.

[18] Q. Chen, D.W.E. Allsopp, Group velocity delay in coupled-cavity waveguides based on ultrahigh-Q cavities with Bragg reflectors, Journal of Optics A: Pure and Applied Optics, 11 (2009) 054010.

[19] T. Baba, T. Kawaaski, H. Sasaki, J. Adachi, D. Mori, Large delay-bandwidth product and tuning of slow light pulse in photonic crystal coupled waveguide, Opt. Express, 16 (2008) 9245-9253.

[20] Q. Xu, P. Dong, M. Lipson, Breaking the delay-bandwidth limit in a photonic structure, Nat Phys, 3 (2007) 406-410

[21] S.M. Hamid, M.M. Tehranchi, A. Bananej, Adjustability of Optical Properties in Complex One-Dimensional Photonic Crystals in: W.L. Dahl (Ed.) Photonic Crystals: Optical Properties, Fabrication and Applications, Nova Publishers, N.Y. New york, 2010, pp. 53-88.

[22] N.C. Panoiu, M. Bahl and R. M. Osgood, All-optical tunability of a nonlinear photonic crystal channel drop filter, Optics Express,12 (2004) 1605-1610.

[23] M. J.Weber, Handbook of optical materials, CRC Press, 2003.

[24] X. Hu, P. Jiang and Q. Gong, Tunable multichannel filter in one-dimensional nonlinear ferroelectric photonic crystals, Journal of Optics A: Pure and Applied Optics, 9 (2007) 108-113.

[25] K. Hosomi, T. Katsuyama, A dispersion compensator using coupled defects in a photonic crystal, IEEE journal of Quantum Electronics, 38 (2002) 825 - 829.

5

Mid-Infrared Surface-Emitting Two Dimensional Photonic Crystal Semiconductor Lasers

Binbin Weng and Zhisheng Shi
School of Electrical and Computer Engineering
University of Oklahoma, Norman Oklahoma
USA

1. Introduction

The mid-infrared (mid-IR) region, covering the electromagnetic spectrum from about 2.5 to 25 μm wavelength, is of unique interest for many applications, especially for molecular spectroscopy, because the vibrational frequencies of almost all target molecules and hydrocarbons are in this spectral region, such as NO, NO_2, CO, CO_2, HF, and CH_4. The specific application areas for molecular spectroscopy include scientific research, vehicle exhaust investigation, atmospheric pollution monitoring, medical diagnostics and biological & chemical weapon detection. Besides molecular spectroscopy applications, there are also some other potential applications, such as free space optical communications, etc. (Tittel, Richter, & Fried, 2003)

For all these applications, typical requirements for mid-IR lasers are manifold. First of all, to reach high spectral selectivity performance, a narrow spectral linewidth is needed. Secondly, optical output power should be larger than 100 μW to lift detector noise limits. (Tacke M. , 2001) Thirdly, high beam quality is favorable for optimum coupling with the gas sampling cell. Also, room temperature or thermoelectric-cooled operation is desirable, considering the additional weight, volumes and costs with the additional cryogenic cooling equipment. Furthermore, a compact set-up is preferable for hand-held *in situ* measurement in the field.

Two dimensional (2D) photonic crystal lasers, potentially have great practical applications in the mid-IR region due to their unique features in, for example, surface emission, circular beams, low threshold operation, miniaturization and simplicity of on-chip monolithic integration, etc. The studies of surface emitting 2D photonic crystal lasers started in and have expanded worldwide since the middle of the 1990s. The first 2D photonic crystal semiconductor laser of 1.55 μm emission peak was demonstrated in 1999. (Painter, et al., 1999) After that, this type of semiconductor lasers has been developed rapidly in near IR region. (Park, Hwang, Huh, Ryu, & Lee, 2001; Loncar, Yoshie, Scherer, Gogna, & Qiu, 2002; Altug, Englund, & Vuckovic, 2006) Unfortunately, mainly restricted by the etch-induced surface recombination, the research development of 2D photonic crystal semiconductor lasers in mid-IR range of electromagnetic spectrum was relatively slow. Even so, due to their intrinsic advantages addressed previously, this type of lasers operating in mid-IR range is still very desirable. Therefore, some major steps of the progress are very important and need to be paid close attention nowadays.

In this chapter, we will review the development of the mid-IR surface emitting 2D photonic crystal semiconductor lasers in recent decade. The performance of the reported mid-IR 2D photonic crystal lasers with different structure design and different material system is going to be introduced and compared. Their major advantages and disadvantages will also be discussed.

2. Mid-IR surface emitting 2D photonic crystal lasers

The researches on the mid-IR semiconductor sources and structures for optoelectronics devices applications have been developed for several decades. Recently, the most competitive mid-IR semiconductor lasers are realized with GaAs- and GaSb-based III-V material system making use of their unique intersubband (Bauer, et al., 2011; Yu, Darvish, Evans, Nguyen, Slivken, & Razeghi, 2006; Slivken, Evans, Zhang, & Razeghi, 2007) and inter-band (Yang R. Q., 1995; Yang, Bradshaw, Bruno, Pham, Wortman, & Tober, 2002) cascade transition mechanisms. Meanwhile, thanks to their suppressed Auger non-radiative loss, (Zhao, Wu, Majumdar, & Shi, 2003) IV-VI lead-salt materials have also been an excellent choice of mid-IR lasers for gas sensing application, and will continue to be so in the future. (Springholz, Schwarzl, & Heiss, 2006) In the following, we will introduce the development of mid-IR surface emitting photonic crystal lasers using all mentioned intersubband cascade, type-II "W" active region III-V semiconductors, and lead-salt materials.

2.1 Quantum cascade surface-emitting photonic crystal laser

Quantum cascade laser (QCL) is a type of mid-IR laser based on intersubband transitions in multiple quantum wells for photon generation as shown in figure 1. As a result, the quantum efficiencies would be greater than 100%. On the other hand, the emission wavelength of QCL is able to be tuned from 4 to 24 μm in the mid-IR range. (Yu, Darvish, Evans, Nguyen, Slivken, & Razeghi, 2006; Diehl, et al., 2006) However, most of the QCL works have been done were focused on edge-emitting lasers, and limited by its intrinsic transverse-magnetic polarization of the intersubband transition.

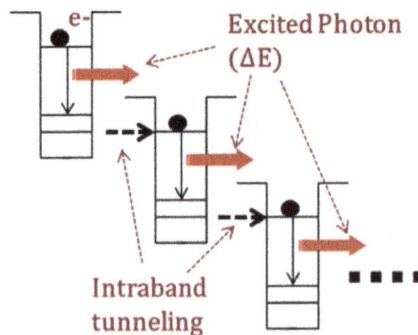

Fig. 1. The schematic of physics mechanism for quantum cascade laser (QCL)

Fortunately, 2D photonic crystal technology allows quantum cascade lasers to achieve surface emission. The first mid-IR surface emitting 2D photonic crystal QCL was realized in 2003. (Colombelli F. , et al., 2003) In this demonstration, the 2D arrays were fabricated with the values of lattice spacing a from 2.69 to 3.00 µm, and the ratio of holes radius r to a from 0.28 to 0.32. As presented in figure 2B and figure 2C, the etched photonic crystal pattern penetrated through the active region down into the bottom substrate, which reduces the diffraction of radiation into the substrate and consequently provides strong in-plane optical feedback. Usually, in diode lasers, deep-etching induced surface recombination would substantially increase the leakage currents. (Colombelli F. , et al., 2003) But this serious obstacle does not affect in this case, because QCL are unipolar devices.

After processing, electroluminescence (EL) measurements were conducted at a temperature of 10 K and the light spectra were acquired using a Nicolet Fourier transform infrared (FTIR) spectrometer and a nitrogen-cooled HgCdTe detector. From the measurement, for the designed structure with a set of values of a and r having the A, B, and C flat-band regions overlapping the gain spectrum, three line-narrowing emission peaks emerge from the EL spectrum, as shown in Figure 3A.

It is worth to mention that, no defect mode emission has been demonstrated, because of the little difference observed between the emission spectra with and without a central defect design. (Colombelli F. , et al., 2003) Therefore, the device should be classified as the band-edge mode photonic crystal lasers. Simply explaining, the optical gain is intensified by reducing the group velocity of light, supported by a basic relation about light amplification in which (H.C. Casey & Panish, 1978)

$$\gamma = gNv_g \, ,$$

where γ is the net stimulated-emission rate, g is the optical gain, N is the photon density, and v_g is the group velocity of light. It has been proven that the gain enhancement is strongly correlated with the group velocity, which decreases rapidly in the vicinity of the photonic band edges. (Nojima, 1998)

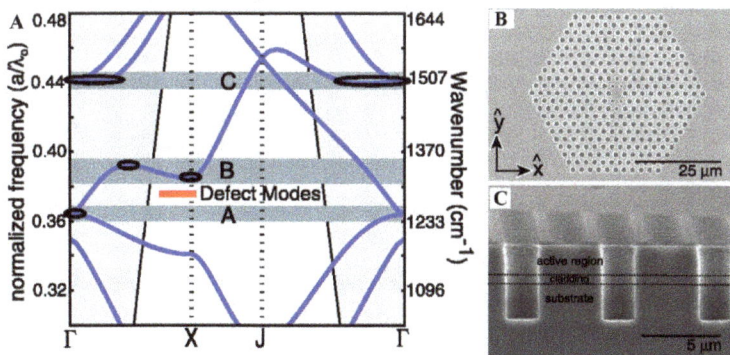

Fig. 2. (A) Theoretical simulation of 2D in-plane transverse magnetic (TM) photonic band structure. The flat-band region (A, B, and C) are indicated by dark grey horizontal bands. The red thick line indicates the highly localized defect modes. (B) Top-view scanning electron microscope (SEM) image of an etched device with a central defect. (C) Cross-section SEM of a portion of a photonic crystal QC device. (Colombelli F. , et al., 2003)

This work also characterized the threshold behavior of the device in figure 4. The light-output versus current characteristic of the device (red line) shows that the device fabrication was successful. However, the current voltage (I-V) relationship (blue line) presents the somewhat soft turn-on effect in the figure. The authors believe that a poor current confinement resulting from the absence of a mesa etched around the device should be the major reason. (Colombelli F. , et al., 2003) In addition, the estimation of its current density at threshold is 9 to 10 kA/cm² which is higher than the conventional edge-emitting QCL.

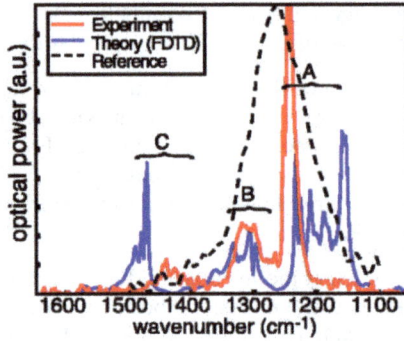

Fig. 3. Electroluminescence and FDTD-simulated spectrum for a device with a = 2.92 µm and r/a = 0.30. Dashed line indicates the reference spectrum collected from non-resonance area of the same device. (Colombelli F. , et al., 2003)

Nevertheless, the electrically pumped demonstration of surface-emitting photonic crystal QCL is an important step in the development of the practical photonic crystal microcavity lasers. It opens up the new horizons in device design and application in the mid-IR ranges using the combination of electronic and photonic bandgap engineering in semiconductor material. But this proof-of-concept demonstration also leaves some significant engineering issues, such as the high laser threshold (9-10 kA/cm²), and extremely low operation temperature (10K), for further research exploration. (Bahriz, Moreau, & Colombelli, 2007; Xu, et al., 2010)

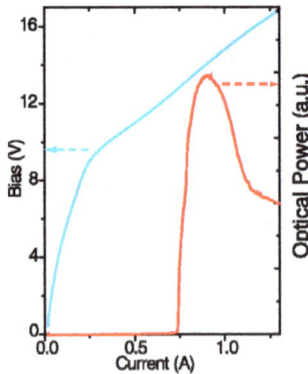

Fig. 4. Voltage versus current (V-I) and light-output versus injected current characteristics of a typical device at 10 K, the peak power with a central defect is from 400 µW to 1 mW, while the ones without defect have less power output. (Colombelli F. , et al., 2003)

2.2 Surface-emitting photonic crystal distributed-feedback (DFB) laser based on III-V type-II "W" active regions

In contrast to the mid-IR photonic crystal QCL lasers discussed previously, reported type-II "W" shape III-V surface-emitting photonic crystal DFB laser has a relatively shallow grating photonic crystal structure etched into the laser waveguide, as shown in figure 5A. In 2003, I. Vurgaftman and J. R. Meyer used a theory based on a time-domain Fourier-Galerkin (TDFG) numerical solution for simulating the surface-emitting photonic crystal DFB operation. They found that a "weak" 2D photonic crystal grating with only refractive-index modulation will yield high-efficiency surface emitting photonic crystal DFB emission into a second order single in-phase symmetric mode (Figure 5B), which is coherent over a wide device area. (Vurgaftman & Meyer, 2003)

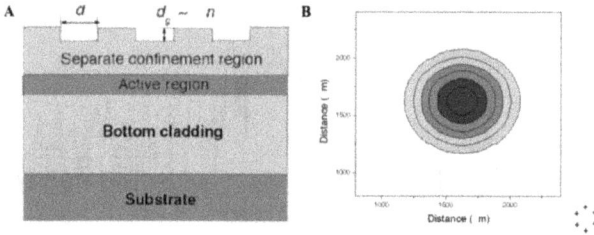

Fig. 5. (A) Cross section of surface emitting photonic crystal distributed feedback (DFB) laser; (B) simulated near-field profile for the single-lobed in-phase mode of hexagonal-lattice with transverse electric (TE) polarization. (Vurgaftman & Meyer, 2003)

Following the theoretical studies guideline, the experimental demonstration of surface emitting photonic crystal DFB lasers with type-II "W" active regions emitting in the mid-IR range was reported by the same group in 2006. (Kim, et al., 2006) The epitaxial structure was grown on a (100) GaSb substrate using molecular beam epitaxy (MBE) method, consisting of a 3.5 μm thick AlAsSb bottom optical cladding layer, an active region of 10 pairs of InAs/GaSb/AlSb quantum wells, and a 500 nm thick GaSb separate-confinement layer where the grating was applied. After growth, two triangular lattice grating patterns with etch depths of 27 and 60 nm were fabricated by using reactive ion etching and chemical wet etching, respectively. Figure 6 shows one titled view SEM image of a 60nm deep photonic crystal DFB grating pattern processed after wet etching.

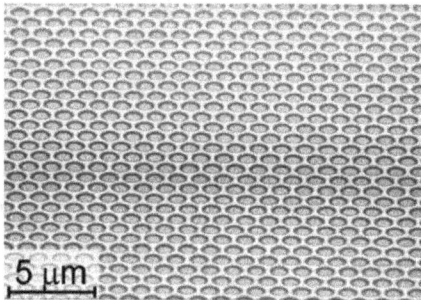

Fig. 6. Tilted view SEM image of the surface emitting photonic crystal DFB grating pattern with 60 nm etching depth. (Kim, et al., 2006)

A high-resolution (~0.05 nm) FTIR emission spectrum for the 27 nm deep surface emitting photonic crystal DFB lasers is demonstrated in Figure 7, with the cw pumping power ~ 6 W, and the pumping spot size ~ 1.4 mm. The emission peak with full width of half maximum (FWHM) 0.13 nm is observed. The confirmed side-mode suppression ratio of >21dB was limited by the apodization artifact in the instrument. (Kim, et al., 2006)

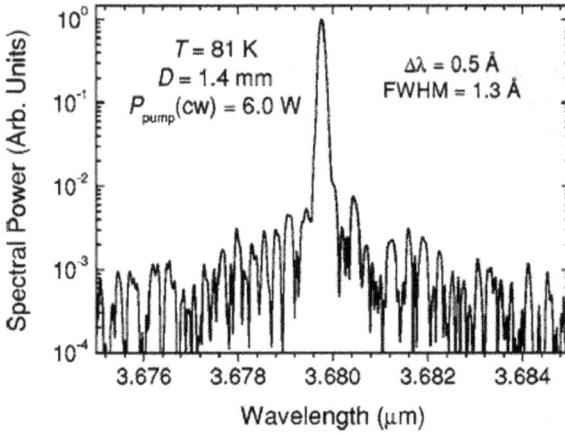

Fig. 7. FTIR spectrum for the surface emitting photonic crystal DFB laser with 27 nm etching depth at 81 K. (Kim, et al., 2006)

Figure 8A shows the temperature dependent pulsed and cw light-light (L-L) characteristics for the photonic crystal DFB lasers with 27 nm etching depth. The slow increased threshold between 100 K and 140 K in contrast to the faster variation from 140 K to 180 K, is due to the peak match between material gain and grating resonance, as presented in figure 8B. As indicated in Fig. 8A, the highest working temperature of the measured sample is 180 K under pulse mode and 81 K under cw mode. (Kim, et al., 2006) Although the authors claim that 180 K is below the maximum lasing temperature, there is no higher temperature value indicated in that letter. The maximum quantum efficiency calculated is only 0.6%. They point out that these weak efficiencies are caused by the low pumping energy absorption in the active region. (Kim, et al., 2006) Besides on this reason, we believe that the weak lateral optical confinement due to the shallow etching grating could somehow enhance the optical leakages in the device. Therefore, the energy conversion rate would be affected.

Although the simulation theory for the surface emitting photonic crystal DFB lasers described above is totally different from the one discussed in the surface emitting photonic crystal QCL section, the physical mechanisms of these two devices are basically quite the same. Both of them use the coupled lasing modes resulting from the 2D photonic crystal distributed feedback mechanism. Therefore, these lasers have the unique coherent lasing mechanism over the wide 2D photonic crystal areas. However, due to the wide area cavity, the threshold current or optical pumping power would be relatively large. For a practical application, this issue should be improved.

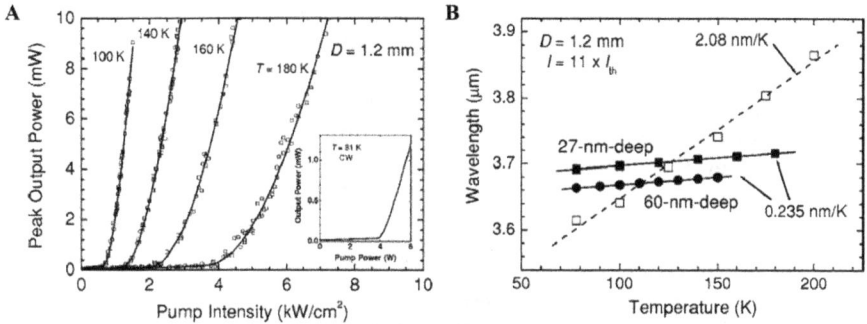

Fig. 8. (A) Temperature dependent pulsed light-light curves for the surface emitting photonic crystal DFB laser with 27 nm etching depth. Inset is the cw light-light curve at 81 K. (B) Peak emission position variation under the different temperature for the regular FP laser (open square), photonic crystal DFB lasers with 27 nm etching depth (solid square), and 60 nm etching depth (solid circle). (Kim, et al., 2006)

2.3 Lead-salt based surface-emitting photonic crystal laser on Si

Lead salt or IV-VI semiconductor system has been used as a major material source for mid-IR diode lasers for a long time, because of its energy bandgap covering the wavelength region from 3 to 30 μm (Preier, 1990; Ishida & Fujiasyu, 2003; Tacke M. , 2000) and also its near two orders of magnitude lower non-radiative Auger recombination rates as compared to other conventional narrow band gap semiconductors (Findlay, et al., 1998). In this section, the realization of surface emitting photonic crystal lasers based on the lead salt compounds is described.

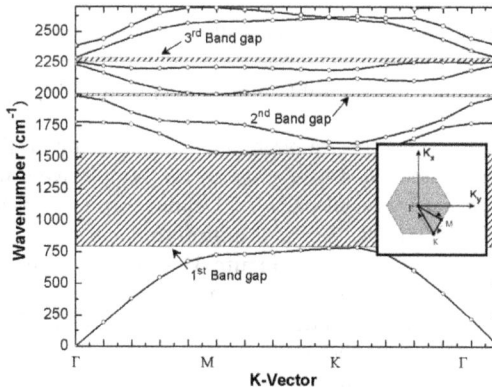

Fig. 9. Dispersion diagram of 2D transverse magnetic (TM) modes of the hexagonal photonic crystal holes array; the first Brillouin zone in reciprocal space is displayed in the inset. (Weng, Ma, Wei, Xu, Bi, & Shi, Mid-infrared surface-emitting photonic crystal microcavity light emitter on silicon, 2010)

In order to achieve the strong in-plane optical feedback, and also avoid etching of epitaxial films which alleviates the problem of surface recombination, the 2D photonic crystal honeycomb structure was fabricated onto the silicon substrate prior to the following IV-VI epitaxial layers growth. Using this modified experimental proposal, B. Weng *et al.* reported a PbSe/PbSrSe multiple quantum wells (MQW) mid-IR photonic crystal coupled light emitter in 2010. (Weng, Ma, Wei, Xu, Bi, & Shi, Mid-infrared surface-emitting photonic crystal microcavity light emitter on silicon, 2010) For the initial proof-of-concept demonstration, they intentionally increased the designed values of interhole spacing and holes radius to micrometer range as to allow high tolerance for possible errors in etching and lateral growth. Therefore, the second and third order photonic bandgaps are moved into the mid-IR spectral region of interest, as presented in figure 9.

In the experiment, Si substrate was patterned by a hexagonal photonic crystal array using electron-beam (E-Beam) lithography. Defect cavity was created in the center of the pattern. RIE system was used to fabricate the photonic crystal patterned substrate with 3 μm etching depth. After that, the epitaxial layer consisted of a seven-pair PbSe/PbSrSe MQW active region and 1μm optical confinement BaF$_2$ top layer. (Weng, Ma, Wei, Xu, Bi, & Shi, Mid-infrared surface-emitting photonic crystal microcavity light emitter on silicon, 2010) Figure 10 shows top view SEM images of both patterned Si substrate after RIE etching process and after lead salt epitaxial growth.

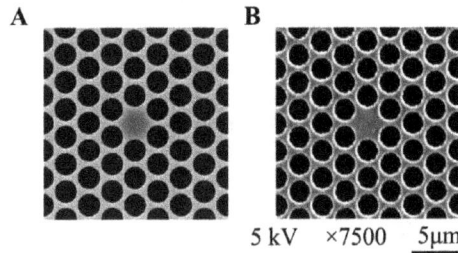

5 kV   ×7500   5μm

Fig. 10. Top view SEM images of (A) patterned Si(111) substrate; (B) PbSe/PbSrSe MQW structure grown on patterned Si(111). (Weng, Ma, Wei, Xu, Bi, & Shi, Mid-infrared surface-emitting photonic crystal microcavity light emitter on silicon, 2010)

The optical pumping emission spectra were characterized by a Bruker IFS 66/S FTIR spectrometer in step-scan mode. Measurements on the sample were taken at the areas both inside and outside the photonic crystal patterns. Detail setup information is discussed in the reference (Weng, Ma, Wei, Xu, Bi, & Shi, Mid-infrared surface-emitting photonic crystal microcavity light emitter on silicon, 2010) Figure 11A presents the temperature dependent photoluminescence spectra of the un-patterned planar area on the sample. As it indicates, the emission peak shifting agrees well with the temperature dependent IV-VI semiconductor QW gain spectra. However, in figure 3B, there are two temperature-independent emission peaks at around 1960 and 2300 cm^{-1}. The emission spectra were taken from the epitaxial layer grown on the 2D photonic crystal patterned Si substrate. Comparing with the theoretical simulation, these two peak positions align well with the calculated 2nd and 3rd band gaps. As we can see, there is only one emission mode emerging in each spectrum at a given temperature. The 2nd order photonic crystal coupled emission mode at 1960 cm^{-1} dominates the range from 77 ~ 160 K which the 3rd one emerges from 180 ~ 270 K. This phenomenon is due to the shift of the IV-VI MQW gain spectra as shown in figure 11A. (Weng, Ma, Wei, Xu, Bi, & Shi, Mid-infrared surface-emitting photonic crystal microcavity light emitter on silicon, 2010)

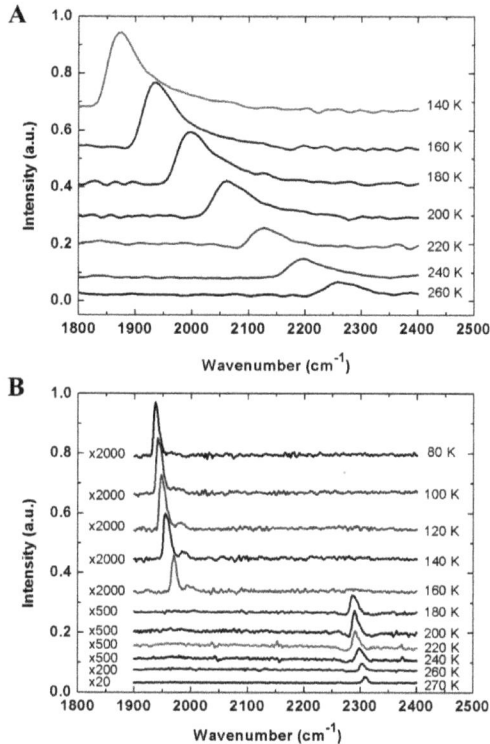

Fig. 11. Temperature-dependent light emission of PbSe/PbSrSe MQW (A) on un-patterned area and (B) patterned area on Si (111). (Weng, Ma, Wei, Xu, Bi, & Shi, Mid-infrared surface-emitting photonic crystal microcavity light emitter on silicon, 2010)

Meanwhile, the output emission power was calibrated by a standard blackbody reference source. The maximum emission peak-power is around 3.99 W at 100 K, under 1kW pumping peak-power excitation. Therefore, the quantum efficiency is 12.8%. As can be seen from figure 12, the collected power from photonic patterned area is over two orders of magnitudes higher than that from planar area. The authors contribute such significant enhancement to several factors: (Weng, Ma, Wei, Xu, Bi, & Shi, Mid-infrared surface-emitting photonic crystal microcavity light emitter on silicon, 2010) First of all, the 2D photonic crystal structure forces all the photons generated to funnel through its given modes and be extracted, where the extraction efficiency of PL emission from the planar epilayer is low under the total reflection limitation. Secondly, the narrow beam divergence of the emission also increases the collected efficiency. Thirdly, great improvement in material quality due to the growth on patterned substrate could significantly reduce the material losses. (Weng, Zhao, Ma, Yu, Xu, & Shi, 2010)

After the initial demonstration of their surface emitting 2D photonic crystal lead-salt light emitter, a new room temperature mid-IR surface emitting 2D photonic crystal laser on Si was successfully realized by B. Weng et al., (Weng, et al., Room temperature mid-infrared surface-emitting photonic crystal laser on silicon, 2011) recently. Up to date, this is the first mid-IR surface emitting photonic crystal semiconductor laser realized at room temperature.

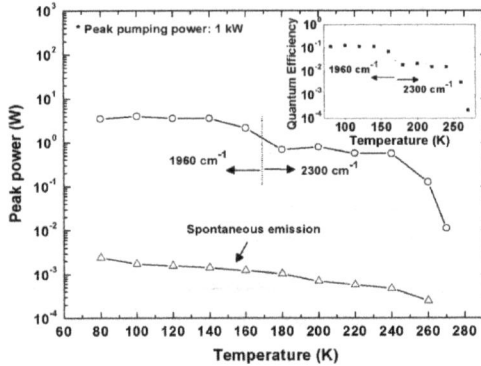

Fig. 12. Calibrated temperature dependent output emission power from PbSe/PbSrSe MQW structure on patterned and un-patterned areas. Quantum efficiency of the emission is show in the inset. (Weng, Ma, Wei, Xu, Bi, & Shi, Mid-infrared surface-emitting photonic crystal microcavity light emitter on silicon, 2010)

As described previously in this section, for the easy proof-of-concept demonstration, the bandgaps for mode emissions aligning with the gain peak of a PbSe/PbSrSe QW were at cryogenic temperatures. Therefore, in order to obtain room temperature photonic crystal modulated mode emission, a modified photonic crystal dispersion diagram was designed which matches the room temperature gain peak of a designed PbSe/PbSrSe QW, presented in figure 13.

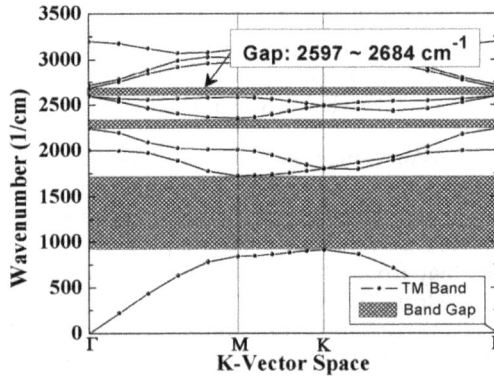

Fig. 13. Dispersion diagram of 2D TM modes of the designed photonic crystal array. (Weng, et al., Room temperature mid-infrared surface-emitting photonic crystal laser on silicon, 2011)

The photonic crystal structure fabrication on Si substrate and also the epitaxial layer growth procedures by MBE system both followed the methods reported in reference (Weng, et al., Room temperature mid-infrared surface-emitting photonic crystal laser on silicon, 2011). Figure 14A shows a top-view SEM image of patterned area with the defect on Si substrate after MBE growth. Later on, without any further processing, the emission spectra of the as grown sample were measured by the same Bruker FTIR system, and the signal was detected by a liquid N_2 cooled InSb detector (Judson J-10D). Within the temperature range of 270 ~ 310 K, there is only one temperature-independent emission peak observed. However, under the same pumping power, the emission spectrum at 310 K is below the threshold and exhibits a much broader emission linewidth. The phenomenon of the fixed peak position under different temperatures is similar to the reported result as shown in figure 11B. (Weng, Ma, Wei, Xu, Bi, & Shi, Mid-infrared surface-emitting photonic crystal microcavity light emitter on silicon, 2010)

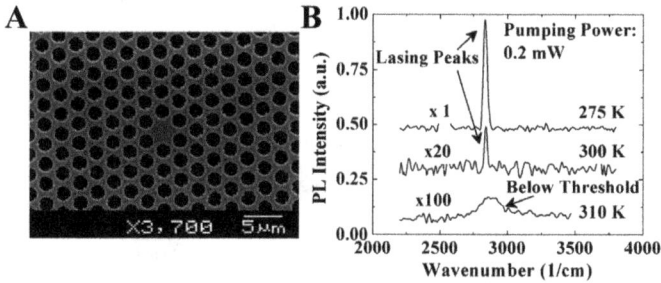

Fig. 14. (A) Top-view SEM image of PbSe/PbSrSe MQWs structure grown on patterned Si(111); (B) Temperature dependent light emission of the same sample. (Weng, et al., Room temperature mid-infrared surface-emitting photonic crystal laser on silicon, 2011)

Figure 15 shows the surface emitting 2D photonic crystal light emission spectrum at 300 K above threshold. As figure 15A indicates, the lasing peak doesn't perfectly match the designed emission wavelength (~3.85 μm) of the QW gain peak position on un-patterned area of the same sample. Comparing with the designed photonic bandgap matching with 3.85μm, the after-processed sample has the lasing peak blue shifted to 3.52 μm. After checked with the dimension the etched pattern, it is found that, the after-etched radius is larger than the designed 1.1 μm, which is 1.14 μm. Simulation using the actual radius parameter shows a higher photonic bandgap which matches the experimental result. As indicated in figure 15B, lasing emission from the photonic crystal structure has multi-mode characteristics. Although the individual peaks could not totally be resolved experimentally due to the restriction by the instrument limitation, using Gaussian fitting function, there are four stimulated peaks resolved mathematically from the spectrum obtained by FTIR spectrometer. Their full width of half maximum (FWHM) values vary from 2.45 nm to 6.02 nm, as presented. According to their design, these peaks are the photonic crystal coupled microcavity modes generated by the defect structure in the center of the array shown in figure 14A. (Weng, et al., Room temperature mid-infrared surface-emitting photonic crystal laser on silicon, 2011) Meanwhile, the authors point out some other mechanisms might contributing the multi-mode emissions, (Weng, et al., Room temperature mid-infrared surface-emitting photonic crystal laser on silicon, 2011) such as degeneracy splitting of dipole mode due to the structure imperfection (Painter, Vuckovic, & Scherer, Defect modes of a two-dimensional photonic crystal in an optically thin dielectric slab, 1999), photonic band-edge mode emission (Colombelli R. , et al., 2004), etc.

The optical output power versus input pumping power density curve of the photonic crystal surface emitting laser was characterized and displayed in figure 16. A clear threshold behavior can be observed. As it indicates, the threshold pumping power density is approximately 24 kW/cm² at 300 K. The relatively low threshold is mainly due to the reasons: (a). IV-VI epitaxial layer grown on holes-patterned Si substrate has better material quality than planar grown structure, therefore crystal defects and dislocations have less impact on the photonic crystal device which ultimately lowered non-radiative recombination rate; (b). with the help of 2D photonic crystal structure, group-velocity anomaly phenomenon could also have played an important role in a large reduction of the threshold.

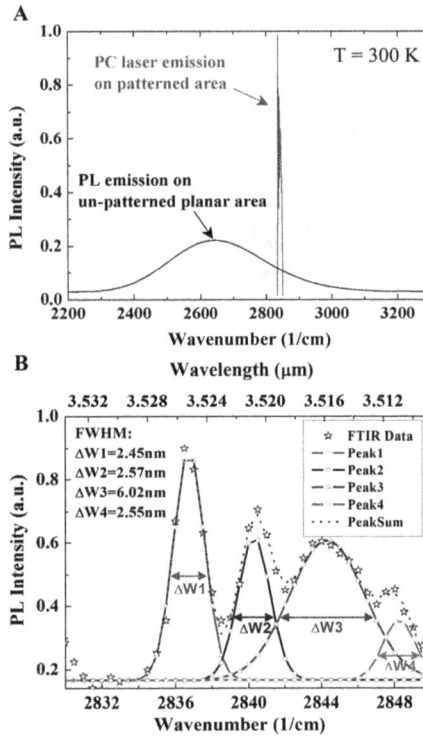

Fig. 15. (A) Comparison of photonic crystal emission spectrum above threshold and the light emission spectrum on un-patterned planar area at room temperature; (B) Multi-mode laser spectrum and linewidths determined by Gaussian function fitting. (Weng, et al., Room temperature mid-infrared surface-emitting photonic crystal laser on silicon, 2011)

As shown in figure 15, the peak emission powers were calibrated using a standard blackbody source. For the output power of 2.74 mW, the quantum efficiency calculated is about 1%. (Weng, et al., Room temperature mid-infrared surface-emitting photonic crystal laser on silicon, 2011) This low efficiency is mainly caused by the peak positions mismatch between photonic crystal coupled mode and the QW optical gain, as indicated in figure 15A. It is also pointed out that with the design modification to realign photonic crystal coupled mode and QW gain peak spectrum, once the peaks match up, significant efficiency increase would be realized.

3. Conclusions

In conclusion, surface emitting 2D photonic crystal lasers offer attractive properties as the mid-IR solid state laser sources. In particular, this group of lasers features intrinsic circular beam emission with small beam divergence, low threshold, single mode operation etc. Since the first

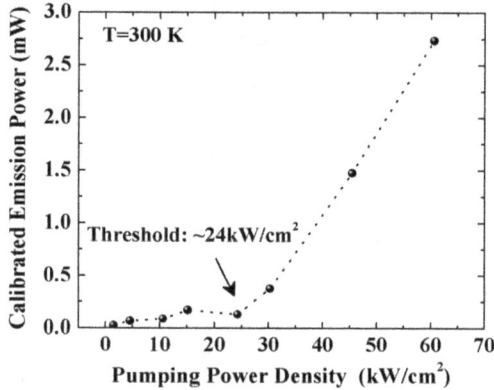

Fig. 16. Calibrated surface emitting photonic crystal laser peak output power as a function of the peak optical pumping power density at room temperature. (Weng, et al., Room temperature mid-infrared surface-emitting photonic crystal laser on silicon, 2011)

demonstration of the surface emitting 2D photonic crystal laser at 10K in mid-IR range of electromagnetic field using intersubband cascade structure based on III-V material system, several attempts have been made in order to bring this type of devices up to the practical application level in the real world. Up to now, room temperature operation of the surface emitting 2D photonic crystal laser is only achieved by using IV-VI lead salt material system under pulsed optical excitation. But with further technological design and fabrication efforts, room temperature electrically pumped CW lasers should be feasible in near future, and this will definitely explore many promising applications in a wide variety of fields.

4. References

Altug, H., Englund, D., & Vuckovic, J. (2006). Ultrafast photonic crystal nanocavity laser . *Nature Physics, 2*, 484.

Bahriz, M., Moreau, V., & Colombelli, R. (2007). Design of mid-IR and Thz quantum cascade laser cavities with complete TM photonic bandgap. *Optics Express, 15*, 5948.

Bauer, A., Roβner, K., Lehnhardt, T., Kamp, M., Hofling, S., Worschech, L., et al. (2011). Mid-infrared semiconductor heterostructure lasers for gas sensing applications. *semiconductor science and technology, 26*, 014032.

Colombelli, F., Srinivasan, K., Troccoli, M., Painter, O., Gmachl, C. F., Tennant, D. M., et al. (2003). Quantum cascade surface-emitting photonic crystal laser. *Science, 302*, 1374.

Colombelli, R., Srinivasan, K., Troccoli, M., Painter, O., Gmachl, C. F., Tennant, D. M., et al. (2004). Quantum cascade photonic-crystal microlasers. *Proceedings of SPIE, 5365*, 228.

Diehl, L., Bour, D., Corzine, S., Zhu, J., Hfler, G., Loncar, M., et al. (2006). High-power quantum cascade lasers grown by low-pressure metal organic vapor-phase epitaxy operating in continuous wave above 400K. *Applied Physics Letters, 88*, 201115.

Findlay, P. C., Pidgon, C. R., Kotitschke, R., Hollingworth, A., Murdin, B. N., Langerak, C. J., et al. (1998). Auger recombination dynamics of lead salts under picosecond free-electron-laser excitation. *Physical Review B, 58*, 12908.

H.C. Casey, J., & Panish, M. B. (1978). *Heterostructure Lasers*. San Diego, USA.

Ishida, A., & Fujiasyu, H. (2003). *Lead Chalcogenides: Physics and Applications*. (Khokhlov, Ed.) New York, USA: Taylor and Francis.

Kim, M., Kim, C. S., Bewley, W. W., Lindle, J. R., Canedy, C. L., Vurgaftman, I., et al. (2006). Surface-emitting photonic-crystal distributed-feedback laser for the midinfrared. *Applied Physics Letters, 88,* 191105.

Loncar, M., Yoshie, T., Scherer, A., Gogna, P., & Qiu, Y. (2002). Low-threshold photonic crystal laser. *Applied Physics Letters, 81,* 2680.

Nojima, S. (1998). Enhancement of optical gain in two-dimensional photonic crystals with active lattice points. *Jpn. J. Appl. Phys., 37,* 565.

Painter, O., Lee, R. K., Scherer, A., Yariv, A., O'Brien, J. D., Dapkus, P. D., et al. (1999). Two-dimensional photonic band-gap defect mode laser. *Science, 284,* 1819.

Painter, O., Vuckovic, J., & Scherer, A. (1999). Defect modes of a two-dimensional photonic crystal in an optically thin dielectric slab. *Journal of the Optical Society of America B , 16,* 275.

Park, H., Hwang, J., Huh, J., Ryu, H., & Lee, Y. (2001). Nondegenerate monopole-mode two-dimensional photonic band gap laser. *Applied Physics Letters, 79,* 3032.

Preier, H. (1990). Physics and applications of IV-VI compound semiconductor lasers . *Semiconductor Science and Technology, 5,* S12.

Slivken, S., Evans, A., Zhang, W., & Razeghi, M. (2007). High-power, continuous-operation intersubband laser for wavelengths greater than 10 μm. *Applied Physics Letters, 90,* 151115.

Springholz, G., Schwarzl, T., & Heiss, W. (2006). Mid-infrared Vertical Cavity Surface Emitting Lasers based on the Lead Salt Compounds. *Mid-infrared Semiconductor Optoelectronics, Springer Series in Optical Sciences, 118,* 265.

Tacke, M. (2000). *Long Wavelength Infrared Emitters Based on Quantum Wells and Superlattices.* (M. Helm, Ed.) Amsterdam, Holand: Gordon and Breach Science Publishers.

Tacke, M. (2001). Lead-salt lasers. *Phil. Trans. R. Soc. Lond. A, 359,* 547.

Tittel, F. K., Richter, D., & Fried, A. (2003). *Solid State Mid-Infrared Lasers Sources.* (S. I. T, & V. K. L, Eds.) Berlin: Springer-Verlag.

Vurgaftman, I., & Meyer, J. R. (2003). Design optimization for high-brightness surface-emitting photonic-crystal distributed-feedback lasers. *IEEE Journal of Quantum electronics, 39,* 689.

Weng, B., Ma, J., Wei, L., Li, L., Qiu, J., Xu, J., et al. (2011). Room temperature mid-infrared surface-emitting photonic crystal laser on silicon. Applied Physics Letters, 99, 221110.

Weng, B., Ma, J., Wei, L., Xu, J., Bi, G., & Shi, Z. (2010). Mid-infrared surface-emitting photonic crystal microcavity light emitter on silicon. *Applied Physics Letters, 97,* 231103.

Weng, B., Zhao, F., Ma, J., Yu, G., Xu, J., & Shi, Z. (2010). Elimination of threading dislocations in as-grown PbSe film on patterned Si(111) substrate using molecular beam epitaxy. *Applied Physics Letters, 96,* 251911.

Xu, G., Colombelli, R., Braive, R., Beaudoin, G., Gratiet, L. L., Talneau, A., et al. (2010). Surface-emitting mid-infrared quantum cascade lasers with high-contrast photonic crystal resonators. *Optics Express, 18,* 11979.

Yang, R. Q. (1995). Infrared laser based on intersubbandtransitions in quantum wells. *Superlattices and Microstructures, 17,* 77.

Yang, R. Q., Bradshaw, J. L., Bruno, J. D., Pham, J. T., Wortman, D. E., & Tober, R. L. (2002). Room temperature type-II interband cascade laser. *Applied Physics Letters, 81,* 397.

Yu, J. S., Darvish, S. R., Evans, A., Nguyen, J., Slivken, S., & Razeghi, M. (2006). Room-temperature continuous-wave operation of quantum-cascade lasers at $\lambda \sim 4$ μm. *Applied Physics Letters, 88,* 041111.

Zhao, F., Wu, H., Majumdar, A., & Shi, Z. (2003). Continuous wave optically pumped lead-salt mid-infrared quantum-well vertical-cavity surface-emitting lasers. *Applied Physics Letters, 83,* 5133.

Part 2

Innovative Materials and Systems

6

Employing Optical Nonlinearity in Photonic Crystals: A Step Towards All-Optical Logic Gates

Mohammad Danaie and Hassan Kaatuzian
Photonics Research Lab., Amirkabir University of Technology
Iran

1. Introduction

Employing nonlinear elements in photonic crystals (PCs) opens up lots of new design opportunities. In comparison to ordinary linear PC structures, using optically nonlinear elements in PCs leads to the observation of many interesting phenomena that can be utilized to design all-optical devices. Optical bistability in PCs is among the mentioned observations, which due to its many applications has attracted the researchers' attention. Many optical devices such as limiters, switches, memories can be implemented when nonlinear elements are embedded in PCs. Kerr type nonlinearity is mostly used for this purpose. In this chapter we are going to discuss the benefits of using nonlinearity in photonic crystal devices. Since most of nonlinear optical devices are either based on directional couplers (DCs) or coupled cavity waveguides (CCWs), we have focused on these two groups to provide a better insight into their future prospects.

The first practical case of employing nonlinearity in PCs was for optical switches. Scholz and his colleagues were among the first ones to use optical Kerr nonlinearity in PCs (Scholz et al., 1998). They designed an all-optical switch using a one dimensional photonic crystal which was placed inside two cross waveguides. In their switch, a strong pump signal lateral to the PC layers made the crystal nonlinear and slightly shifted the position of the bandgap. The data signal wavelength was chosen on the bandgap edge such that it could not be placed in the shifted bandgap region (when the probe signal was present). Therefore; when no probe signal existed, the data signal could not pass through the PC layers, while when the probe was present, the shifted photonic bandgap would allow its transmission. Later the optical bistability in two dimensional nonlinear photonic crystal waveguides coupled to a micro-cavity was discussed in (Centeno & Felbacq, 2000) and it was suggested that the mentioned phenomena could be use to design all-optical switches. Mingaleev and Kivshar placed nonlinear elements in waveguides and bends to obtain optical limiters (Mingaleev & Kivshar, 2002). Meanwhile other structures were suggested by (Soljac˘ic´ et al., 2002) and (Fan, 2002) for optical switching. Later in 2004 (Locatelli et al., 2004) and (Cuesta-Soto et al, 2004) suggested optical switches using directional couplers. Thereafter, some logic components were reported in literature such as: An optical AND gate by (Zhu et al., 2006), an all-optical PC on-chip memory implemented in (Shinya et. al., 2008), a PC half-adder structure designed in (Liu et al., 2008) and so forth.

In this chapter, first different types of optical nonlinearities are briefly explained in electromagnetics' terms. Thereafter, numerical and analytic methods for modeling optical nonlinearity in photonic crystals are reviewed. As a case study nonlinear optical Kerr effect is used to design a directional coupler switch and later as the second case study an structure for all-optical AND gate operation is proposed.

2. Optical nonlinearity in PCs

Nonlinear optics field came to existence when (Franken et al., 1961) published their achievements on second harmonic generation. It was mainly due to acquiring the technology to produce high intensity coherent lights that could trigger the nonlinearity in materials. Such ability was facilitated by the introduction of laser by (Maiman, 1960). Since then many other optical phenomena such as parametric oscillation, four wave mixing, stimulated Raman and Brillouin scattering, phase conjugation etc. have been observed and studied (Boyd, 2003).

Nonlinear optical materials can be considered as the basic elements for all-optical processing systems. Although optical bistability in nonlinear materials has been extensively studied since 1980 (Saleh, 1991 and Kaatuzian, 2008), but since most materials exhibit pretty weak nonlinear characteristics, the large size and high operational power of such devices makes them unsuitable for all-optical integrated devices. Introduction of PCs, due to its unique characteristics (Yablonovitch, 1987 and John, 1987), helped to reduce both the operational power and size of nonlinear optical components and presented a new opportunity for their integration.

In a linear material the dipole moment density (or equally known as polarization) P(t) and electric field E(t) have a linear relationship:

$$P(t) = \varepsilon_0 \chi \ E(t). \tag{1}$$

The notation "χ" is usually referred to as susceptibility. In a nonlinear medium (1) can be generalized as follows (Boyd, 2003):

$$
\begin{aligned}
P(t) &= \varepsilon_0 [\chi^{(1)} E(t) + \chi^{(2)} E^2(t) + \chi^{(3)} E^3(t) + ...] \\
&= P^1(t) + P^2(t) + P^3(t) + ... \\
&= P^L + P^{NL},
\end{aligned}
\tag{2}
$$

where, the notations $\chi^{(1)}$, $\chi^{(2)}$, $\chi^{(3)}$ are called the first, second and third order susceptibility coefficients respectively. Also $P^1(t)$, $P^2(t)$, $P^3(t)$ are referred to as the first, second and third order polarizations. Therefore, P(t) can be considered to have a linear term P^L and a nonlinear term P^{NL}. In nonlinear optics, most of nonlinear observations are usually due to $\chi^{(2)}$ and $\chi^{(3)}$ which is mainly due to the fact that the higher order terms need much higher input intensities to show significant impact. In addition it is shown (Boyd, 2003) that $\chi^{(2)}$ can be only observed in non-centrosymmetric crystals; therefore liquids, gas and many solid crystals do not display those phenomena that are originated by $\chi^{(2)}$.

The ith component for the vector P can be expressed as (3) (Saleh, 1991):

$$P_i = \varepsilon_0 \sum_{j=1}^{3} \chi_{ij}^{(1)} E_j + \varepsilon_0 \sum_{j=1}^{3}\sum_{k=1}^{3} \chi_{ijk}^{(2)} E_j E_k + \varepsilon_0 \sum_{j=1}^{3}\sum_{k=1}^{3}\sum_{l=1}^{3} \chi_{ijkl}^{(3)} E_j E_k E_l + ...,$$ (3)

where, i=1,2,3 (corresponding to x,y,z). For a linear material, only the first term of this equation is significant. One of the most important nonlinear optical effects which originates from the $\chi^{(3)}$ coefficients is the Kerr effect (Weinberger, 2008). It was discovered in 1875 by John Kerr. Almost all materials show a Kerr effect. The Kerr effect is a change in the refractive index of a material in response to an external electric field. For materials that have a non-negligible Kerr effect, the third, χ(3) term is significant, with the even-order terms typically dropping out due to inversion symmetry of the Kerr medium. In the optical Kerr effect, an intense beam of light in a medium can itself provide the modulating electric field, without the need for an external field to be applied. In this case, the electric field is given by:

$$E = E_\omega Cos(\omega t).$$ (4)

where E_ω is the amplitude of the wave. Combining this with the equation for the polarization will yield:

$$P \cong \varepsilon_0 \left(\chi^{(1)} + \frac{3}{4}\chi^{(3)} |E_\omega|^2 \right) E_\omega Cos(\omega t).$$ (5)

As before:

$$\chi = \chi^L + \chi^{NL} = \chi^{(1)} + \frac{3}{4}\chi^{(3)} |E_\omega|^2.$$ (6)

and since:

$$n = \sqrt{1+\chi} = \sqrt{1 + \chi^L + \chi^{NL}} \cong n_0 \left(1 + \frac{1}{2n_0^2} \chi^{NL} \right),$$ (7)

where $n_0 = (1+\chi^L)^{1/2}$ is the linear refractive index. Therefore:

$$n = n_0 + \frac{3\chi^{(3)}}{8n_0} |E_\omega|^2 = n_0 + n_2 I,$$ (8)

where n_2 is the second-order nonlinear refractive index, and I is the intensity of the wave. The refractive index change is thus proportional to the intensity of the light travelling through the medium.

2.1 Modeling optical nonlinearity in PCs

In order to numerically analyze linear PCs, various methods have been proposed in the literature. Among these methods, the most popular are: plane wave expansion (PWE), finite difference in time domain (FDTD), finite element method (FEM), multiple multimode method (MMP) and Wannier function method (WFM). In the nonlinear regime, in addition to the mentioned methods, complementary strategies have to be used to analyze the effect of the nonlinear elements on the linear system. Transfer matrix method (TMM), perturbation theory, coupled mode theory (CMT) and FDTD can be used for this purpose. In order to model the Kerr effect using FDTD, several methods have been proposed. Here FDTD and perturbation theory are briefly reviewed. Also In section 4 CMT is used to analyze a PC limiter.

From the Maxwell equations, the wave equation for the nonlinear medium can be written as:

$$\nabla^2 E - \mu_0 \varepsilon_0 \varepsilon_r(x,y,z,\omega,E)\frac{\partial^2 E}{\partial t^2} = 0. \tag{9}$$

It is shown in (Joseph & A Taflov, 1997) that (9) can be rewritten as:

$$\nabla^2 E - \mu_0 \varepsilon_0 n_0^2 \frac{\partial^2 E}{\partial t^2} = \mu_0 \frac{\partial^2 P^{NL}}{\partial t^2}. \tag{10}$$

Based on the type of nonlinearity different approaches have to be made to solve the mentioned equation. The FDTD model for Kerr-type materials assumes an instantaneous nonlinear response. The nonlinearity is modelled in the relation $D = \varepsilon E$ where:

$$\varepsilon = n^2 = (n_0 + \frac{3\chi^{(3)}}{8n_0}|E_\omega|^2)^2 \approx n_0^2 + 2n_0 \frac{3\chi^{(3)}}{8n_0}|E|^2. \tag{11}$$

Therefore the relation ship between E and D can be iteratively determined using:

$$E = \frac{D}{n_0^2 + \frac{3\chi^{(3)}}{4}|E|^2}. \tag{12}$$

Perturbation theory is also a useful tool in engineering for analyzing systems with small nonlinearities. Using Maxwell equations, the following eigenvalue problem can be derived for linear time invariant PC systems (A waveguide is assumed in our case.). The Dirac notation $|\vec{E}_0\rangle$ specifies the Bloch eigenmode for the electric field.

$$\nabla \times \nabla \times |\vec{E}_0\rangle = \left(\frac{\omega_0}{c}\right)^2 \varepsilon_0(\vec{r})|\vec{E}_0\rangle. \tag{13}$$

It is shown in (Bravo-Abad et al., 2007) that a $\Delta\varepsilon$ change in the dielectric constant can result in a $\Delta\omega$ variation in the original eigenvalue ω_0 as:

$$\Delta\omega = -\frac{\omega_0}{2}\frac{\langle\vec{E}_0|\Delta\varepsilon|\vec{E}_0\rangle}{\langle\vec{E}_0|\varepsilon_0|\vec{E}_0\rangle} = -\frac{\omega_0}{2}\frac{\int d^3r \Delta\varepsilon(\vec{r})|\vec{E}_0(\vec{r})|^2}{\int d^3r \varepsilon_0(\vec{r})|\vec{E}_0(\vec{r})|^2}. \tag{14}$$

But since for a waveguide $v_g \approx \Delta\omega / \Delta k$ then it can be shown that (Bravo-Abad et al., 2007) the following approximation is valid:

$$\Delta k = -\frac{\omega_0}{v_g}\frac{\Delta n}{n_r}\frac{\displaystyle\int_{HIGH-\varepsilon} d^3r\varepsilon_0(\vec{r})\left|\vec{E}_0(\vec{r})\right|^2}{\displaystyle\int_{ALLSPACE} d^3r\varepsilon_0(\vec{r})\left|\vec{E}_0(\vec{r})\right|^2}. \tag{15}$$

The mentioned change in the wavenumber can cause an optical phase difference between the linear and nonlinear states of a system. Many optical devices such as directional couplers or Mach-Zehnder interferometers are sensitive to the induced phase. Assuming that L is the waveguide length, $\Delta\phi = L\Delta k$ shows the phase difference between the linear and nonlinear cases. According to (15) if the waveguide is designed to have a low group velocity, for a fixed amount of refractive index variation a larger Δk is obtained. It means that for a fixed $\Delta\phi$ a smaller device size L is needed; or equally for a fixed L a smaller Δn (which is proportional to operational power) is required. Since photonic crystals can be used to design waveguides with very low group velocities (Solja˘cic´ & Joannopoulos, 2004); therefore they provide the opportunity to reduce both the device size and the operational power; making such devices able to be integrated on a single chip.

3. Case study 1: A PC all-optical switch

The idea of a one dimensional PC all-optical switch was first proposed in (Scholz et al., 1998). Coupled cavity waveguides (CCW) with Kerr nonlinearity, were there after suggested for all-optical switching. The main drawback of CCW switches is that when the switch is in the OFF state, all the data signal is reflected back to the input port. Since the backscattered signal can affect other optical devices on an optical chip, it makes them unsuitable for all optical integrated circuit applications. A combination of directional couplers and nonlinear optical elements can be used to solve the mentioned problem (Yamamoto et al., 2006, Cuesta-Soto et al., 2004, Rahmati & N. Granpayeh, 2009).

In a directional coupler based switch, according to the ON or OFF state of the switch, most of the data signal power is guided to either of the two output ports. Usually only a very small amount leaks to the input port. It gives the designer the ability of using the switch in sequential optical circuits.

Here a PC directional coupler is designed first.The PC lattice used for this design is a two dimensional array of GaAs rods which is known to have Kerr type optical nonlinearity.

The band diagram of the PC, Which is obtained using Plane Wave Expansion (PWE) method, is shown in Fig. 1. The shaded region is the optical bandgap. No optical signal within the normalized frequency range of $0.28a/\lambda$ to $0.45a/\lambda$; where a is the lattice constant and λ is free space wavelength, can propagate through the PC lattice. The value of a is chosen 635nm in our simulations. The radii of rods are equal to 0.2a and their refractive index is chosen equal to 3.4, which is equal to the refractive index of GaAs at the 1550nm wavelength.

Introduction of defects into PC lattice is the first step in designing PC devices. These are usually classified in two different categories of point defects and line defects, which can create resonators and waveguides respectively. As an Instance, in the mention PC, removing a column of rods (Fig. 2a) creates a waveguide mode in the bandgap region between the normalized frequencies 0.340a/λ and 0.447a/λ (See Fig. 3a). A simple directional coupler can be obtained by removing two rows of rods adjacent to a central row (See Fig. 2b.) (Zimmermann et al., 2004, Nagpal, 2004). As discussed in (Nagpal, 2004), when two PC waveguides are placed close to each other, light propagating in one of the waveguides can be coupled to the neighboring waveguide.

TM Band Structure

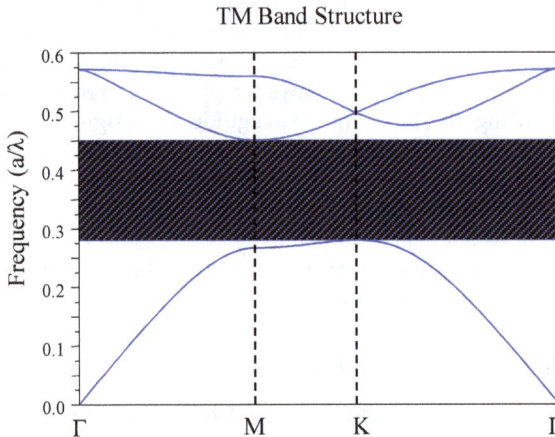

Fig. 1. The band diagram for a 2D hexagonal array of GaAs rods, where the ratio of the rods radius to the lattice constant is 0.2.

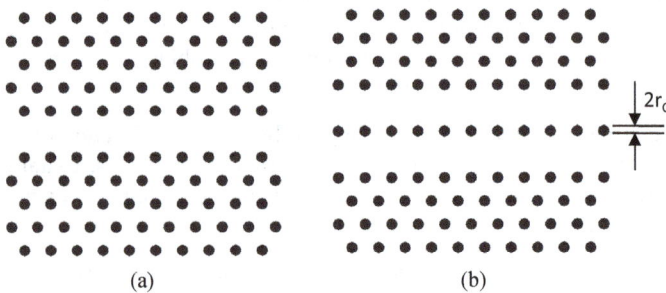

Fig. 2. Different PC structures for TM modes: The dark circles are assumed to be dielectric and the background is air (a) a PC W1 waveguide (b) A PC directional coupler.

PWE method can also be used to analyze the structure in Fig. 2b. In this case, using a super-cell, the two adjacent waveguides can be treated as a single symmetric line defect in the PC structure. The super-cell which is used for this purpose and the band diagram obtained in this case are shown is Fig. 3b. Here, the radii of the central rods, r_C, is assumed 0.2a. An odd and an even defect mode are produced in this case.

It is shown (Zimmermann et al., 2004, Nagpal, 2004)] that the light that is travelling in one of these waveguides, can be periodically coupled to the other one after passing a certain distance referred to as the coupling length (L_C). The coupling length is related to the propagation constants of the odd (k_o) and even (k_e) modes as follows:

$$L_C = \frac{\pi}{ke - ko}. \tag{16}$$

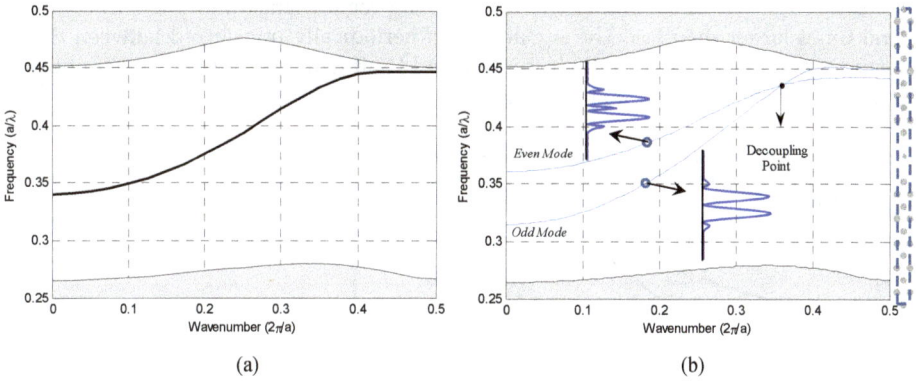

Fig. 3. The waveguide modes for (a) the W1 waveguide shown in Fig. 2(a) and (b) the directional coupler shown in Fig. 2(b). The super-cell for obtaining the directional coupler is depicted in the dashed box.

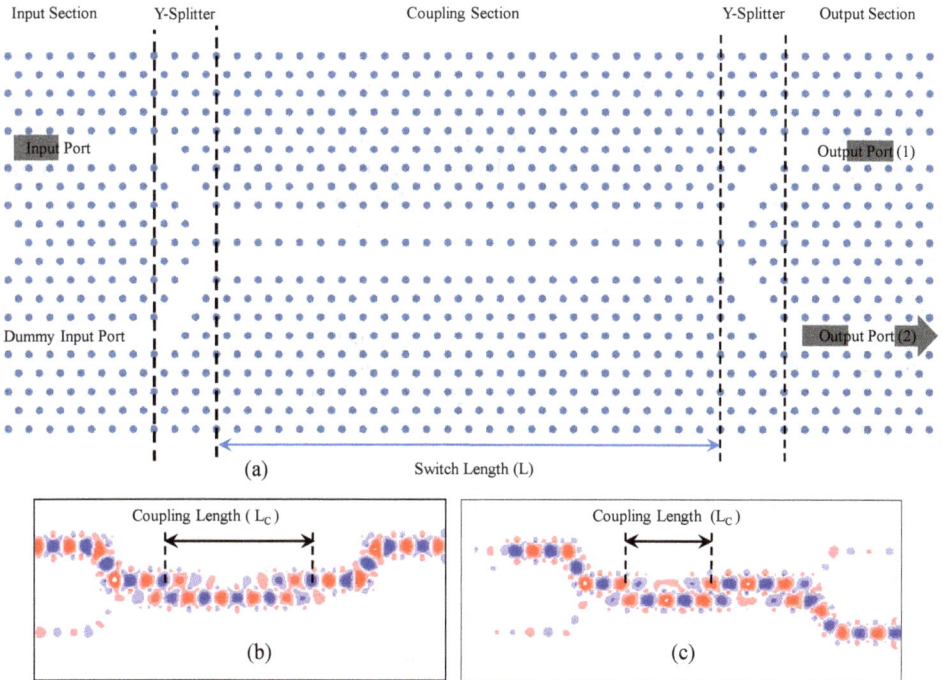

Fig. 4. (a) A conventional PC directional coupler (b) f= 410a/λ (c) f= 0.396a/λ.

Directional couplers can be used as wavelength selective devices, in their linear regime, or switches when optical nonlinearity is introduced to their structure. A typical symmetrical directional coupler switch is shown in Fig. 4a. It consists of two input ports, two output ports and a central coupler. Assuming that a signal has entered from one of the input ports, it passes through the bends and enters the coupling region which has a length equal to L (several times larger than L_C). The signal is then periodically transferred between the two waveguides and regarding the ratio between L and L_C, it will be directed to either one of the output ports (or even both for a poor design).

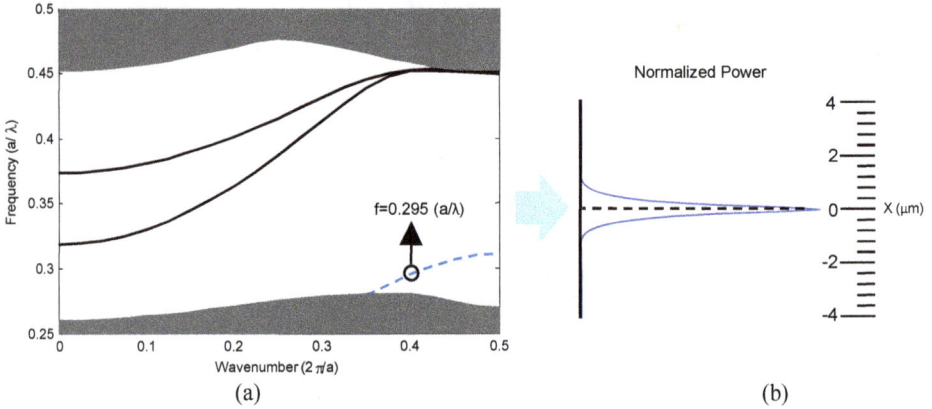

(a)   (b)

Fig. 5. (a) The band diagram for the directional coupler shown in Fig. 2(b) when r_C is reduced to 0.15a, (b) The spatial profile for the control mode at f = 0.295 (a/λ).

Fig. 4b and Fig.4c demonstrate the selective behavior of the device for two different input wavelengths. Since from (16) the coupling length is inversely related to the difference of ke and ko; then according to Fig. 3b, it is obvious that for each input signal frequency a different coupling length is perceived. This phenomenon can be used to separate different wavelengths (Fig. 4b and Fig. 4c).

Reduction of r_C create a mode that can be used as the control signal. Fig. 5a shows the band diagram for the directional coupler in Fig. 2b when r_C is 0.15 (a/λ). As seen in Fig. 5b the mentioned mode is highly confined. Since the rods are nonlinear, if the control signal power is large enough the refractive index of the middle row can be changed which will affect the system behaviour in regard to the linear case.

In a conventional directional coupler, if the refractive index of the central region between the two waveguides is somehow changed such as: n→n+δn, then the wavenumber of the odd and even modes change accordingly to ke + δke and ko + δko. From (16), it is obvious that the new coupling length for the modified structure will be slightly different. Assuming that the difference in the coupling length be δL_C, and that m = L/L_C, then if (m+1)δL_C equals a coupling length, the light will be transferred to the other output port. It means that a change in the refractive index of the central row can provide a switching mechanism. From the previous equations, it can be concluded that (m+1)δL_C = L_C is the switching condition. Since m = L/L_C then:

$$L = L_C^2 \times \frac{1}{\delta L_c} - L_C.$$

(17)

Also:

$$\delta L_C = \pi \left(\frac{1}{ke - ko} - \frac{1}{ke + \delta ke - ko - \delta ko} \right) \cong \pi \frac{\delta ke - \delta ko}{(ke - ko)^2}.$$

(18)

Neglecting the L_C term with regard to $L^2_C/\delta L_C$ in (17) leads to:

$$L \cong L_C^2 \times \frac{(ke - ko)^2}{\pi(\delta k_e - \delta k_o)} = L_C^2 \times \frac{(ke - ko)^2}{\pi \delta k}.$$

(19)

Using (16), equation (19) can be simplified to:

$$L \cong \pi \times \frac{1}{\delta k}.$$

(20)

According to (20), the switch length is inversely proportional to δk; which is itself a function of δn. The relationship between δk and δn is dependant upon the structure geometry and is usually determined numerically using PWE method. In order to minimize the switch length and operational power, it is very important that for a slight variation in n, a large δk be obtained. In (Yamamoto et al., 2006) some methods are suggested to improve the δk and δn relation for a PC directional coupler for TE modes.

The refractive index can be modified using electro-optic, thermo-optic or Kerr optical effect. From the mentioned optical phenomena, only Kerr effect can be used for all-optical applications. In materials that possess the Kerr effect, the refractive index can be linearly changed using an optical pump signal as follows:

$$n = n_0 + n_2 I,$$

(21)

where, n_0 is the refractive index in the linear regime, I is the optical field intensity and n_2 is the nonlinear Kerr coefficient. Since δk is a function of δn, it can be seen from (20) that there is a trade-off between switch size and power, i.e. in order to reduce the operating power of the switch, longer device size is needed, or vice versa. The relationship between δk and δn should be calculated numerically (Danaie & Kaatuzian, 2011). Afterwards according to (20) the required switch length can be estimated for different signal frequencies. PWE method can be used for this purpose. We have depicted the relationship between signal frequency (f_s) and δk in Fig. 6.

In order to be able to optimize the switch length, we first try to fit an analytical expression on the curves shown in Fig. 6. Each of the five curves in Fig. 6 can be considered to be the bottom black curve multiplied by a γ factor. If γ is calculated for the four upper curves, it is seen that these factors are neatly placed on a line. On the other hand, the first glance at the curves shown in Fig. 6 suggests an exponential relation between δk and frequency; while a more accurate study reveals second order polynomial relation.

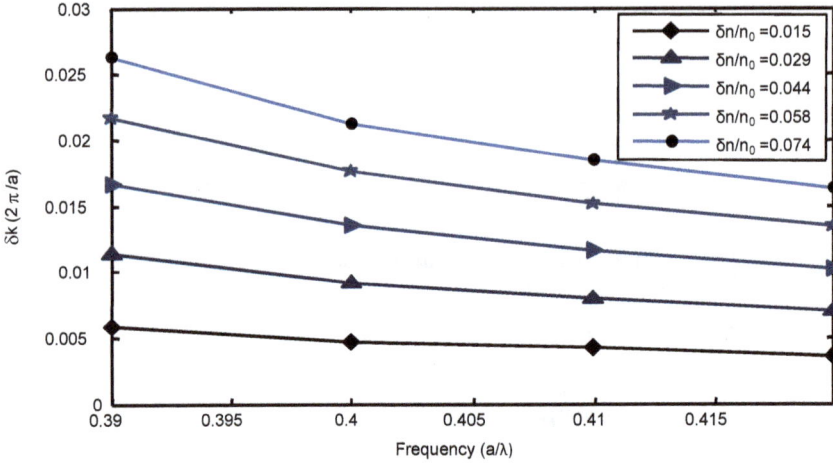

Fig. 6. The relationship between signal frequency and δk of the directional coupler depicted in Fig. 2(b) with r_C = 0.15a, obtained using PWE method for different values of central row's refractive index change.

Based on the previous observations the relationship between δk and δn can be assumed as:

$$\delta k(\delta n, f_s) = h(\delta n).g(f_s),\qquad(22)$$

where f_s is the signal frequency and h and g are first and second order polynomials dertermined as:

$$h(\delta n) = a.\delta n + b.$$

$$g(f) = c.f_s^2 + d.f_s + e.\qquad(23)$$

For our structure (r_C = 0.15a), the factors a-f are determined as follows using curve fitting techniques:

$$\delta k(\delta n, f) = (17.5\delta n + 0.17).\left(1.25 f_s^2 - 1.08 f_s + 0.24\right).$$

The previous equation can be used for optimization purpose; in which the f is considered per (a/λ) and the δk is per $(2\pi/a)$. Since δn = n_2I, then the relationship between the switch length (per lattice constant) and optical power intensity (on the rods) becomes as follows:

$$L \cong \pi \times \frac{1}{(\frac{2\pi}{a})(a.\delta n + b)(c.f_s^2 + d.f_s + e)} = \frac{a}{2(a.n_2 I + b)(c.f_s^2 + d.f_s + e)}.\qquad(24)$$

In the above equation a is the latice constant. If the designer wishes to choose the input signal for the 1550nm wavelength, then the the lattice constant must be chosen equal to f_s.1550nm; therefore the actual size of the switch becomes equal to:

$$L \cong \frac{f_s . 1550\text{nm}}{2(a.n_2 I + b)(c.f_s^2 + d.f_s + e)}.$$ (25)

When an optical pulse enters a medium with a low group velocity, it becomes squeezed in the time domain. In order for the pulse energy to remain constant its amplitude should increase (Solja˘cic´ & Joannopoulos, 2004). The term $n_2 I$ in (25) can be therefore rewritten as $a.n_2.I_{in}(c_0/v_g)$; where I_{in} is the input power intensity of the control signal; v_g is the group velocity, c_0 is the free space light speed and a is a correction factor which is related to the control signal mode profile. It determines what ratio of the input power is concentrated on the nonlinear rods. The a factor should be calculated numerically for each frequency. Since v_g and a are functions of the control frequency (f_c), therefore equation (25) is then rewritten as:

$$L \cong \frac{f_s . 1550\text{nm}}{2 \left[a.n_2 I_{in} (\frac{c_0}{v_g(f_c)}) \alpha(f_c) + b \right](c.f_s^2 + d.f_s + e)}.$$ (26)

The relation between the group velocity and f_c can be obtained using PWE. For $f_c = 0.295$ (a/λ) a can be obtained from integration of control mode profile from $x = -95\text{nm}$ to $+95\text{nm}$ (the central rods section). Using a simple integration it can be seen that 0.35 percent of the control power is located on the rods (Fig. 5b).

Fig. 7. The all-optical switch designed using a directional coupler.

As a case study, we have designed a switch in this section. The control signal frequency is chosen $f_c = 0.295$ (a/λ), so as to minimum the reflection in the control signal path. The signal frequency is chosen $f_s = 0.41$ (a/λ). Since:

$$L \cong \frac{a}{2(17.5\delta n + 0.17).(1.25 f_s^2 - 1.08 f_s + 0.24)},$$ (27)

then if we decide the switch length L to be equal to 25a, then according to (27) the required δn will be equal to 0.204. We use 2D FDTD method to analyze the time domain behavior of the directional-coupler-based switch. The rods are assumed to have Kerr type nonlinearity and the nonlinear Kerr coefficient n_2 is assumed $1.5 \times 10^{-17} m^2/W$. Each unit cell is meshed 16×16. The final switch structure is shown in Fig. 7.

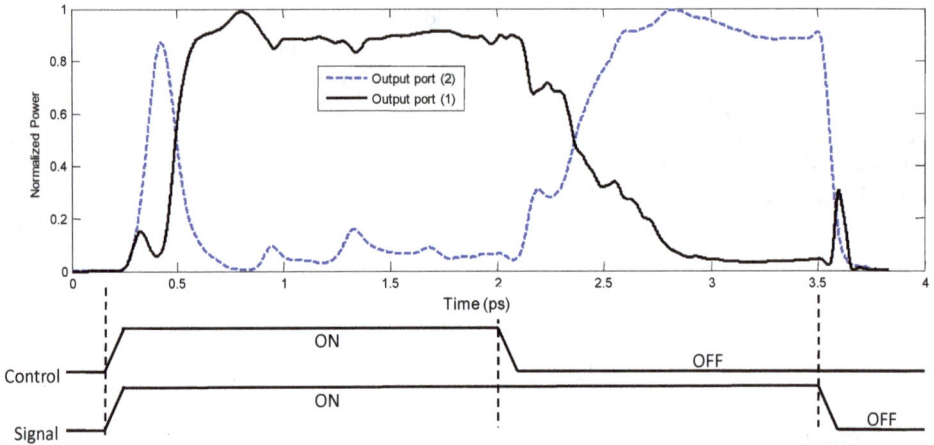

Fig. 8. FDTD time domain simulation results for the directional coupler shown in Fig. 7. The solid line represents the optical power directed to the first output port and the dashed line shows the optical power which goes to the second output port.

In order to be able to observe the time-domain characteristics of the switch, the control and signal inputs are turned on simultaneously. After 2ps the control signal is turned off and after 3.5ps the data signal is turned off as well. Fig. 8 show the power density versus time for the first and second output ports for the structure shown in Fig. 7. Since the group velocity of the data signal is higher than the control signal, it reaches the output section faster. Thus for a brief period of time (t = 0.4ps), it goes to the second output port. At t = 0.5ps the control signal reaches the output section (turns the rods nonlinear) and forces the data signal to travel to the first output port. When the control signal is turned off (t = 2ps), the rods are tuned linear again and the data signal travels back to the first output port.

4. Case study 2: Design of an optical gate using CCWs

Recently more research attention has been directed towards photonic crystal logic gates (Andalib & Granpayeh, 2009, Bai eta al., 2009). In (Zhu et al., 2006) a structures for All-optical AND gate is proposed. The main problem of the structures is that the AND gate's inputs have different input wavelengths. In an ideal two-input AND gate, the wavelength of the inputs should be the same or the mentioned AND gate cannot be used in all-optical large-scale circuits. Here a photonic crystal AND gate with identical input characteristics is proposed. Most PC optical devices reported in the literature use TM photonic crystal structures, such as a square lattice of dielectric rods to create a switching mechanism; while here a triangular lattice of holes in a dielectric substrate is used instead (which can be used for TE modes). It can provide a considerably large bandgap and can easily be fabricated using integrated circuit manufacturing technology. The radius of the holes is chosen 0.3a. Choosing larger holes will reduce the guiding strength in vertical direction. The dielectric is assumed to be GaAs as before. In this section, first a sharp PC limiter is designed and then by combining it with a Y-junction, a PC gate is obtained.

4.1 PC limiters

One of the best methods to be able to observe a strong switching mechanism in a PC is to create a nonlinear PC cavity that is coupled to a PC waveguide. Optical nonlinearity can shift the position of the cavity, which can result in creating a hysteresis loop in the path of signal (i.e. an optical limiter can be created.). In order to do so, the PC structure, waveguide topology and cavity specifications each play an important role.

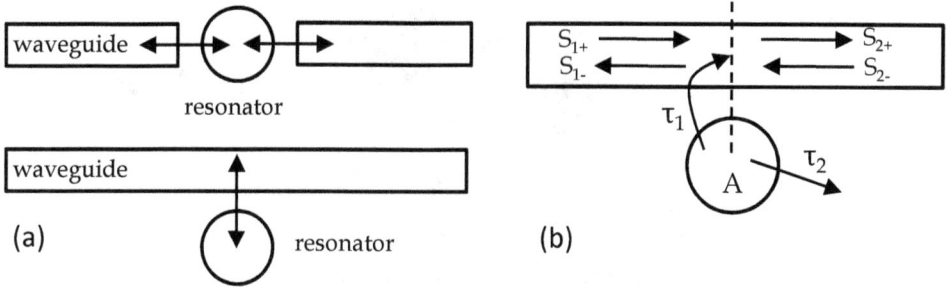

Fig. 9. (a) Two different possible coupling schemes between a cavity and a waveguide (b) Schematic description of CMT model parameters for the side coupled cavity waveguide.

It was briefly mentioned that FDTD and perturbation methods can be used to analyze PCs which have optical nonlinearity. In this section another popular method is used to explain the behaviour of a nonlinear PC component. CMT is a general method that can be used for analyzing the behaviour of cavities that are weakly coupled to a waveguide (Saleh, 1991). Generally a cavity can be either placed at the end of a waveguide (directly coupled) or be side-coupled to one (Yanik et al., 2003) (See Fig. 9a).

In the linear case, if we assume that the EM fields inside the cavity are proportional to the parameter A (Fig. 9b), this system can be modelled using CMT by the following set of equations (Bravo-Abad et. al, 2007):

$$\frac{dA}{dt} = i\omega_c A - \frac{1}{\tau_1}A - \frac{1}{\tau_2}A + \kappa s_{1+} + \kappa s_{2-}$$

$$s_{1-} = s_{2-} - \kappa A$$

$$s_{2+} = s_{1+} - \kappa A,$$

(28)

where ω_c is the resonant frequency of the cavity. τ_1 is the decay constant of the cavity into the waveguide modes propagating to the right and to the left; while the magnitude τ_2 relates to the decay rate due to cavity losses. The parameters s_{1+} and s_{1-} (s_{2+} and s_{2-}) denote the complex amplitudes of the fields propagating to the left and right at the input (output) of the waveguide. The parameter κ governs the input coupling between the resonant cavity and the propagating modes inside the waveguide. From (Bravo-Abad et. al, 2007) this magnitude is given by $\kappa = (1/\tau_1)^{(1/2)}$.

Now, if we assume that the input signal launched only from the left ($s_{2-}=0$) and that external losses can be neglected ($\tau_2 \to \infty$), (29) can be expressed in terms of just s_{1+} and s_{1-} as:

$$\frac{ds_{1-}}{dt} = i\omega_c s_{1-} - \frac{1}{\tau_1}s_{1-} - \frac{1}{\tau_1}s_{1+}. \qquad (29)$$

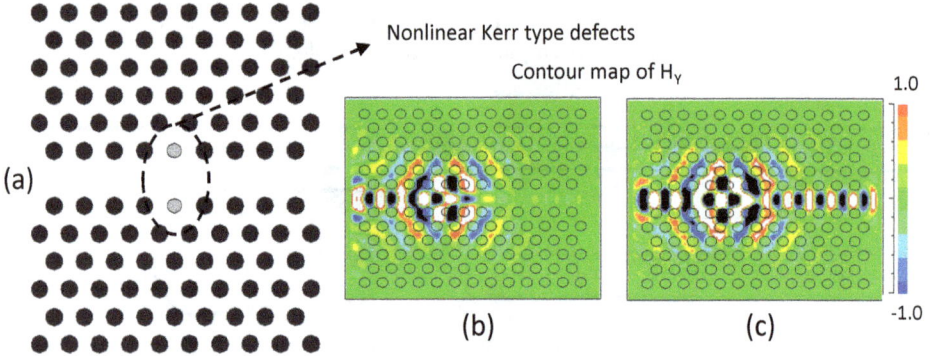

Fig. 10. (a) A simple symmetric side-coupled cavity waveguide. The dark circles are assumed to be air holes in a GaAs substrate and the grey circles are assumed to contain the nonlinear elements. (b) Steady state response of the limiter for the linear case i.e. low input intensity (c) Transmission for high input intensity.

It can be shown that for Kerr type nonlinearity:

$$\frac{ds_{1-}}{dt} = i\omega_c\left(1 - \frac{1}{2Q}\frac{|s_{1-}|^2}{P_0}\right)s_{1-} - \frac{1}{\tau_1}s_{1-} - \frac{1}{\tau_1}s_{1+}, \qquad (30)$$

where Q is the quality factor of the cavity and P_0 is characteristic power of the system (See (Bravo-Abad et. al, 2007) for its definition.).

The PC used in this section to create a waveguide is a triangular lattice of holes in a GaAs substrate. The radius of the holes is chosen 0.3a. The triangular lattice of air holes can be more easily fabricated than the rod lattice version which is used for TM modes. To create a waveguide a row of holes is removed from the PC.

In order to create a cavity mode in the bandgap region of a PC, a defect has to be added to the structure of the waveguide. The defect can later be doped with a material which exhibits strong Kerr effect (Fushman et al., 2008, Nakamura et al., 2004) so as to be able to create the desirable hysteresis effect. Doping quantum dots in to photonic crystals has long been known to create large Kerr type nonlinearity coefficients. Here, the defect is assumed to be doped using the method presented in (Nakamura et al., 2004) resulting in a refractive index equal to 2.6 and a nonlinear Kerr effect equal to $n_2 = 2.7\times10^{-9}$ m²/W. The defect radius is $r_d = 0.25a$. First the structure shown in Fig. 10a is chosen as the limiter. The grey circles are assumed to be the nonlinear elements. FDTD method is used for time domain analysis of such a device. First we assume that the PC device is comprised of linear element ($n_2 = 0$); in such a case the input frequency is tuned at the cavity's notch. No signal can pass the waveguide in this case. Then we assume that the defect has Kerr effect which causes a shift in cavity's central frequency and allows the signal to pass. The switching mechanism of this limiter is shown in Fig. 10b and Fig. 10c.

The mentioned structure cannot provide a sharp limiting operation. Based on our observations, if a number of cavities are used in cascade a sharper curve can be obtained. It should be noted that the distance between the cavities should be chosen long enough so that they cannot have loading effect on each other. The proposed structure is shown in Fig. 11a. In order to be able to observe the switching behaviour more accurately, the transmission ratio of the switch versus the input power has been depicted in Fig. 11b. We need that by doubling the input power a switching mechanism be observed in our design; therefore the triple cascade combination can be used for our design.

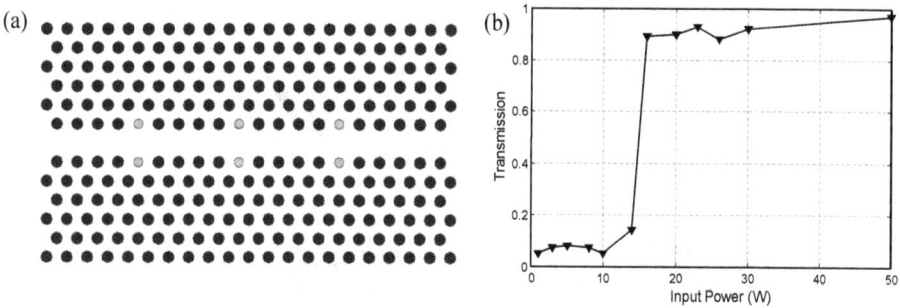

Fig. 11. (a) The limiter topology used to design an optical gate, (b) The transmission curve for the limiter in fig. 11(a).

4.2 A PC Y-splitter

As it was mentioned before, combination of a Y-junction and a limiter is proposed as the AND gate. Using methods presented in the literature (Wilson et al., 2003, Frandsen et. al, 2004, Yang et al, 2010, Danaie et. al, 2008) we have optimized the Y-splitter. A simple Y-junction and the optimized junction are shown in Fig. 12a and Fig. 12b respectively. By introducing modifications to the structure of the bends and branches the transmission spectrum can be improved (Wilson et al., 2003, Frandsen et. al, 2004, Yang et al., 2010). There have been some detailed topologies reported in literature which enhance the bandwidth (Borel et al., 2005, Têtu et al., 2005). (Yang et al., 2010) used a triangular lattice of dielectric rods to design a Y-Branch. They showed that the Y-branch can be treated as a cavity that couples with the input and output waveguides.

In (Yang et al., 2010) to enhance the transmittance, additional rods are added to the junction area and the corner rods are displaced. The movement of the corner rods increases the volume of their cavity and make the cavity mode resonant with the waveguide modes. It is shown in (Yang et al., 2010) that using the coupled mode theory the reflection coefficient can be expressed as (31).

$$R = \frac{\left| -j(\omega - \omega_0) + \dfrac{1}{\tau_1} - \dfrac{1}{\tau_2} - \dfrac{1}{\tau_3} \right|^2}{\left| j(\omega - \omega_0) + \dfrac{1}{\tau_1} + \dfrac{1}{\tau_2} + \dfrac{1}{\tau_3} \right|}, \tag{31}$$

where, ω_0 is the resonance frequency, τ_i is the time constant regarding the amplitude decay of the resonance into the ith port. It is seen that if the (32) is satisfied, the reflection will be zero for $\omega = \omega_0$.

$$\frac{1}{\tau_1} = \frac{1}{\tau_2} + \frac{1}{\tau_3}. \tag{32}$$

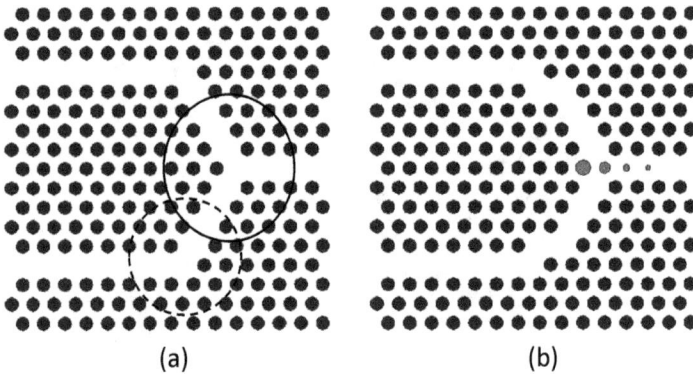

(a)   (b)

Fig. 12. (a) A simple photonic crystal Y-splitter consisting of a Y- junction (solid line) and two 60-degree bends (dashed line) (b) a Y-junction optimized to decrease the input reflection. The grey hole have radii equal to 0.4a, 0.31a, 0.15a, 0.11a respectively from left to right.

We already know that for our junction due to symmetry $\tau_2 = \tau_3$; therefore the mentioned condition can be simplified as: $\tau_2 = 2\tau_1$. It means that the decay rate of resonance to the output ports should be half the decay rate of resonance to the input port. In a simple Y-junction all the three time constants are equal. The tapered structure of the holes inserted in our proposed design, improves the coupling strength between the resonator and the input waveguide. It results in the reduction of τ_1. If the tapering is designed so that τ_1 is reduced to half τ_2, then the zero reflection condition can be satisfied. Numerical optimization was used to design such a tapering.

4.3 Design of a PC AND gate

In order to design a PC AND gate the structure depicted in Fig. 13 is proposed. It is comprised of the proposed Y-junction in Fig. 12b and the sharp limiter in Fig. 11a. Each signal is assumed to have a power equal to 10W. The junction section adds the signals together.

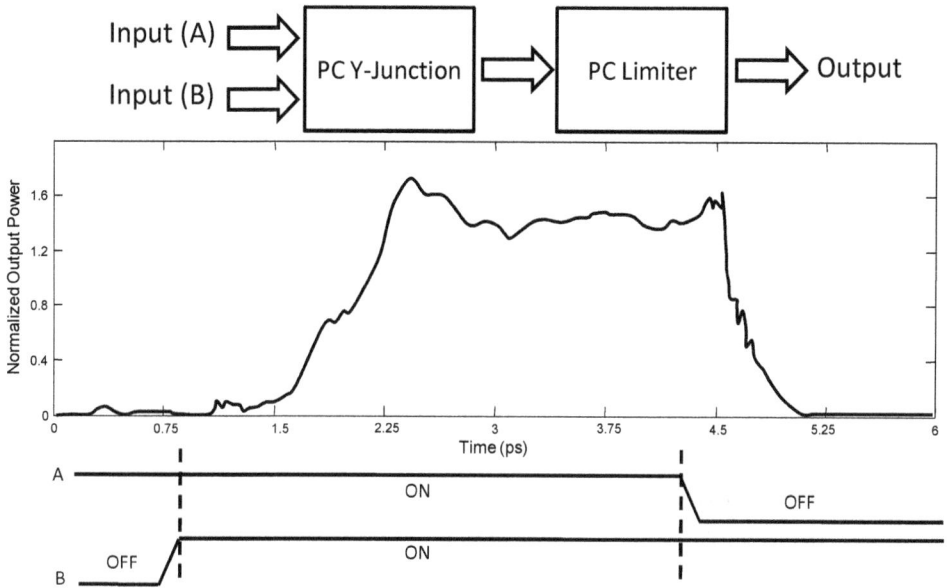

Fig. 13. The structure for the proposed AND gate.

If only one signal is sent into the junction the power will be bellow the limiter threshold and the signal will not be able to reach the output port. However when the two signals are present, the output power of the splitter will be high enough to trigger the switching mechanism of the limiter. Therefore the signal will be directed to the output port. In order to have a better insight on the performance of the device; the time domain response is also depicted. First it is assumed that only one of the input signals is ON. Then both signals are turned ON for a while and one of them is turned OFF afterwards. The normalised output power is measured for each case. It can be seen that the transition time is less than 1ps.

4.4 Implementation issues

Many researchers have worked on methods to efficiently couple an optical fiber to a photonic crystal waveguide. The issue of designing a coupling scheme becomes more important when photonic crystal waveguides with low group velocities are used. However, in our case since the group velocity of the W1 waveguides in inputs and outputs is not much low, the coupling problem to the W1 waveguides does not seem to be acute. In order to design the input and output couplers for the AND gate, the same efficient coupling structure implemented in (Ayre et al., 2005) can be used. The mentioned paper suggests a wideband Y-splitter in a GaAs substrate for TE modes. A 5μm ridge waveguide is coupled to the photonic crystal waveguide with period = 430 nm and 36.2% air filling factor (corresponding to a radius over period (r/a) ratio of 0.30) via a 9μm injector tapering to provide a bandwidth of the order of 300 nm.

Due to nearly exact same prameters, for the structure proposed as the AND gate and the Y-splitter in (Ayre et al., 2005), the fabrication procedure described in (Ayre et al., 2005) can also be employed for design and implementation of the gate. The AlGaAs/GaAs heterostructure can be grown by metal-organic chemical vapor-phase deposition (MOCVD); where a 300 nm silicon oxide is deposited over the wafer using plasma enhanced chemical vapor-phase deposition (PECVD), followed by a 200-nm-thick layer of PMMA as electron-beam resist. The pattern is then transferred into the oxide hard mask via reactive ion etching (RIE) using a fluorine process. The deep etching of the PC holes can be achieved using chemically assisted ion beam etching (CAIBE) in a Chlorine-Argon process (Ayre et al., 2005). The holes would thus be approximately 605nm deep. The input wavelength is assumed 1550nm which can be obtained using an Erbium doped fiber laser which results in the device size to be approximately equal to 5.95µm×12.90µm.

5. References

Andalib, P.; Granpayeh, N. (2009). All-optical ultra-compact photonic crystal NOR gate based on nonlinear ring resonators, J. of Optics A: Pure and Applied Optics, vol. 11, no. 8, (Aug. 2009) 08523.

Ayre, M.; Karle, T. J.; Wu, L.; Davies, T.; Krauss, T. F. (2005) Experimental verification of numerically optimized photonic crystal injector, Y-splitter, and bend, IEEE J. Selected Areas in Communications, vol. 23, no. 7 (Jul. 2005) pp. 1390-1395.

Bai, J.; Wang, J.; Jiang, J.; Chen, X.; Li, H.; Qiu, Y.; Qiang, Z. (2009). Photonic crystal NOT and NOR gates based on a single compact photonic crystal ring resonator, Applied Optics, vol. 48, no. 36 (Dec. 2009) pp.6923-6927.

Borel, P.I.; Frandsen, L.H.; Harpøth, A.; Kristensen, M.; Jensen, J. S.; Sigmund, O. (2005). Topology optimised broadband photonic crystal Y-splitter, IEE Electron. Let., vol. 41, no. 2 (Jan. 2005) pp. 69-71.

Boyd, R. W. (2003) Nonlinear Optics, Academic Press, San Diego, USA, ISBN: 0121216829.

Bravo-Abad, J.; Fan, S.; Johnson, S. G.; Joannopoulos, J. D.; Solja˘cic´, M. (2007). Modeling nonlinear optical phenomena in nanophotonics, J. of Lightwave Technol., vol. 25, no. 9, (Sep. 2007) pp. 2539-2546.

Centeno, E.; Felbacq, D; (2000). Optical bistability in finite-size nonlinear bidimensional photonic crystals doped by a microcavity, Phys. Rev. B, Condens. Matter, vol. 62, no. 12, (Sep. 2000) pp. R7683-R7686.

Cuesta-Soto, F; Martinez, A.; Garcia, J.; Ramos, F.; Sanchis, P.; Blasco, J.; Marti, J. (2004). All-optical switching structure based on a photonic crystal directional coupler, Optics Express, vol. 12, no. 1, (Jan. 2004) pp. 161-167.

Cuesta-Soto, F; Martinez, A; Garcia, J; Ramos, F; Sanchis, P; Blasco, J; Marti, J. (2004). All-optical switching structure based on a photonic crystal directional coupler, Optics Express, vol. 12, no. 1, (Dec. 2003) pp. 161-167.

Danaie, M.; Attari, A. R.; Mirsalehi, M. M.; Naseh, S. (2008). Design of a high efficiency wide-band 60° bend for TE polarization, Photonics and Nanostructures: Fundamentals and Applications, vol. 6, no. 3-4, (Dec. 2008) pp. 188–193.

Danaie, M., Kaatuzian, H. (2011). Bandwidth Improvement for a Photonic Crystal Optical Y-splitter, Journal of the Optical Society of Korea, vol. 15, no. 3 (Sep. 2011) pp. 283-288.

Danaie, M.; Kaatuzian, H. (2011). Improvement of power coupling in a nonlinear photonic crystal directional coupler switch, Photonics and Nanostructures – Fundamentals and Applications, vol. 9, no. 1 (Feb. 2011) pp. 70–81.

Fan, S. (2002). Sharp asymmetric line shapes in side-coupled waveguide-cavity systems, Appl. Phys. Lett., vol. 80, no. 6, (Feb. 2002) pp. 908-910.

Frandsen, L. H.; Borel, P. I.; Zhuang, Y. X.; Harpøth, A.; Thorhauge, M.; Kristensen, M. (2004). Ultralow-loss 3dB photonic crystal waveguide splitter, Opt. Lett., vol. 29, no. 14 (Jul. 2004) pp. 1623-1625.

Franken, P. A.; Hill, A. E.; Peters, C. W.; Weinreich, G. (1961). Generation of optical harmonics, Phys. Rev. Lett., vol. 7, no. 4, (Aug. 1961) pp. 118-199.

Fushman, I.; Englud, D.; Faraon, A.; Stoltz, N.; Petroff, P.; Vuckovic, J. (2008) Controlled phase shifts with a single quantum dot, Science, vol. 320, no. 5877 (May 2008) pp. 769-772.

John, S. (1987). Strong localization of photons in certain disordered dielectric superlattices, Phys. Rev. Lett., vol. 58, no. 23, (Jun. 1987) pp. 2486-2489.

Joseph, R. M.; Taflov, A. (1997). FDTD Maxwell's equations models for nonlinear electrodynamics and optics, IEEE Tran. on Anten. and Prop., vol. 45, no. 3, (Mar 1997) pp. 364-374.

Kaatuzian, H. (2008) photonics, vol. 1, 2nd edition (in Persian), Amirkabir University Press, Tehran, Iran, ISBN: 9644632710.

Liu, Q; Ouyang, Z; Wu, C. J.; Liu, C. P.; Wang, J. C. (2008). All-optical half adder based on cross structures in two-dimensional photonic crystals, Optics Express, vol. 16, no. 26. (Nov. 2008) pp.18992-19000.

Locatelli, A; Modotto, D; Paloschi, D; Angelis, C. D. (2004). All optical switching in ultrashort photonic crystal couplers, Optics Commun., vol. 237, (March) pp. 97-102.

Maiman, T. H. (1960). Stimulated optical radiation in ruby. Nature, vol. 187 (4736): (Aug. 1960) pp. 493-494.

Mingaleev, S. F.; Kivshar, Y. S. (2002). Nonlinear transmission and light localization in photonic-crystal waveguides, J. Opt. Soc. Amer. A, Opt. Image Sci., vol. 19, no. 9, (Sep. 2002) pp. 2241-2249, ISSN: 0740-3224.

Nagpal, Y.; Sinha, R. K. (2004) Modelling of photonic band gap waveguide couplers, Microwave and Opt. Technol., Lett., vol. 43, no. 1 (Oct. 2004) pp. 47-50.

Nakamura, H.; Sugimoto, Y.; Kanamoto, K.; Ikeda, N.; Tanaka, Y.; Nakamura, Y.; Ohkouchi, S.; Watanabe, Y.; Inoue, K.; Ishikawa, H.; Asakawa, K. (2004) Ultrafast photonic crystal/quantum dot all-optical switch for future photonic networks, Optics Express, vol. 12, no. 26 (Dec. 2004) pp. 6606-6614.

Rahmati, A. T.; Granpayeh, N. (2010). Design and simulation of a switch based on nonlinear directional coupler, Optik -Internat. J. for Light and Electron. Optics, vol. 121, no. 18, (Oct. 2010) pp. 1631-1634.

Saleh, B. E. A.; Teich, M. C. (1991). Fundamentals of Photonics, John Wiley & Sons, New York, USA, ISBN: 0471839655.

Scholz, S.; Hess, O.; Rüuhle, R. (1998). Dynamic cross-waveguide optical switching with a nonlinear photonic band-gap structure, Optics Express, vol. 3, no. 1, (Jul. 1998) pp. 28-34.

Shinya, A.; Matsuo, Yosia, S; Tanabe, T.; Kuramochi, E; Sato, T.; Kakitsuka, T.; Notomi, M. (2008). All-optical on-chip bit memory based on ultra high Q InGaAsP photonic crystals, Optics Express, vol. 16, no. 23, (Nov. 2008) pp. 19382-19387.

Soljaˇcic´, M.; Ibanescu, M.; Johnson, S. G.; Fink, Y.; Joannopoulos, J. D. (2002). Optimal bistable switching in nonlinear photonic crystals, Phys. Rev. E, vol. 66, no. 5, (Nov. 2002) pp. 055601R.

Soljaˇcic´, M.; Joannopoulos, J. D. (2004) Enhancement of nonlinear effects using photonic crystals, Nat. Mater., vol. 3, no. 4, (Apr. 2004) pp. 211–219.

Têtu, A.; Kristensen, M.; Frandsen, L.; Harpøth, A.; Borel, P.; Jensen, J.; Sigmund, O. (2005) Broadband topology-optimized photonic crystal components for both TE and TM polarizations, Optics Express, vol. 13, no. 21 (Oct. 2005) pp. 8606-8611.

Weinberger, P. (2008). John Kerr and his Effects Found in 1877 and 1878, Philosophical Magazine Letters, vol. 88, no. 12, (Dec 2008) pp.897–907.

Wilson, R.; Karle, T. J.; Moerman, I.; Krauss, T. F. (2003). Efficient photonic crystal Y-junctions, J. Opt. A: Pure Appl. Opt., vol. 5, no. 4 (Jul. 2003) pp.S76–S80.

Yablonovitch, E. (1987). Inhibited spontaneous emission in solid-state physics and electronics, Phys. Rev. Lett., vol. 58, no. 20, (May 1987) pp. 2059–2062.

Yamamoto, N.; Ogawa, T; Komori, K. (2006). Photonic crystal directional coupler switch with small switching length and wide bandwidth, Optics. Express, vol. 14 (Feb. 2006) pp. 1223-1229.

Yang, W.; Chen, X.; Shi, X.; Lu, W. (2010) Design of high transmission Y-junction in photonic crystal waveguides, Physica B: Condensed Matter., vol. 405, no. 7 (Apr. 2010) pp. 1832-1835.

Yanik, M. F.; Fan, S.; Soljaˇcic´, M. (2003). High-contrast all-optical switching in photonic crystal microcavites, App. Phys. Lett., vol. 83, no. 14, (Oct. 2003), pp. 2739-2741.

Zhu, Z.-H.; Ye, W.-M.; Ji, J.-R.; Yuan, X.-D. ; Zen, C. (2006). High-contrast light-by-light switching and AND gate based on nonlinear photonic crystals, Optics Express, vol.14, no. 5, (Mar. 2006)pp. 1783-1788.

Zimmermann, J.; Kamp, M.; Forchel, A.; Marz, R. (2004). Photonic crystal waveguide directional couplers as wavelength selective optical filters, Opt. Communications, vol. 230 no. 4-6, (Feb. 2004) pp. 387-392.

7

On the Applicability of Photonic Crystal Membranes to Multi-Channel Propagation

Bartłomiej Salski[1], Kamila Leśniewska-Matys[1] and Paweł Szczepański[1,2]
[1]Warsaw University of Technology
[2]National Institute of Telecommunications
Poland

1. Introduction

The most common 2D geometrical arrangements of photonic crystals (PhC) are square and triangular (hexagonal) lattices as shown in Fig.1. Assuming that a PhC structure is expanded to infinity along the x-axis, the problem belongs to a so-called vector 2D class (Gwarek et al., 1993). However, it may frequently be simplified even further to a scalar 2D class, restricting a wave vector k to a PhC plane (yz-plane in Fig.1). In such a case, any electromagnetic field propagating in the PhC plane can be decomposed into two orthogonal modes, usually denoted as transverse magnetic (TM) and transverse electric (TE) with respect to the x-axis.

Although performance of PhC-based devices relies, in most cases, on the confinement of light within a photonic bandgap (PBG), photonic crystals also exhibit remarkable dispersion properties in their transmission bands, thus opening the perspective for new optical functionalities.

A lot of research activities have been undertaken in the development of planar PhC passive optical devices, like waveguides (Loncar et al., 2000; Chow et al., 2001), filters (Ren et al., 2006; Fan et al., 1998), couplers (Yamamoto et al., 2005; Tanaka et al., 2005), power splitters (Park et al., 2004; Liu et al., 2004) or, recently, active devices for laser beam generation operating as a surface-emitting microcavity laser (Srinivasan et al., 2004), a photonic band-edge laser (Vecchi et al., 2007) or an edge-emitting laser (Shih et al., 2006; Lu et al., 2009). However, PhC devices in practical realizations are of a finite thickness (see Fig.2), thus, limiting applicability of the approximate 2D modelling approach to those scenarios where the PhC's thickness is large enough with respect to wavelength. Otherwise, the problem becomes 3D and a complete full-wave EM approach is essential.

Similarly to 2D waveguiding slabs, optical confinement of light in thin membranes depends primarily on a contrast between the membrane's and cladding's refractive indices. Most of all, a propagating mode has to be located beyond a light cone of the cladding, if energy leakage wants to be suppressed. Secondly, the mode has to be confined within a channel processed between the surrounding photonic crystal boundaries. The photonic bandgap exists only for those modes that are totally internally reflected at the interface between the channel and the photonic crystal. Furthermore, if the membrane is deposited on a low-index dielectric film, instead of being symmetrically surrounded with air, additional complications of a design process are introduced.

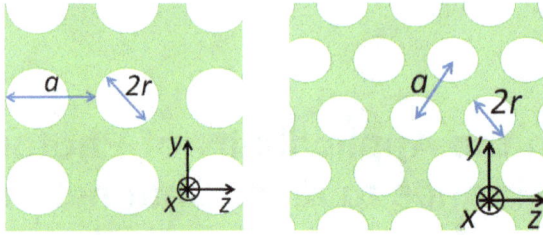

Fig. 1. The definition of square (left) and triangular (right) air-hole lattices.

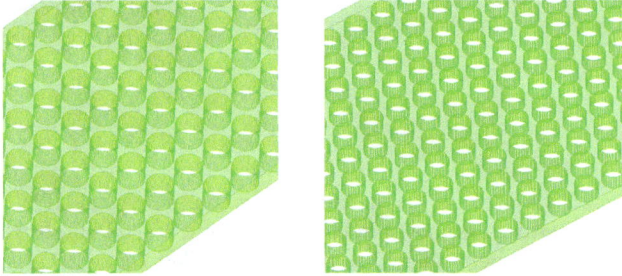

Fig. 2. A perspective view of dielectric membranes with square (left) and triangular (right) PhC air-hole lattices.

In the next Section, a brief overview of the developments of mutli-channel laser generation techniques is given, especially in the context of so-called supermode multi-channel propagation.

2. State-of-the-art in multi-channel laser generation techniques

A phase-locked operation of multi-channel waveguide devices supporting propagation of lateral modes (also known as supermodes) was studied mostly in 80's. The main goal of theoretical and experimental research was to achieve higher power density of the coherent laser beams generated in semiconductors. Phased array lasers, consisting of N single-mode waveguides, can guide, in total, N array modes. In practice, the most likely excited mode is of the highest order (Yariv, 1997). Consequently, relatively broad far-field patterns as well as broad spectral linewidths are obtained. To solve or at least alleviate that disadvantageous property, it is essential to distinguish appropriate supermodes.

Fig.3 shows near-field patterns of five supermodes supported by index-guided arrays consisting of five identical and non-identical channels. The supermode patterns were calculated with a numerical solver of Maxwell's equations (Kapon et al., 1984a). In particular, the excitation of a fundamental supermode results in a single-lobe radiation beam aligned with the array channels. However, as it has been shown in Fig.3a, in uniform arrays with identical channels, intensity patterns of the fundamental and the highest order supermodes are similar to each other, so their discrimination becomes difficult. Moreover, as it has been shown in (Kapon et al., 1984a), since inter-channel regions are usually lossy, the highest order supermode, with a two-lobe far field pattern, is often favoured over the other modes. Subsequently, variation in the channels' width (known as chirped arrays) results in significantly different near field envelope patterns of the fundamental and higher-order supermodes, in contrast to the case of uniform arrays (Kapon et al., 1984a). In such arrays, higher order supermodes can be suppressed by employing a proper gain distribution.

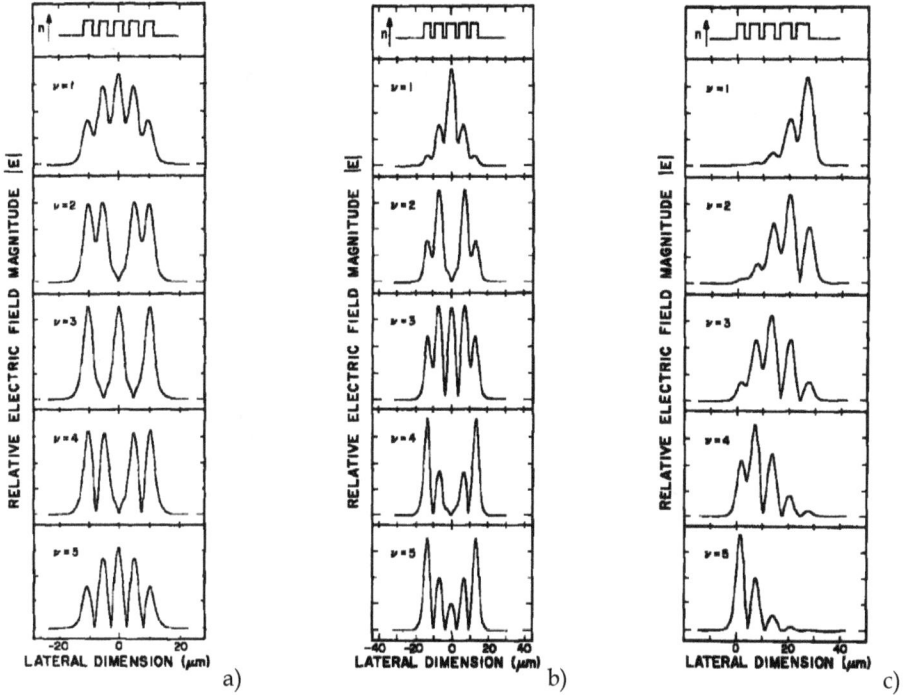

Fig. 3. Near-field patterns of the supermodes in a five-element a) uniform array b) inverted-V chirped array, c) linearly chirped array (Kapon et al., 1984a).

Next, in an inverted-V chirped array (Fig.3b), the power of the fundamental supermode is concentrated in central channels, whereas the higher order supermodes are more localized in the outermost channels. Since gain in the active region is larger when the laser channels are wider, the fundamental supermode is expected to have a higher modal gain (near threshold) and, in consequence, is more likely to oscillate (Kapon et al., 1984b).

In (Kapon et al., 1986), a buried ridge array has been proposed. In such arrays, a small refractive index contrast between the channels and inter-channel regions is applied. That soft index profile ensures effective coupling between the adjacent array laser channels via their evanescent optical fields. Since the channels in those arrays are defined by a built-in distribution of the refractive index, it is possible to achieve a uniform gain distribution across the array, while maintaining the channel definition. Such an approach makes the buried ridge arrays different from the gain guided arrays, in which inter-channel regions are inherently more lossy. Moreover, buried ridge arrays operate mainly with the fundamental supermode, thus, producing a single-lobe radiation beam.

The first analytical interpretation of supermodes behaviour in the phased array lasers was proposed in (Scifres et al., 1979). Experimental data was interpreted by considering a diffraction pattern of a structure with equally-spaced slits corresponding to individual laser array elements. Such an approach is usually known as a simple diffraction theory. Although the simple diffraction theory has been proved useful to interpret some experimental results (Scifres et al., 1979; AcMey & Engelmann, 1981; van der Ziel et al., 1984), it provides no means to describe the allowed oscillating modes in the array of coupled emitters.

In the early 70's, an alternative method, known as a coupled mode theory, was intensively investigated (Yariv, 1973; Yariv & Taylor, 1981; Kogelnik, 1979). It has been successfully applied to the modelling and analysis of various guided-wave optoelectronic and fibre optical devices, such as optical directional couplers (Taylor, 1973; Kogelnik & Schmidt, 1976), optical fibres (Digonnet & Shaw, 1982; Zhang & Garmire, 1987), phase-locked laser arrays (Kapon et al., 1984c; Mukai et al., 1984; Hardy et al., 1988), distributed feedback lasers (Kogelnik & Shank, 1972) and distributed Bragg reflectors (Schmidt et al., 1974).

One of major assumptions made in the conventional coupled mode theory is that the modes of uncoupled systems are orthogonal to each other. In coupled systems, however, one often chooses the modes of isolated systems as the basis for the mode expansion and these modes may not be orthogonal. Therefore, the orthogonal coupled mode theory (OCMT) is not suitable for the description of the mode-coupling process in that case. Non-orthogonality of modes in optical couplers, due to crosstalk between the waveguide modes, was first recognized in (Chen & Wang, 1984). Later on, several formulations of the non-orthogonal coupled mode theory (NCMT) were developed by several authors (Hardy & Streifer, 1985; Chuang, 1987a; Chuang, 1987b; Chuang, 1987c). It has been shown that NCMT yields more accurate dispersion characteristics and field patterns for the modes in the coupled waveguides. Better accuracy is even more essential to the modelling of coupling between non-identical waveguides. It is evident for weak coupling, though the new formulation extends the applicability of the coupled mode theory to geometries with more strongly coupled waveguides. However, NCMT becomes inaccurate when considering very strongly coupled waveguide modes (Hardy & Streifer, 1985).

To the best of authors' knowledge, edge-emitting multi-channel membrane lasers have not been manufactured so far, although single-channel membrane lasers processed on a GaAs photonic crystal membrane were already presented (Yang, et al., 2005; Yang, et al., 2007; Lu, et al. 2009). One of the major reasons lies in technological challenges in achieving acceptable repeatability of the photonic crystal structure manufacturing process (Massaro, et al., 2008). However, with the advent of new technology nodes those challenges will likely be overcome or at least substantially alleviated, opening a wide range of applications to the methodology addressed below.

In this Chapter, a complete design cycle of a new type of phased array laser structures processed in photonic crystal membranes is presented. Due to a very strong coupling between the adjacent channels in the array, a non-orthogonal coupled mode theory was applied in order to maintain the rigidity of the analysis.

3. Numerical modelling of optical channels in photonic crystal membranes

A complete electromagnetic design cycle of single- and multi-channel optical propagation in PhC membranes is presented in this Section, together with the computational methods and tools applied. First, dispersive properties of an infinitely large PhC membrane with no defects are investigated to exemplify general rules for the photonic bandgap (PBG) generation as a function of PhC membrane geometry and incident light wavelength. Once a PBG dispersion diagram is achieved, a defect channel is processed in the PhC membrane and dispersive properties of such an optical waveguide are considered. For the purpose of this Chapter, propagation of transverse–electric (TE) modes in defect PhC membrane channels based on the square lattice type is studied only. However, the introduced methodology may be easily extended to other lattice types with either TE or TM polarisation. The obtained PBG diagrams will help detecting the supermodes within a photonic bandgap. Eventually, electric field patterns of those modes are computed to assess their applicability to the laser beam generation. As it is shown in Section 4.3, those field distributions are useful to calculate laser characteristics of the single- and multi-channel photonic crystal membrane lasers.

3.1 Bandgaps in photonic crystal membranes

Two common lattice types processed in a photonic crystal membrane are investigated, namely square and triangular (see Fig.4). The lattices are cut with air holes in an indium gallium arsenide phosphide (InGaAsP) layer with a refractive index of $n = 3.4$. At this stage, the goal is to specify design rules for the photonic bandgap generation as a function of the most critical parameters of those structures, that is, a membrane's thickness d, a lattice constant a and an air holes' radius r.

Numerical computations are performed using a full-wave electromagnetic approach with a finite-difference time-domain (FDTD) method implemented in a QuickWave-3D simulator (Taflove & Hagness, 2005; QWED). Since the structure is periodic in two dimensions, the computation with FDTD is enhanced with the Floquet's theorem (Collin, 1960), also known as the Bloch's one, which allows us to reduce a computational domain to a single period of the lattice (Salski, 2010), as exemplified in Fig.4. Considering periodicity along the z-axis, the following periodic boundary conditions (PBCs), derived from the Floquet's theorem, are enforced at periodic faces of the structure:

$$\vec{E}_\perp \left(x,y,z+L,t \right) = \vec{E}_\perp \left(x,y,z,t \right) e^{j\psi} \tag{1}$$

$$\vec{H}_\perp \left(x,y,z,t \right) = \vec{H}_\perp \left(x,y,z+L,t \right) e^{-j\psi} \tag{2}$$

where L is the period of the structure along the z-axis, $\perp$ denotes the components transverse to periodicity (in this case x- and y- components), and ψ is a fundamental Floquet phase shift per period L understood as a user-defined parameter.

As it has been shown in (Celuch-Marcysiak & Gwarek, 1995; Salski, 2010), incorporation of the Floquet's theorem into FDTD schemes results in a complex notation of time-domain electromagnetic fields with the real and imaginary FDTD grids computed simultaneously at the same structure's mesh and coupled via PBCs in each iteration cycle. The method is known as Complex-Looped FDTD (CL-FDTD) and is implemented in the QuickWave-3D simulator (QWED). Additionally, due to conformal meshing implemented in QuickWave-3D (Gwarek, 1985), curvature of the air holes, as shown in Fig.4, is accurately represented on the FDTD mesh with no deteriorating effect on memory storage and computing time. A vertical cross-section of a unit membrane lattice cell sketched in Fig.4 (right) indicates that the structure is situated in air which, in order to reduce the computational volume, is truncated with absorbing boundary conditions, usually known as Mur superabsorption (Mei et al., 1992).

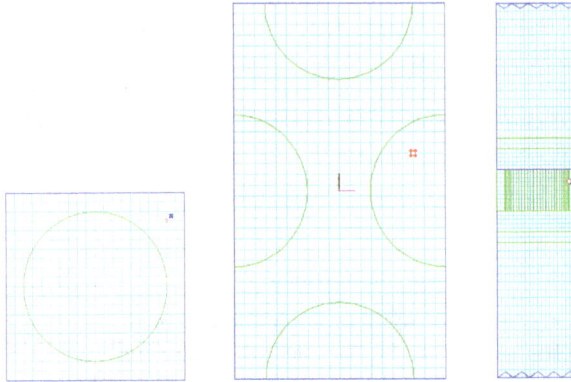

Fig. 4. A horizontal cross-section view of FDTD models of a unit cell of square (left) and triangular (centre) air-hole lattices and a vertical cross-section (right) with absorbing boundary conditions truncating the air regions below and above the membrane.

In each simulation run, for a particular set of Floquet's phase shifts per y- and z- periods, a point excitation located somewhere inside a unit cell is driven with a wideband pulse (e.g. a Dirac's delta), injecting energy into the structure. As the simulation continues, the Fourier transform is iteratively calculated until a convergent state is achieved. Fig.5 shows an exemplary spectrum of an injected electric current for the lattice shown in Fig.4 (left). Resonances indicate the eigenvalues (frequencies) of the detected modes satisfying Floquet's phase shifts imposed by periodic boundary conditions.

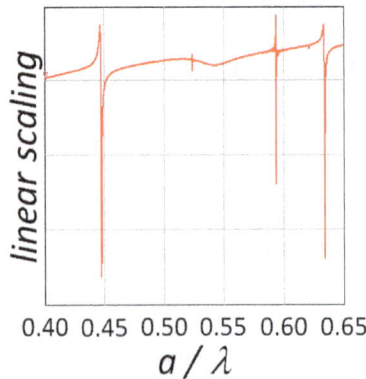

Fig. 5. The spectrum of an electric current injected into a unit cell of a square PhC membrane as shown in Fig.4 (left) with the imposed Floquet's phase shifts along y- and z-axis $\psi_y = \psi_z = 0$ radians ($n = 3.4$, $r/a = 0.4$, $d/a = 0.4$).

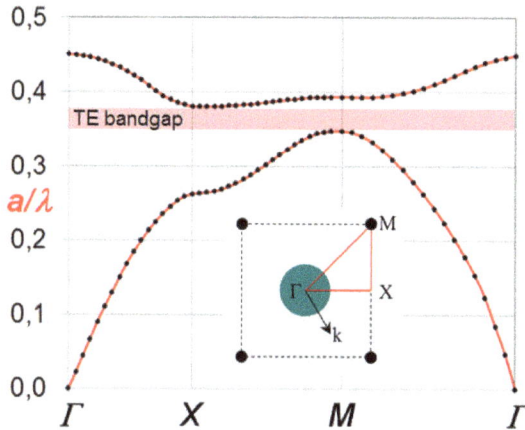

Fig. 6. A TE mode photonic bandgap diagram for an air-hole square lattice cut in an InGaAsP membrane ($n = 3.4$, $r/a = 0.4$, $d/a = 0.4$).

Owing to the wideband spectral properties of the CL-FDTD method, a single simulation provides information about all the modal frequencies within the spectrum of our interest satisfying the imposed Floquet's phase shifts. Thus, the simulator has to be invoked as many times as the number of wave vector points chosen to collect a PBG diagram along a whole contour of an irreducible Brillouin zone (Salski, 2010). In this case, a single simulation of the model consisting of 18 400 FDTD cells (ca. 4MB RAM) takes 11 seconds on Intel Core i7 CPU 950 with the speed of 1785 iter/sec. An FDTD cell size is set to $a/20$, leading to at least 40 FDTD cells per wavelength in free space and ca. 12 in the membrane. Calculation of the whole PBG diagram shown in Fig.6 with 55 wave vector points takes, in total, ca. 55 x 11 sec = 10 minutes.

In the case of an air-hole square lattice cut in an InGaAsP membrane, the PBG diagram of which is shown in Fig.6, an 8.8% wide indirect X-M TE bandgap for the normalised frequency $a/\lambda = 0.348 \ldots 0.380$ is found. Although it is not exemplified in Fig.6, a TM bandgap is not present in that spectrum range, what may be considered as a potential disadvantage in applications when the precise control of beam propagation is necessary. It can be solved using a triangular lattice, where both TE and TM bandgaps may coincide within the same spectrum range. However, this issue extends beyond the scope of the Chapter and is not considered here.

Consider now the impact of geometrical settings on a TE bandgap in the investigated square PhC membrane. Fig.7 shows the modal dispersion as a function of the membrane's thickness d/a. It can be seen that, the TE bandgap decreases with the increasing membrane's thickness d/a, while covering the spectrum width of 8.8%, 9.2% and 10.4% for $d/a = 0.4$, 0.5 and 0.6, respectively. It shows that the membrane's thickness d/a has a relatively minor impact on the photonic bandgap width. Next, Fig.8 presents the computation results for a variable air-holes' radius r/a, and it is evident that the radius, in contrast to the membrane's thickness, has a substantial impact on the bandgap width, which reaches zero below ca. $r/a = 0.25$. Later in this paper, those PhC structures are applied for single- and multi-channel propagation of optical pulses.

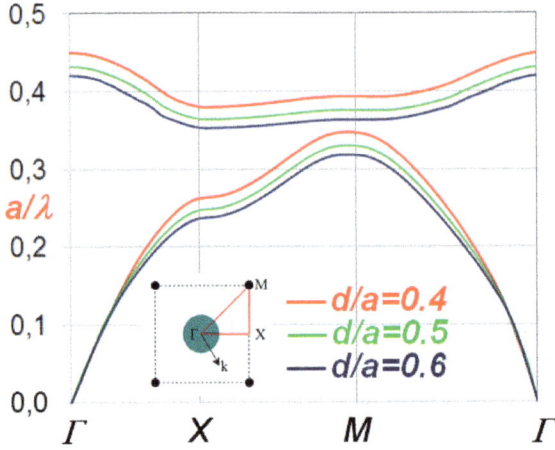

Fig. 7. TE mode photonic bandgap diagrams for air-hole square lattices cut in an InGaAsP membrane in function of a membrane's thickness d/a (n = 3.4, r/a = 0.4).

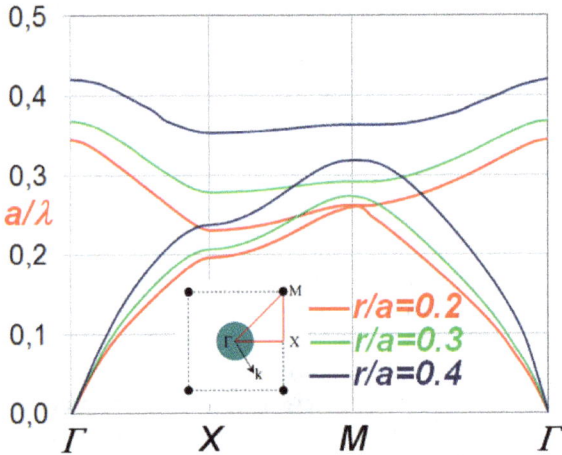

Fig. 8. TE mode photonic bandgap diagrams for air-hole square lattices cut in an InGaAsP membrane in function of an air-holes' radius r/a (n = 3.4, d/a = 0.4).

Similar computations were carried out for a membrane with a triangular air-hole lattice. First of all, as it may be inferred from Fig.9, a direct TE bandgap is achieved at a Γ critical point. Secondly, the achieved TE bandgap spectra are much wider when compared to their counterparts computed for the square lattice. Fig.9 shows that the spectrum width amounts to 40.7%, 42.6% and 43.8% for d/a = 0.4, 0.5 and 0.6, respectively. Next, Fig.10 depicts the impact of the air-holes' radius r/a on the TE bandgap spectrum width, which amounts to 16.6%, 40.7% and 36.9% for r/a = 0.3, 0.35 and 0.4, respectively. The simulations show that if a narrow bandgap is favourable in a considered application, the square lattice is a better option, while the triangular one allows creation of a wider bandgap.

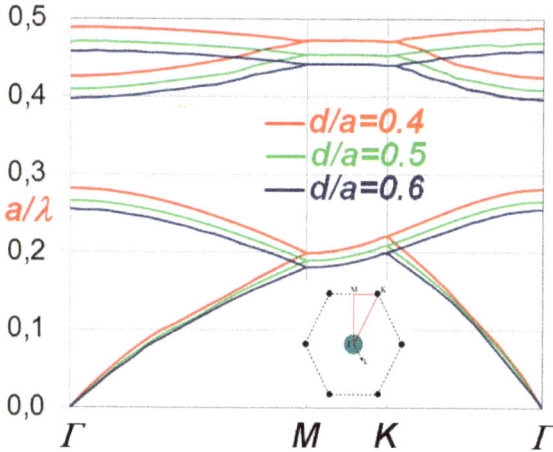

Fig. 9. TE mode photonic bandgap diagrams for air-hole triangular lattices cut in an InGaAsP membrane in function of a membrane's thickness d/a (n = 3.4, r/a = 0.35).

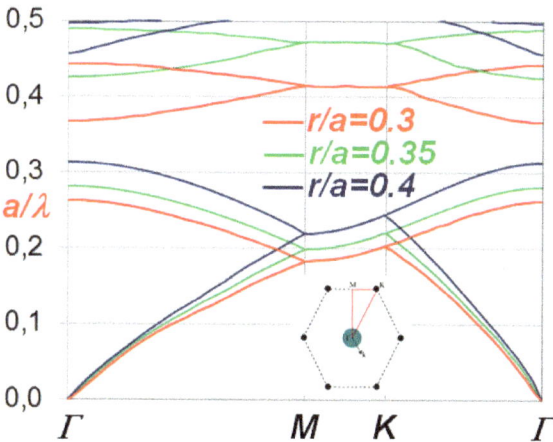

Fig. 10. TE mode photonic bandgap diagrams for air-hole triangular lattices cut in an InGaAsP membrane in function of an air-holes' radius r/a (n = 3.4, d/a = 0.4).

Concluding, it may be noticed from the investigation given in this Section and from the literature as well, that two-dimensional air-hole photonic crystals processed in thin dielectric membranes exhibit the following properties:

- PBG for TE modes shifts up in frequency and broadens with the increasing air-holes' radius r/a for both square and triangular air-hole lattices,
- PBG for TE modes shifts up in frequency with the increasing membrane's thickness d/a with no meaningful influence on its bandwidth for both square and triangular air-hole lattices,
- a PBG bandwidth for TE modes is much wider when a triangular air-hole lattice is used,

- PBG cannot be created for TM modes in a two-dimensional square air-hole photonic crystal processed in a dielectric membrane, while it is feasible in a triangular one when the membrane is thick enough (Joannopoulos et al., 2008).

3.2 Bandgaps in defect channels processed in photonic crystal membranes

Spectral properties of TE modes propagating in defect channels, as exemplified in Fig.11, processed in PhC membranes are investigated below. PBG diagrams are computed with the aid of the QuickWave-3D electromagnetic FDTD simulator (QWED), in the same way as in the case of non-defect PhC membranes as shown in Fig.4. This time, however, an FDTD model consists of a single PhC row, as marked with a red dashed line in Fig.11. Since it is assumed that the waveguide is infinitely long, the Floquet's periodic boundary conditions are enforced only along the channel's axis, while lateral dimensions are truncated with the absorbing boundary conditions (Mei, 1992). PBG diagrams for the square lattice channels are computed for phase shifts within a range designated by Γ and X critical points of the first irreducible Brillouin zone of the corresponding non-defect PhC membranes.

Fig.12 depicts the modes computed for a single square channel in an air-hole square lattice cut in an InGaAsP membrane (n = 3.4, r/a = 0.4, d/a = 0.4, b/a = 0.3). Black curves indicate the propagating modes with one distinguished by a green colour, while the red curves depict the modes of the non-defect PhC membrane surrounding the channel. It can be seen that a single defect mode is achieved (green) within a photonic bandgap (red semi-transparent zone) of the surrounding PhC. The mode has a uniquely defined phase constant within the a/λ = 0.356... 0.369 spectrum range, that is 3.6% wide, although a light cone additionally limits the allowed spectrum range to a/λ = 0.356... 0.365 (2.5% wide). Since single-mode propagation is achievable in that spectral range, it may be useful to design edge-emitting lasers based on 2D photonic crystal membranes.

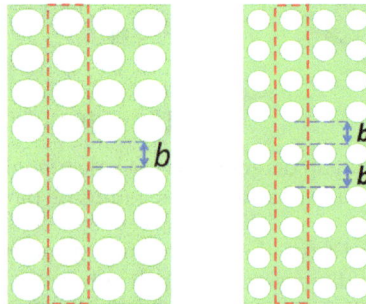

Fig. 11. The definition of square single- (left) and dual-channel (right) air-hole lattices with regions chosen for an FDTD simulation (see red dashed line).

A single simulation of a model consisting of 220 320 FDTD cells (ca. 26MB RAM) takes 140 seconds on Intel Core i7 CPU 950 with the speed of 220 iter/sec. An FDTD cell size is set to $a/20$, leading to at least 40 FDTD cells per wavelength in free space and ca. 12 in the membrane. Thus, calculation of the whole PBG diagram with the step of $\pi/10$ takes about 19 x 140 sec = 45 minutes.

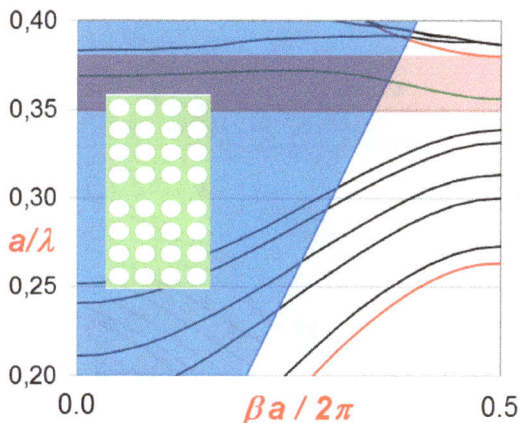

Fig. 12. A TE mode photonic bandgap diagram for a single-channel in an air-hole square lattice cut in an InGaAsP membrane ($n = 3.4$, $r/a = 0.4$, $d/a = 0.4$, $b/a = 0.3$). The light cone is shown with a blue semi-transparent colour.

Fig.13 shows a PBG diagram for TE polarisation computed for a square dual-channel ($n = 3.4$, $r/a = 0.4$, $d/a = 0.6$, $b/a = 0.3$). Two supermodes are distinguished with blue and green colours. However, only the 1st order supermode (green) has a uniquely defined phase constant within a photonic bandgap (red semi-transparent zone), additionally reduced by a light cone to the $a/\lambda = 0.337... 0.342$ spectrum range (1.5% wide). Comparing the results shown in Fig.14 with those in Fig.13, it can be seen that an increase in the channel's width b/a from 0.3 to 0.4 results in a decrease of the supermode's frequency. Most of the 1st order supermode's unique phase constant range shown in Fig.14 is within the photonic bandgap (red semi-transparent zone). Unfortunately, the light cone limits the choice to the $a/\lambda = 0.318... 0.323$ spectrum range (1.5% wide). In this case, however, the allowed spectrum is more distant from the 2nd order supermode (blue), reducing the risk of its unintended oscillation. On the other hand, excitation of the modes in the photonic crystal surrounding the channel is more likely to happen. Concluding, it can be seen that an appropriate adjustment of the light cone, photonic bandgap and channel's width gives a lot of possibilities to modify the allowed supermode's spectrum range (Lesniewska-Matys, 2011).

In the next Section, electric field distributions of a few exemplary supermodes obtained in photonic crystal membrane channels are given.

3.3 Electromagnetic field distribution in photonic crystal membrane channels

The calculation of laser characteristics of above-threshold generation in the considered PhC membrane channels requires quantitative knowledge of a field distribution of an undisturbed travelling wave propagating along the channel at one of selected modes (see Section 4.3). Therefore, envelopes of electric field components within a unit row of the photonic crystal waveguides have to be computed. For that purpose, an FDTD computational model as shown in Fig.15 is used to generate a travelling wave in the channel(s), which may be then integrated in time to obtain the envelopes. The photonic crystal is equipped on the left with an additional input section, where an appropriate mode

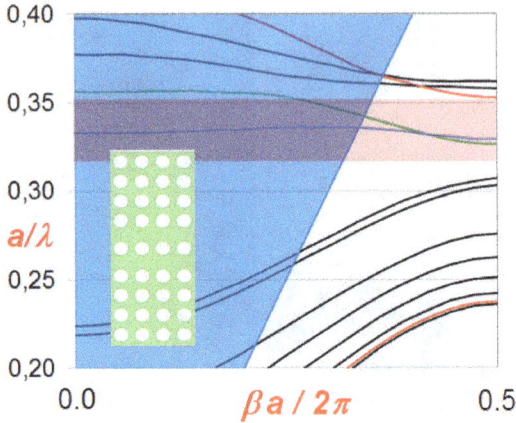

Fig. 13. A TE mode photonic bandgap diagram for a dual-channel in an air-hole square lattice cut in an InGaAsP membrane (n = 3.4, r/a = 0.4, d/a = 0.6, b/a = 0.3). The light cone is shown with a blue semi-transparent colour.

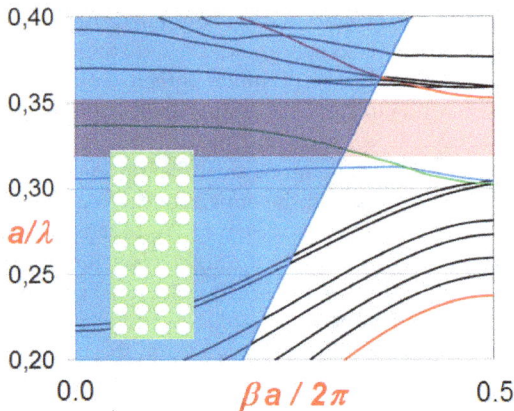

Fig. 14. A TE mode photonic bandgap diagram for a dual-channel in an air-hole square lattice cut in an InGaAsP membrane (n = 3.4, r/a = 0.4, d/a = 0.6, b/a = 0.4). The light cone is shown with a blue semi-transparent colour.

is excited using a mode template generation technique (Celuch-Marcysiak et al., 1996). The end of the waveguide on the right is truncated with a perfectly matched layer (PML) (Berenger, 1994) to avoid any reflections that would disturb the travelling wave.

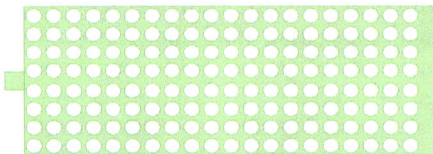

Fig. 15. The view of an FDTD model of a photonic crystal waveguide.

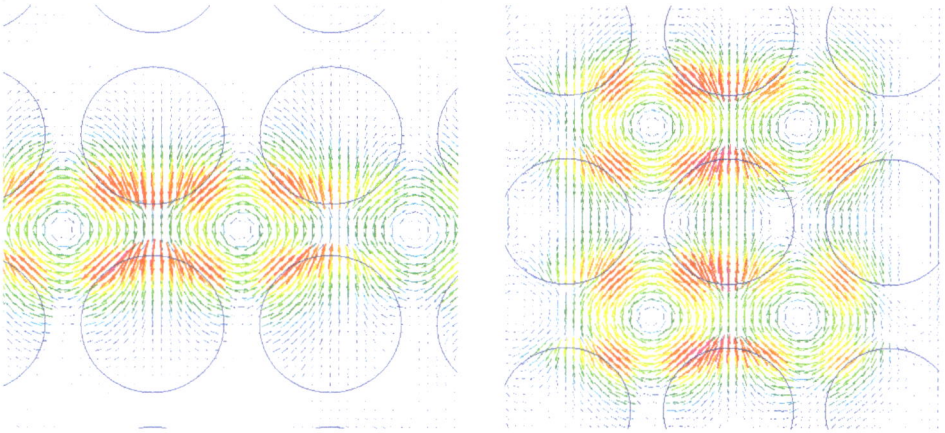

Fig. 16. Vector views of an instantaneous electric field for $a/\lambda = 0.356$ (left) and for $a/\lambda = 0.321$ (right) for the single-channel (left) and dual-channel (right) waveguides, the PBG diagrams of which are shown in Fig.12 and Fig.14, respectively.

For instance, Fig.16 shows the distribution of an instantaneous electric field vector in the single-channel waveguide for $a/\lambda = 0.356$. It can be seen that the field is mostly concentrated within the channel near the hole and has its minimum in the middle between the rows, where the longitudinal electric component dominates over the transverse one. Similarly, Fig.17 presents the distribution of an instantaneous electric field vector in the dual-channel waveguide for $a/\lambda = 0.321$. This time, it is crucial to determine whether the fields in both channels oscillate in-phase or not. A thorough look onto the picture reveals that vectors in the adjacent channels have the same direction prompting the conclusion that both modes creating the supermode are in-phase polarised. Thus, the gain of a far-field radiation pattern increases leading to higher laser beam intensity.

Instantaneous electric field distributions like those shown in Fig.16 are, afterwards, integrated in time and in a whole volume of a single row of a channel. As it is shown in the subsequent Section, those envelopes are used to compute gain characteristics of phased array lasers based on photonic crystal membranes.

4. Supermode laser generation in photonic crystal membranes

4.1 The model of an effective planar waveguide

Taking advantage of the already computed PBG diagrams of the photonic crystal membranes with one and two waveguide channels (see Section 3), the phase constant of the supermodes may be easily determined. That knowledge is essential to build an equivalent effective waveguide model, which enables an approximate analytical representation of a field distribution of the guided modes in passive structures (Lesniewska-Matys, 2011). As it can be seen in Fig.17, in the proposed model, a photonic crystal waveguide is replaced with a two dimensional planar one with the same membrane's thickness but the channel's width adjusted so as to obtain the same phase constant.

Refractive indices in the planar equivalent structure are chosen in the following way: the refractive index of a waveguide core (n_1 in Fig.17b) is the same as in the photonic crystal structure, while n_4 and n_5 are equal to the value of the material filling the holes in the photonic crystal structure (air in the examples considered in this Chapter).

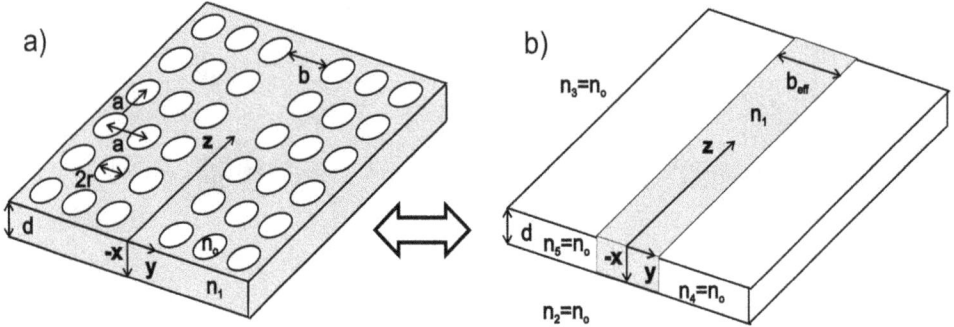

Fig. 17. A perspective view of a) a single-channel in a square-PhC membrane and b) an effective planar waveguide equivalent.

Next, to evaluate the field distribution in the already defined effective planar waveguide, a method proposed in (Marcuse, 1974) is applied. Those analytically derived waveguide modes are used, afterwards, to describe the operation of laser modes in such effective planar waveguides. Subsequently, the field distribution in a multichannel structure with the propagating supermodes is obtained using a non-orthogonal (strongly) coupled mode theory as proposed in (Chuang, 1987a; Chuang, 1987b; Chuang, 1987c). The overlapping integrals between the modes and their coupling coefficients in planar dual-channel waveguides are derived using formulae describing an EM field distribution in the single-channel planar structure. Eventually, the obtained field distributions for the single- and dual-channel structures may be used to estimate approximate operation conditions of laser structures above a generation threshold. (see Section 4.3).

4.2 Mode propagation in an effective N-waveguide structure

Fig.18 shows a top view of a dual-channel defect waveguide, where shaded regions indicate the equivalent effective planar structure. The channels' widths in the equivalent model are adjusted so as to provide the same phase constant of the fundamental supermode as in the corresponding PhC channels.

A total EM field distribution in the coupled planar waveguides may be represented as a weighted sum of the modes propagating in each of the waveguides separately:

$$\vec{E}_t(x,y) = \sum_{p=1}^{N} \vec{a}_p(z)\, \vec{E}_t^{(p)}(x,y) \tag{3}$$

$$\vec{H}_t(x,y) = \sum_{p=1}^{N} \vec{a}_p(z)\, \vec{H}_t^{(p)}(x,y) \tag{4}$$

$$\vec{E}_z(x,y) = \sum_{p=1}^{N} \vec{a}_p(z) \frac{\varepsilon^{(p)}}{\varepsilon} \vec{E}_z^{(p)}(x,y) \tag{5}$$

$$\vec{H}_z(x,y) = \sum_{p=1}^{N} \vec{a}_p(z) \vec{H}_z^{(p)}(x,y) \tag{6}$$

where: $\vec{E}_t$, $\vec{H}_t$, $(\vec{E}_z, \vec{H}_z)$ denote transverse (longitudinal) electric and magnetic field components, respectively.

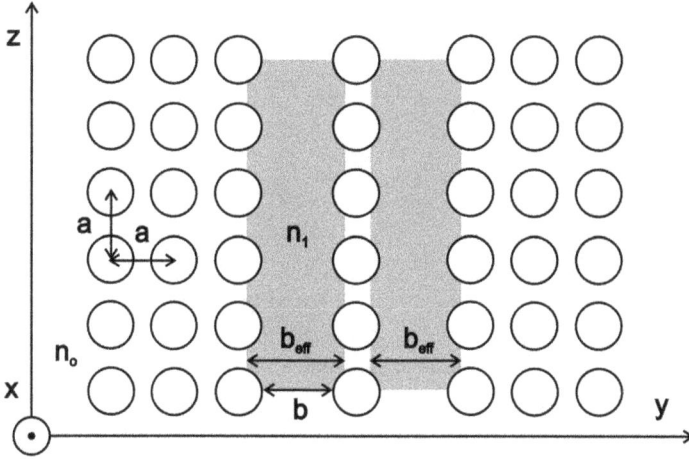

Fig. 18. A top view of a dual-channel defect waveguide processed in a membrane with a photonic crystal square lattice. Shaded regions indicate an equivalent effective waveguide structure.

According to (Kogelnik & Shank, 1972; Schmidt et al., 1974; Chen & Wang, 1984), the amplitudes of the modes guided in the coupled waveguides satisfy the following conditions:

$$\bar{C}\frac{d}{dz}\vec{a}(z) = i\,S\,\vec{a}(z) \quad \text{or} \quad \frac{d}{dz}\vec{a}(z) = i\,M\,\vec{a}(z) \tag{7}$$

where an N-element vector $\vec{a}(z)$ consists of $a_p(z)$ amplitudes of the modes propagating in consecutive waveguide channels.

The matrices S and M are defined as follows:

$$S = \bar{C}B + \tilde{K} \quad \text{or} \quad M = \bar{C}^{-1}S = B + \bar{C}^{-1}\tilde{K} \tag{8}$$

where $\bar{C}$ is an NxN square matrix, where each element of the overlapping integrals $\bar{C}_{pq}$ is defined as follows:

$$\bar{C}_{pq} = \bar{C}_{qp} = \frac{C_{pq} + C_{qp}}{2} \tag{9}$$

where

$$C_{pq} = \frac{1}{2} \int\limits_{-\infty}^{\infty} \int\limits_{-\infty}^{\infty} \left(\vec{E}_t^{(q)} \times \vec{H}_t^{(p)} \right) \hat{z} \, dx \, dy \tag{10}$$

It should be emphasized that the electromagnetic field was normalised so as to:

$$C_{pp} = C_{qq} = 1 \tag{11}$$

Another NxN matrix $\tilde{K}$, applied in Eq.8, consists of coupling coefficients between all the N waveguides, which are defined in the following way:

$$\tilde{K}_{pq} = \frac{\omega}{4} \int\limits_{-\infty}^{\infty} \int\limits_{-\infty}^{\infty} \Delta\varepsilon^{(q)} \left[\vec{E}_t^{(p)} \cdot \vec{E}_t^{(q)} - \frac{\varepsilon^{(p)}}{\varepsilon} \vec{E}_z^{(p)} \cdot \vec{E}_z^{(q)} \right] dx \, dy \tag{12}$$

where: $\Delta\varepsilon^{(q)} = \varepsilon - \varepsilon^{(q)}$, $\varepsilon^{(q)} = \left(n^{(q)} \right)^2$.

The matrix B is an NxN diagonal matrix with phase constants β_i $(i = 1...N)$ of the modes propagating in all the waveguide channels. It should be noted that matrices $\overline{C}$ and S are symmetric, what is very important to prove the orthogonality of the supermodes, whereas the matrix M, in general, is not necessarily symmetric.

The solution of the coupled mode equations given by Eq.8 leads to the modal field distribution in the entire array for a given propagation constant γ_p. In the system consisting of N coupled waveguides, P supermodes are generated $(N = P)$, which may be written as follows:

$$\tilde{E}_1 = A_1 \cdot \left[a_1^{(1)}(z) E_1^{(1)}(x,y) + ... + a_N^{(1)}(z) E_N^{(1)}(x,y) \right] \cdot e^{i\gamma_1 z}$$
$$\vdots \tag{13}$$
$$\tilde{E}_P = A_P \cdot \left[a_1^{(P)}(z) E_1^{(P)}(x,y) + ... + a_N^{(P)}(z) E_N^{(P)}(x,y) \right] \cdot e^{i\gamma_P z}$$

where A_k $(k = 1...P)$ are scaling coefficients, P is the order of a supermode, and $a_N^{(1)}$ indicates the amplitude of a field distribution of the 1st supermode in the Nth waveguide.

4.3 The model of light generation in planar multi-channel photonic crystal membrane lasers

In this Section, an approximate model of laser generation in planar multi-channel PhC membranes is described. Fig.19 shows phased array lasers with mirrors made of 1D and 2D photonic crystals processed in a membrane. The effective values of reflection coefficients are denoted with r_1 and r_2. It is assumed hereafter that the reflection coefficient of an input mirror is $r_1 = 1.0$.

In the proposed model of light generation, a field distribution in the single- and multi-channel photonic crystal membrane lasers is substituted with a field in the equivalent effective planar waveguides (see Section 4.2). To achieve laser characteristics of those structures, energy theorem is used (Szczepanski et al., 1989; Szczepanski, 1988). It allows us to represent a normalised small-signal gain saturation of the laser as a function of a saturation power, a distributed losses coefficient and a laser's geometry. The field distribution of the modes generated in the membrane was obtained in two ways: applying the FDTD method (Taflove & Hagness, 2005; QWED) and using analytical formulas derived for the effective planar waveguides. As it is shown in the next Section, discrepancy between lasers characteristics achieved with both methods remains below ca. 10 %.

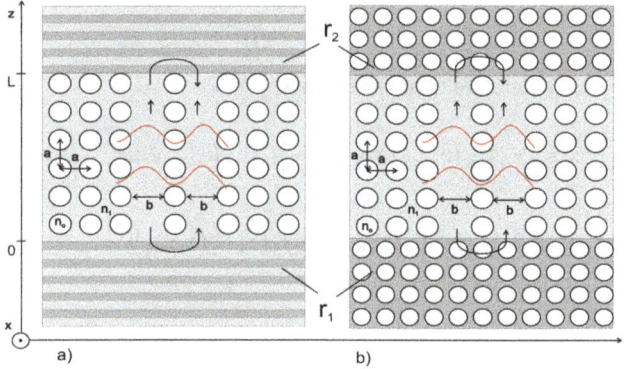

Fig. 19. A schematic view of phased array photonic crystal lasers processed in a membrane with a square photonic crystal lattice with mirrors made of a) a 1D photonic crystal and b) a 2D photonic crystal.

The field distribution in the laser is written in the following way:

$$
\begin{aligned}
E(x,y,z) = R(x,y,z) + S(x,y,z) &= f_R(z)\,E_R(x,y,z)\,e^{i\beta z} + \\
&+ f_S(z)\,E_S(x,y,z)\,e^{-i\beta z}
\end{aligned}
\tag{14}
$$

where β is a phase constant of a laser mode, $f_R(z)$ and $f_S(z)$ denote complex amplitudes of forward and backward propagating waves, whereas $E_R(x,y,z)$ and $E_S(x,y,z)$ represent a transverse distribution of the laser mode.

The coupled-mode equations for the considered laser structures may be written as (Szczepanski, 1994):

$$
\left\{
\begin{aligned}
&\frac{df_R(z)}{dz} + (\alpha_L - i\delta)\,f_R(z) = \frac{1}{N_R}\iint g_o\,\gamma\,D\,d_\alpha\,f_R(z)\,\big|E_R(x,y,z)\big|^2\,dxdy + \\
&+\frac{1}{2N_R}\iint g_o\,\gamma\,D\,d_\kappa\,f_S(z)\,E_S(x,y,z)\,E_R^*(x,y,z)\,dxdy, \\
&-\frac{df_S(z)}{dz} + (\alpha_L - i\delta)\,f_S(z) = \frac{1}{N_S}\iint g_o\,\gamma\,D\,d_\alpha\,f_S(z)\,\big|E_S(x,y,z)\big|^2\,dxdy + \\
&+\frac{1}{2N_S}\iint g_o\,\gamma\,D\,d_\kappa^*\,f_R(z)\,E_R(x,y,z)\,E_S^*(x,y,z)\,dxdy,
\end{aligned}
\right.
\tag{15}
$$

where g_o is a small-signal gain coefficient, α_L stands for laser's distributed losses, δ denotes a frequency-shift parameter understood as the discrepancy between an oscillating frequency in passive and active resonators. The normalisation factors N_R and N_S may be calculated as:

$$N_R = \iint \left| E_R\left(x,y,z\right) \right|^2 dxdy , \ N_S = \iint \left| E_S\left(x,y,z\right) \right|^2 dxdy \tag{16}$$

The shape of a gain spectral line applied in Eq.15 is given by:

$$D = \left(\gamma + i\left(\omega - \upsilon\right) \right)^{-1} \tag{17}$$

whereas the other parameters are given as follows:

$$C_i = 1 + \left[\left| f_R\left(z\right) E_R\left(x,y,z\right) \right|^2 + \left| f_S\left(z\right) E_S\left(x,y,z\right) \right|^2 \right] \cdot \frac{L}{I_S} \tag{18}$$

$$C_c = 2 \left[\left| f_R\left(z\right) E_R\left(x,y,z\right) \right|^2 \cdot \left| f_S\left(z\right) E_S\left(x,y,z\right) \right|^2 \right] \cdot \frac{L}{I_S} \tag{19}$$

$$d_\alpha = \left(C_i^2 - C_c^2 \right)^{-0.5} \tag{20}$$

$$d_\kappa = -C_c \ e^{i\theta} \left[\sqrt{C_i^2 - C_c^2} \left(C_i + \sqrt{C_i^2 - C_c^2} \right) \right]^{-1} \tag{21}$$

where L denotes the length of the laser (see Fig.19).

The saturation power in the active region can be written as:

$$P_S = I_S \cdot A_l \tag{22}$$

where I_S is a saturation intensity

$$I_S = \frac{h\upsilon}{\sigma\tau} \tag{23}$$

and A_l denotes a cross-section of the laser, h is the Planck constant, υ is the frequency of a laser mode, σ is an emission cross-section, τ represents recombination lifetime in the active region.

Operations on Eq.15 lead to (Szczepanski et al., 1989; Szczepanski, 1988):

$$\frac{d}{dz}\left(\left| f_R\left(z\right) \right|^2 - \left| f_S\left(z\right) \right|^2 \right) = -2\alpha_L \left(\left| f_R\left(z\right) \right|^2 + \left| f_S\left(z\right) \right|^2 \right) + \frac{2L\left(\omega - \upsilon\right)}{N^R} \int \frac{g_o \left| f_R\left(z\right) \right|^2 \left| E^R\left(x,y,z\right) \right|^2}{\sqrt{C_i^2 - C_c^2}} dxdy +$$

$$+ \frac{2L\left(\omega - \upsilon\right)}{N^S} \int \frac{g_o \left| f_S\left(z\right) \right|^2 \left| E^S\left(x,y,z\right) \right|^2}{\sqrt{C_i^2 - C_c^2}} dxdy - \frac{2L\left(\omega - \upsilon\right)}{N^R} \int \frac{g_o \left| f_S\left(z\right) E^S\left(x,y,z\right) \right| \left| f_R\left(z\right) E^R\left(x,y,z\right) \right|}{\sqrt{C_i^2 - C_c^2} \left(C_i + \sqrt{C_i^2 - C_c^2} \right)} dxdy + \tag{24}$$

$$- \frac{2L\left(\omega - \upsilon\right)}{N^S} \int \frac{g_o \left| f_R\left(z\right) E^R\left(x,y,z\right) \right| \left| f_S\left(z\right) E^S\left(x,y,z\right) \right|}{\sqrt{C_i^2 - C_c^2} \left(C_i + \sqrt{C_i^2 - C_c^2} \right)} dxdy.$$

The solution of Eq.24 requires boundary conditions to be specified:

$$|f_R(0)| = r_2|f_S(0)| \tag{25}$$

$$P_{S,out} = |f_S(0)|^2 (1 - r_2^2) \iint |E_S(x,y,0)|^2 dxdy \tag{26}$$

$$|f_S(L)| = r_1|f_R(L)| \tag{27}$$

$$P_{R,out} = |f_R(L)|^2 (1 - r_1^2) \iint |E^R(x,y,L)|^2 dxdy \tag{28}$$

where $P_{out} = P_{R,out} + P_{S,out}$ is a total power generated by the laser.

In a threshold approximation, $f_R(z)$ and $f_S(z)$ are equal to:

$$f_R(z) = |A|\exp(\gamma_f z) \tag{29}$$

$$f_S(z) = |A|r_2^{-1}\exp(-\gamma_f z) \tag{30}$$

where

$$\gamma_f = \frac{1}{2L}\ln\frac{1}{r_1 r_2} \tag{31}$$

An approximate expression relating the normalised small-signal gain coefficient $g_o L$ to the output power and the parameters of the planar laser is given as follows:

$$g_o L = \frac{C\int_{-d}^{0}\int_{0}^{b_{eff}} |E_t(x,y,z)|^2 dxdy + 2\alpha_L \int_{-d}^{0}\int_{0}^{b_{eff}}\int_{0}^{L} |E_t(x,y,z)|^2 \cdot \left(|f_R(z)|^2 + |f_S(z)|^2\right) dxdydz}{2\int_{-d}^{0}\int_{0}^{b_{eff}}\int_{0}^{L} \frac{|E_t(x,y,z)|^2 \cdot \left(|f_R(z)|^2 + |f_S(z)|^2\right)}{1 + \frac{P_{out}}{P_S}\frac{1}{K}|E_t(x,y,z)|^2 \cdot \left(|f_R(z)|^2 + |f_S(z)|^2\right)} dxdydz}, \tag{32}$$

where:

$$K = |f_R(L)|^2 (1 - r_1^2) + |f_S(0)|^2 (1 - r_2^2) \tag{33}$$

The transverse electric field distribution of the laser mode in the photonic crystal membrane $E_R = E_S = E_t$ was calculated numerically (see Section 3.3) and analytically using the effective planar waveguide model (see Section 4.2).

4.4 Laser gain characteristics

In this Section, exemplary gain characteristics of phased array lasers processed in defect photonic crystal membranes are given. The transverse field distribution of the laser supermode is calculated numerically with the FDTD method (Taflove & Hagness, 2005; QWED) and analytically, using the non-orthogonal mode theory applied to the calculation of the effective waveguide structure. It is assumed that the distributed losses coefficient is equal to α_L = 200 cm⁻¹ (Zielinski et al. 1989; Lu et al., 2008), whereas the output power to saturation power ratio is P_{out}/P_s = 10⁻⁶ (Lu et al., 2009; Susaki et al., 2008; van den Hoven, 1996).

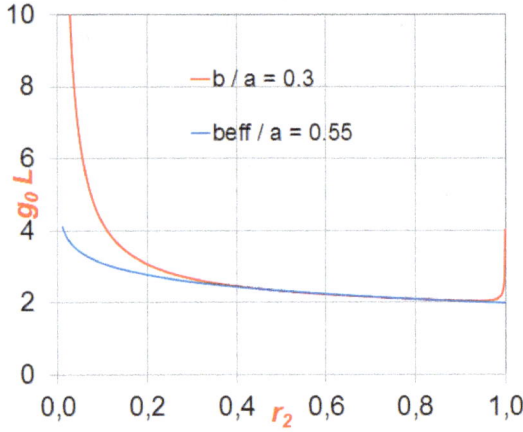

Fig. 20. A normalised small-signal gain at λ = 1.55μm as a function of cavity mirror reflectivity for a single-channel square-PhC membrane laser (r/a = 0.4, d/a = 0.4, b/a = 0.3, a = 0.55μm). Red curve: computed with the aid of FDTD for (a/λ, βa) = (0.356, π); blue curve: effective waveguide model with b_{eff}/a = 0.55.

Fig.20 presents a normalised small-signal gain at λ = 1.55μm as a function of output mirror reflectivity r_2 for a single-channel square-PhC membrane laser (r/a = 0.4, d/a = 0.4, b/a = 0.3, a = 0.55μm), the PBG diagram of which is shown in Fig.12. The red curve depicts the gain characteristics calculated for (a/λ, βa) = (0.356, π) with the aid of FDTD, whereas the blue one indicates the result of analytical computation with Eq.32 for the corresponding effective waveguide model with the channel width b_{eff}/a = 0.55. In principle, the minimum of the calculated characteristics indicates an optimum value of the mirror reflection coefficient r_2 of an output mirror, for which maximum output power efficiency is achieved. It can be seen from Fig.20 that, although the shape of both curves is substantially different, their minima are in a similar position and the optimum reflectivity r_2 amounts to 0.93 and 0.997 for the red and blue curves, respectively. Consequently, it leads to ca. 7.2% of a relative discrepancy between the optimum values computed with the two approaches.

Similar computations were carried out for dual-channel scenarios with r/a = 0.4, d/a = 0.6, b/a = 0.3 and 0.4. Fig.21,22 show the corresponding laser characteristics for (a/λ, βa) = (0.340, 13π/18) and (0.321, 12π/18) with a = 0.53μm and 0.50μm, respectively. In both cases, the reflection coefficient r_2 is equal to 0.91 and 0.996 for numerical and analytical (effective) computations, respectively, leading to ca. 9.4% of the relative discrepancy.

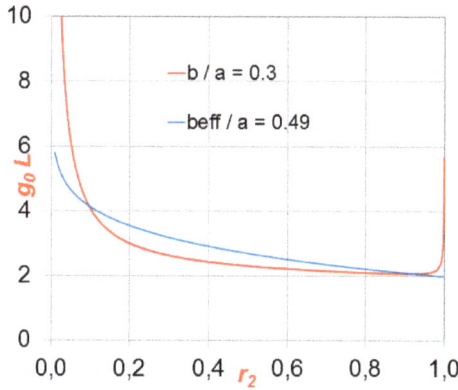

Fig. 21. A normalised small-signal gain at $\lambda = 1.55\mu m$ as a function of cavity mirror reflectivity for a single-channel square-PhC membrane laser ($r/a = 0.4$, $d/a = 0.6$, $b/a = 0.3$, $a = 0.53\mu m$). Red curve: computed with the aid of FDTD for (a/λ, βa) = (0.340, $13\pi/18$); blue curve: effective waveguide model with $b_{eff}/a = 0.49$.

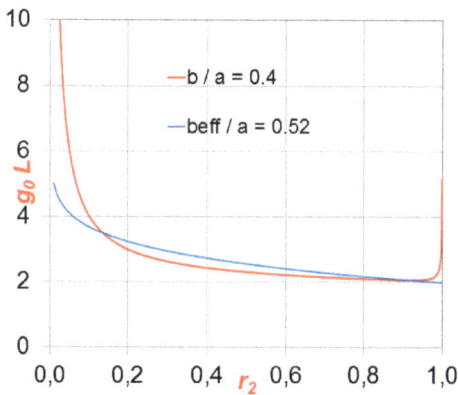

Fig. 22. A normalised small-signal gain at $\lambda = 1.55\mu m$ as a function of cavity mirror reflectivity for a single-channel square-PhC membrane laser ($r/a = 0.4$, $d/a = 0.6$, $b/a = 0.4$, $a = 0.50\mu m$). Red curve: computed with the aid of FDTD for (a/λ, βa) = (0.321, $12\pi/18$); blue curve: effective waveguide model with $b_{eff}/a = 0.52$.

Concluding, exemplary laser small-signal gain characteristics have been shown, which enable the generation of a laser single-mode both in the single- and dual-channel structures in two-dimensional photonic crystal lattices processed in dielectric membranes. It has also been shown that both rigid full-wave and approximate computations of the modal field distributions provide the values of the optimum reflection coefficient of the output mirror, which are in less than 10% agreement.

5. Conclusions

In this Chapter, a complete design cycle of phased array lasers based on photonic crystals processed in dielectric membranes has been given. First, full-wave electromagnetic computations with the FDTD method allow us to determine a photonic bandgap of the selected passive photonic crystal lattices processed in a dielectric membrane. Second, a single- either multi-channel waveguide array is introduced into the lattice and dispersive properties of the modes within the corresponding photonic bandgap are computed. The goal is to evaluate the spectrum, where a single-mode propagation of the supermodes is possible along the channels. Third, for given geometry settings and the mode's wavelength spectrum, the above-threshold laser small-signal gain characteristic is computed with the non-orthogonal coupled mode theory. Gain computations are two-fold. In the first approach, numerical computations of an electric field envelope within a passive structure are executed with the aid of the FDTD method, while the second method is based on an equivalent effective waveguide structure. Both methods provide similar values of the optimum reflection coefficient of the output mirror.

6. References

AcMey, D.E. & Engelmann, R.W.H., High power leaky mode multiple stripe laser, *Appl. Phys. Lett.*, vol. 39, no. 7, pp. 27-29, 1981

Berenger, J.P., A perfectly matched layer for the absorption of electromagnetic waves, *J. Comput. Phys.*, vol. 114, no. 2, pp. 185-200, 1994

Celuch-Marcysiak, M. & Gwarek, W.K., Spatially looped algorithms for time-domain analysis of periodic structures, *IEEE Trans. Microwave Theory Tech.*, vol. MTT-43, No. 4, pp. 860-865, 1995

Celuch-Marcysiak, M., Kozak, A. & Gwarek, W.K., A new efficient excitation scheme for the FDTD method based on the field and impedance template, *IEEE Antennas and Propag. Symp.*, Baltimore, 1996

Chen, K.-L. & Wang, S., Cross-talk problems in optical directional couplers, *Appl. Phys. Lett.*, vol. 44, no. 2, pp. 166-168, 1984

Chow, E., Lin, S.Y., Wendt, J.R., Johnson, S.G. & Joannopoulos, J.D., Quantitative analysis of bending efficiency in photonic-crystal waveguide bends at λ=1.55μm wavelengths, *Optics Letters*, vol. 26, no. 5, pp. 286-288, 2001

Chuang, S.-L., A coupled mode formulation by reciprocity and a variational principle, *J. Lightwave. Technol.*, vol. 5, no. 1, pp. 5-15, 1987

Chuang, S.-L., A coupled mode theory for multiwaveguide systems satisfying the reciprocity theorem and power conservation, *J. Lightwave. Technol.*, vol. 5, no. 1, pp. 174-183, 1987

Chuang, S.-L., Application of the strongly coupled mode theory to integrated optical devices, *J. Quant. Electron.*, vol. 23, no. 5, pp. 499-509, 1987

Collin, R.E., *Field Theory of Guided Waves*, McGraw-Hill Inc., New York, 1960

Digonnet, M.J.F. & Shaw, H.J., Analysis of a tunable single mode optical fiber coupler, *J. Quant. Electron.*, vol. 18, no. 4, pp. 746-754, 1982

Fan, S., Villeneuve, P.R. & Joannopoulos, J.D., Channel drop filters in photonic crystals, *Optics Letters*, vol. 3, no. 1, pp. 4-11, 1998

Gwarek, W.K., Analysis of an arbitrarily-shaped planar circuit - a time-domain approach, *IEEE Trans. Microwave Theory Tech.*, vol. MTT-33, No.10, pp.1067-1072, 1985

Gwarek, W.K., Morawski, T. & Mroczkowski, C., Application of the FDTD Method to the Analysis of the Circuits Described by the Two-Dimensional Vector Wave Equation, *IEEE Trans. Microwave Theory Tech.*, vol. 41, no. 2, pp. 311-316, Feb. 1993

Hardy, A. & Streifer, W., Coupled mode theory of parallel waveguides, *J. Lightwave Technol.*, vol. 3, no. 5, pp. 1135-1146, 1985

Hardy, A., Streifer, W. & Osinski, M., Chirping effects in phase-coupled laser arrays, *Proc. IEEE*, vol. 135, no. 6, pp. 443-450, 1988

Joannopoulos, J.D., Johnson, S.G., Winn, J.N. & Meade, R.D., *Photonic Crystals. Molding the flow of light*, Second Edition, Princeton University Press, ch. 8, pp. 144, 2008

Kapon, E., Lindsey, P., Katz, J., Margalit, S. & Yariv, A., Chirped arrays of diode lasers for supermode control, *Appl. Phys. Lett.*, vol. 45, no. 3, pp. 200-202, 1984

Kapon, E., Lindsey, P., Smith, J.S., Margalit, S. & Yariv, A., Inverted-V chirped phased arrays of gain-guided GaAs/GaAlAs diode lasers, *Appl. Phys. Lett.*, vol. 45, no. 12, pp. 1257-1259, 1984

Kapon, E., Katz, J., Margalit, S. & Yariv, A., Controlled fundamental supermode operation of phase-locked arrays of gain-guided diode lasers, *Appl. Phys. Lett.*, vol. 45, no. 6, pp. 600-602, 1984

Kapon, E., Rav-Noy, Z., Margalit, S. & Yariv, A., Phase-Locked Arrays of Buried-Ridge InP/InGaAsP Diode Lasers, *J. Lightwave Technol.*, vol. 4, no. 7, pp. 919-925, 1986

Kogelnik, H. & Shank, C.V., Coupled-wave theory of distributed feedback lasers, *J. Appl. Phys.*, vol. 43, no. 5, pp. 2327-2335, 1972

Kogelnik, H. & Schmidt, R.V., Switched directional couplers with alternating $\Delta\beta''$, *J. Quant. Electron.*, vol. 12, no. 7, pp. 396-401, 1976

Kogelnik, H., *Theory of dielectric waveguides*, ch. 2, T.Tamir Edition, New York: Springer-Verlag, 1979

Lesniewska-Matys, K., *Modelowanie generacji promieniowania w planarnym wielokanałowym laserze sprzężonym fazowo zbudowanym na bazie dwuwymiarowego kryształu fotonicznego*, Ph.D. Thesis, Warsaw University of Technology, 2011

Liu, T., Zakharian, A.R., Fallahi, M., Moloney, J.V. & Mansuripur, M., Multimode Interference-Based Photonic Crystal Waveguide Power Splitter, *J. Lightwave Technol.*, vol. 22, no. 12, pp. 2842-2846, 2004

Loncar, M., Doll, T., Vuckovic, J. & Scherer, A., Design and Fabrication of Silicon Photonic Crystal Optical Waveguides, *J. Lightwave Technol.*, vol. 18, no. 10, pp. 1402-1411, 2000

Lu, L., Mock, A., Bagheri, M., Hwang, E.H., O'Brien, J. & Dapkus, P.D., Double-heterostructure photonic crystal lasers with lower thresholds and higher slope efficiencies obtained by quantum well intermixing, *Opt. Express*, vol. 16, no. 22, pp. 17342-17347, 2008

Lu, L., Mock, A., Yang, T., Shih, M.H., Hwang, E.H., Bagheri, M., Stapleton, A., Farrell, S., O'Brien, J. & Dapkus, P.D., 120µW peak output power from edge-emitting photonic crystal doubleheterostructure nanocavity lasers, *Appl. Phys. Lett.*, vol. 94, pp. 111101, 2009

Marcuse, D., *Theory of dielectric optical waveguides*, ch. 1.7, Academic Press, New York and London, 1974

Massaro, A., Errico, V., Stomeo, T., Cingolani, R., Salhi, A., Passaseo, A. & De Vittorio, M., 3-D FEM Modeling and Fabrication of Circular Photonic Crystal Microcavity, *J. Lightwave. Technol.*, vol. 26, no. 16, pp. 2960-2968, 2008

Mei, K.K. & Fana, J., Superabsorption - A Method to Improve Absorbing Boundary Conditions, *IEEE Trans. Antennas & Propagat.*, vol. AP-40, pp. 1001-1010, 1992

Mukai, S., Lindsey, C., Katz, J., Kapon, E., Rav-Noy, Z., Margalit, S. & Yariv, A., Fundamental mode oscillation of a buried ridge waveguide laser array, *Appl. Phys. Lett.*, vol. 45, no. 8, pp. 834-835, 1984

Park, I., Lee, H-S., Kim, H-J., Moon, K-M., Lee, S-G., O, B-H., Park, S-G. & Lee, E-H., Photonic crystal power-splitter based on directional coupling, *Opt. Express*, vol. 12, no. 15, pp. 3599-3604, 2004

QuickWave-3D, QWED Sp. z o.o. Available: http://www.qwed.com.pl

Ren, C., Tian, J., Feng, S., Tao, H., Liu, Y., Ren, K., Li, Z., Cheng, B. & Zhang, D., High resolution three-port filter in two dimensional photonic crystal slabs, *Optics Letters*, vol. 12, no. 21, pp. 10014-10020, 2006

Salski, B., *Application of semi-analytical algorithms in the finite-difference time-domain modeling of electromagnetic radiation and scattering problems*, Ph.D. Thesis, Institute of Radioelectronics, Warsaw University of Technology, 2010

Schmidt, R.V., Flanders, D.C., Shank, C.V. & Standley, R.D., Narrow-band grating filters for thin-film optical waveguides, *Appl. Phys. Lett.*, vol. 25, no. 11, pp. 651-652, 1974

Scifres, D.R., Streifer, W. & Burnham, R.D., Experimental and analytic studies of coupled multiple stripecdiode lasers, *J.Quant. Electron.*, vol. 15, no. 9, pp. 917-922, 1979

Shih, M.H., Kuang, W., Mock, A., Bagheri, M., Hwang, E.H., O'Brien, J.D. & Dapkus, P.D., High-quality factor photonic crystal heterostructure laser, *Appl. Phys. Lett.*, vol. 89, pp. 101104, 2006

Srinivasan, K., Painter, O., Colombelli, R., Gmachl, C., Tennant, D.M., Sergent, A.M., Sivco, D.L., Cho, A.Y., Troccoli, M. & Capasso, F., Lasing mode pattern of a quantum cascade photonic crystal surface-emitting microcavity laser, *Appl. Phys. Lett.*, vol. 84, no. 21, pp. 4164-4166, 2004

Susaki, W., Kakuda, S., Tanaka, M., Nishimura, H. & Tomioka, A., Influence of Band Offsets on Carrier Overflow and Recombination Lifetime in Quantum Well Lasers Grown on GaAs and InP, *20th International Conference on Indium Phosphide and Related Materials*, France, 2008

Szczepanski, P., Sikorski, D. & Wolinski, W., Nonlinear operation of a planar distributed feedback laser: energy approach, *J. Quant. Electron.*, vol. 25, no. 5, pp. 871-877, 1989

Szczepanski , P., Semiclassical theory of multimode operation of a distributed feedback laser, *J. Quant. Electron.*, vol. 24, no. 7, pp. 1248-1257, 1988

Szczepanski, P., *Rola przestrzennego rozkładu pola w generacji promieniowania w laserze*, Postdoctoral Thesis, Warsaw University of Technology, 1994

Taflove, A. & Hagness, S.C., *Computational Electrodynamics: The Finite-Difference Time-Domain Method*, Artech House Publishers, 2005

Tanaka, Y., Nakamura, H., Sugimoto, Y., Ikeda, N., Asakawa, K. & Inoue, K., Coupling properties in a 2-D photonic crystal slab directional coupler with a triangular lattice of air holes, *J. Quant. Electron.*, vol. 41, no. 1, pp. 76-84, 2005

Taylor, H.F., Optical switching and modulation in parallel dielectric waveguides, *J. Appl. Phys.*, vol. 44, no. 7, pp. 3257-3262, 1973

van den Hoven, G.N., Koper, R.J.I.M., Polman, A., van Dam, C., van Uffelen, J.W.M. & Smit, M.K., Net optical gain at 1.53 mm in Er-doped Al_2O_3 waveguides on silicon, *Appl. Phys. Lett.*, vol. 68, no. 14, pp. 1886-1888, 1996

van der Ziel, J.P., Mikulyak, R.M., Temkin, H., Logan, R.A. & Dupuis, W.D., Optical beam characteristics of Schottky barrier confined arrays of phase-coupled multiquantum well GaAs lasers, *J. Quant. Electron.*, vol. 20, no. 10, pp. 1259-1266, 1984

Vecchi, G., Raineri, F., Sagnes, I., Yacomotti, A., Monnier, P., Karle, T.J., Lee, K-H., Braive, R., Le Gratiet, L., Guilet, S., Beaudoin, G., Talneau, A., Bouchoule, S., Levenson, A. & Raj, R., Continuous-wave operation of photonic bandedge laser near 1.55 μm on silicon wafer, *Opt. Express*, vol. 15, no. 12, pp. 7551-7556, 2007

Yamamoto, N., Watanabe, Y. & Komori, K., Design of photonic crystal directional coupler with high extinction ratio and small coupling length, *Jpn. J. Appl. Phys.*, vol. 44, no. 4B, pp. 2575-2578, 2005

Yang, T., Lipson, S., O'Brien, J.D. & Deppe, D.G., InAs Quantum Dot Photonic Crystal Lasers and Their Temperature Dependence, *Photon. Technol. Lett.*, vol. 17, no. 11, pp. 2244-2246, 2005

Yang, T., Mock, A. & O'Brien, J.D., Edge-emitting photonic crystal double heterostructure nanocavity lasers with InAs quantum dot active material, *Opt. Lett.*, vol. 32, no. 9, pp. 1153-1155, 2007

Yariv, A., Coupled mode theory for guided-wave optics, *J. Quantum Electron.*, vol. QE-9, no. 9, pp. 919-933, 1973

Yariv, A. & Taylor, H. F., Guided-wave optics, *Proc. IEEE*, vol. 62, no. 8, pp. 131-134, 1981

Yariv, A., *Optical Electronics in Modern Communications*, Ch. 13, Oxford University Press, 1997

Zhang, M. & Garmire, E., Single-mode fiber-film directional coupler, *J. Lightwave Technol.*, vol. 5, no. 2, pp. 260-267, 1987

Zielinski, E., Keppler, F., Hausser, S., Pilkuhn, M.H., Sauer, R. & Tsang, W.T., Optical Gain and Loss Processes in GaInAs/InP MQW Laser Structures, *J. Quant. Electron.*, vol. 25, no. 6, pp. 1407-1416, 1989

8

A Novel Compact Photonic Crystal Fibre Surface Plasmon Resonance Biosensor for an Aqueous Environment

Emmanuel K. Akowuah, Terry Gorman, Huseyin Ademgil,
Shyqyri Haxha, Gary Robinson and Jenny Oliver
University of Kent, Canterbury
United Kingdom

1. Introduction

Surface plasmon resonance (SPR) sensors provide high sensitivity without the use of molecular labels (Homola, Yee, and Gauglitz 1999). They been widely used in the analysis of biomolecular interactions (BIA) and detection of chemical and biological analytes (Homola, Yee, and Gauglitz 1999), where they provide benefits of real-time, sensitive and label-free technology. They have also been used for the detection of various chemical and biological compounds in areas such as environmental protection, food safety and medical diagnostics (Mouvet et al. 1997; Nooke et al. 2010).

Most commercial SPR biosensors are based on the simple, robust and highly sensitive traditional prism-coupled configuration. However, they are not amenable to miniaturization and integration (Jha and Sharma 2009). There is therefore a growing interest in the development of robust, portable and highly sensitive SPR sensing devices capable of out of laboratory measurements (Akowuah et al. 2010; Piliarik et al. 2009; Wang et al. 2010).

Several compact configurations, enabling coupling between optical waveguide modes and surface plasmon waves have been investigated over the last decade. Among these inlcude metalized single-mode, polarization maintaining, and multimode waveguides, metalized tapered fibers, metalized fiber Bragg gratings and lapped D-shaped fiber sensors have been studied(Jorgenson and Yee 1993; Monzon-Hernandez and Villatoro 2006). Fibre optic SPR biosensors offer miniaturization, a high degree of integration, and remote sensing capabilities (Patskovsky et al. 2010; Hoa, Kirk, and Tabrizian 2007).

There is currently interest in the design of photonic crystal fibre (PCF) SPR sensors, which are based on the coupling of a leaky core mode to the SPP mode along a metalized fiber micro-structure (Dhawan, Gerhold, and Muth 2008). This has resulted in the proposal of many types of PCF and micro structured optical fiber (MOF) SPR biosensors, some of which include a three-hole MOF SPR biosensor with a gold layer deposited on the holes, PCF SPR biosensor with enhanced micro fluids and photonic bandgap SPR biosensors (Gauvreau et al. 2007; Hassani et al. 2008; Hassani and Skorobogatiy 2006; Hautakorpi, Mattinen, and Ludvigsen 2008). PCFs are thin silica glass fibers possessing a regular array of microscopic holes that extend along the whole fiber length (Ferrando et al. 2000). The discovery of PCFs has led to several possibilities, ranging from guidance of light in vacuum, to achieving unusual dispersion properties, from enhancing non-linear effects to high confinement of light and minimizing the same non-linear effects through very large mode area single mode fibers (Ademgil et al. 2009). These unusual properties of PCFs have led to an increasing interest in their application in areas such as sensing, signal processing and optical communication systems (Ferrando et al. 2000; Ademgil et al. 2009).

This chapter presents the numerical analysis of a novel sensitive PCF SPR biosensor optimised for operation in aqueous environments. The proposed sensor, shown in Fig. 1, consists of two metalized micro fluidic slots, air holes for light guidance and a small central air hole to facilitate phase matching between guided and plasmon modes. The proposed PCF SPR sensor incorporates extra air holes between the main air holes as a means of reducing the propagation losses whilst ensuring efficient coupling between the core guided and plasma modes.

It will be shown that the proposed PCF SPR sensor can be optimised to achieve a sensitivity of 4000 nm/RIU with regards to spectral interrogation, which is much higher than the 1000 nm/RIU and 3000 nm/RIU reported by (Hautakorpi, Mattinen, and Ludvigsen 2008) and (Hassani and Skorobogatiy 2006) respectively.

With regards to fabrication, the proposed structure should be relatively easy to fabricate due to the notably large micro fluidic slots. Deposition of metal layers inside of the micro fluidic slots can be performed either with the high-pressure chemical vapour deposition technique (Sazio 2006) or electroless plating techniques used in fabrication of metalized hollow waveguides and microstructures (Harrington 2000; Takeyasu, Tanaka, and Kawata 2005).

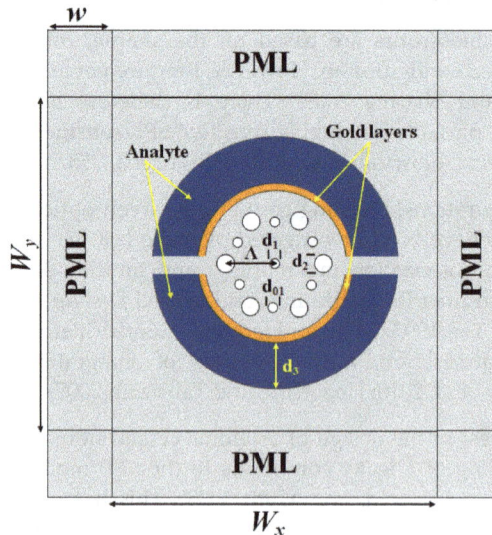

Fig. 1. Cross section of the proposed PCF SPR biosensor showing various sections.

A full – vectorial Finite Element Method (FEM) with perfectly matched layers (PML) is employed to investigate the variation of key performance parameters such as sensitivity and confinement losses with structural parameters such as air hole diameter and gold thickness in the next section of the chapter, followed by conclusion in the final section.

2. Full Vectorial FEM

The finite element method (FEM) can be applied to waveguides in general and PCFs in particular to investigate propagation characteristics of modes. In the case of PCFs, the fibre cross section is divided into homogeneous subspaces where Maxwell's equations are solved by accounting for the adjacent subspaces. These subspaces are triangles which give good approximation of circular structures (Saitoh and Koshiba 2001).

In order to allow the study of fibers with arbitrary air filling fraction and refractive index contrast, a full vector formulation is required. A full vector FEM formulation based on anisotropic perfectly matched layers (PMLs) is able to calculate as many modes as desired in a single run without setting any iterative procedure (Saitoh and Koshiba 2001; Koshiba and Saitoh 2001).

2.1 Implementation of FEM on PCFs

The full – vectorial wave equation can be derived from Maxwell's equations for an optical waveguide with an arbitrary cross section as (Koshiba 1992):

$$\nabla \times \left([p] \nabla \times \phi \right) - k_0^2 [q] \phi = 0 \tag{1}$$

where ϕ represents the electric **E** or magnetic **H** field. The relative permittivity and permeability tensors [p] and [q] can be written as (Koshiba 1992);

$$[p] = \begin{bmatrix} p_x & 0 & 0 \\ 0 & p_y & 0 \\ 0 & 0 & p_z \end{bmatrix} \tag{2}$$

$$[q] = \begin{bmatrix} q_x & 0 & 0 \\ 0 & q_y & 0 \\ 0 & 0 & q_z \end{bmatrix} \tag{3}$$

where $p_x = p_y = p_z = 1$, $q_x = n_x^2$, $q_y = n_y^2, q_z = n_z^2$ for electric field (ϕ = E) and $q_x = q_y = q_z = 1$, $p_x = \dfrac{1}{n_x^2}$, $p_y = \dfrac{1}{n_y^2}$, $p_z = \dfrac{1}{n_z^2}$ for magnetic field ($\phi = H$). In the above expressions, n_x, n_y and n_x represent the refractive indices in the x, y and z directions respectively.

2.2 Analysis of FEM with anisotropic PML

Technically, a PML is not a boundary condition but an additional domain that absorbs the incident radiation without producing reflections. It can have arbitrary thickness and is specified to be made of an artificial absorbing material. The material has anisotropic permittivity and permeability that matches that of the physical medium outside the PML in such a way that there are no reflections regardless of the angle of incidence, polarisation and frequency of the incoming electromagnetic radiation (Koshiba and Saitoh 2001; Buksas 2001). The PML formulation can be deduced from Maxwell's equations by introducing a complex-valued coordinate transformation under the additional requirement that the wave impedance should remain unaffected (Buksas 2001).

If one considers a PML which is parallel to one of the Cartesian coordinate planes, an s matrix of the form (Koshiba 1992; Koshiba and Saitoh 2001):

$$[s] = \begin{bmatrix} \dfrac{s_y s_z}{s_x} & 0 & 0 \\ 0 & \dfrac{s_x s_z}{s_y} & 0 \\ 0 & 0 & \dfrac{s_x s_y}{s_z} \end{bmatrix} \tag{4}$$

$$[s]^{-1} = \begin{bmatrix} \dfrac{s_x}{s_y s_z} & 0 & 0 \\ 0 & \dfrac{s_y}{s_x s_z} & 0 \\ 0 & 0 & \dfrac{s_z}{s_x s_y} \end{bmatrix} \tag{5}$$

can be substituted into Eqn. (1) to permit the use of anisotropic PML. The modified equation thus becomes;

$$\nabla \times \left([p][s]^{-1} \nabla \times \phi \right) - k_0^2 [q][s]\phi = 0 \tag{6}$$

The parameters s_x, s_y, and s_z are complex valued scaling parameters. These parameters are set to a ($a = 1 - a_j$ for leaky mode analysis), when we want to absorb the travelling wave in that direction and unity when no absorption is need. Thus, the absorption by the PML can be controlled by appropriate choice of a_j. A parabolic profile is assumed for a_j such that:

$$\alpha_j = \alpha_{j max}\left(\dfrac{\rho}{\omega}\right)^2 \tag{7}$$

where ρ is the distance from the beginning of the PML and ω, the thickness of the PML.

3. Sensor design and numerical modelling

The proposed PCF SPR biosensor (Fig. 1) consists of circular air holes arranged in a hexagonal lattice, with a small circular air hole at the center. The air hole to air hole spacing is denoted by Λ, whilst d_2 represents the diameters of the circular air holes in the first ring.

Extra air holes of diameter d_{01} are inserted between the main air holes as a means of reducing the propagation losses whilst ensuring efficient coupling between the core guided and plasma modes. A small air hole of diameter d_1 is introduced into the core to facilitate phase matching with a plasmon by lowering the refractive index of the core-guided mode. The first layer of holes work as a low refractive index cladding, enabling mode guidance in the fiber core. The second ring has two slots of uniform thickness d_3, which houses the analyte. Theses slots are coated with gold of thickness t_{Au}. The gap between slots is set to be equal to d_2 throughout this study.

The background material is made of silica which is modeled using the Sellmeier equation (Sellmeier 1871);

$$n(\lambda) = \sqrt{1 + \frac{B_1 \lambda^2}{\lambda^2 - C_1} + \frac{B_2 \lambda^2}{\lambda^2 - C_2} + \frac{B_3 \lambda^2}{\lambda^2 - C_3}} \qquad (8)$$

where n is the refractive index, λ, the wavelength in µm, B(i = 1,2,3) and C(i = 1,2,3) are Sellmeier coefficients. The Sellmeier coefficients used for the background material are B_1=0.696166300, B_2=0.407942600, B_3=0.897479400, C_1=4.67914826×10⁻³µm², C_2=1.35120631×10⁻² µm² and C_3= 97.9340025 µm² (Sellmeier 1871). The permittivity of gold and silver is modeled using data from Johnson and Christy(Johnson and Christy 1972).

Simulations were carried out using a full-vectorial finite element method (FEM) with perfectly matched layers (PMLs). The cross section of the proposed PCF SPR biosensor is divided into many sub-domains with triangular shaped elements in such a way that the step index profiles can be exactly represented. Due to the symmetrical nature of the PCF structure, only one-quarter of the sensor cross section is divided into curvilinear hybrid elements. This results in a computational window area of   (5 µm × 5 µm) terminated by a PML of width = 1 µm. The mesh size is 19,062 elements. Modal analysis of the fundamental mode has been performed on the cross section in the x-y plane of the PCF as the wave is propagating in the z – direction.

4. Simulation results

We begin our analysis by investigating the potential of the proposed PCF for sensing. The structural parameters used are Λ = 1.5 µm, d_1/Λ= 0.2, d_2/Λ= 0.35, d_{01}/Λ = 0.15, d_3= 1.5 µm and t_{Au} = 40 nm. The slots in the second ring are first filled with an analyte whose refractive index, n_a = 1.33 (water) after which the confinement loss of the fundamental mode is calculated. The process is repeated with an analyte of refractive index of 1.34 and the calculated loss spectra for both cases are plotted in Fig 2. It can be observed from Fig. 2 that there are two major attenuation peaks which correspond to the excitation of plasmonic modes on the surface of the metalized channels filled with aqueous analyte , n_a=1.33. It is important to note that the shape of a metallized surface can have a significant effect on the plasmonic excitation spectrum. Hence, a planar metallized surface supports only one plasmonic peak, while a cylindrical metal layer can support several plasmonic peaks.

By changing the analyte refractive index from 1.33 to 1.34, it is observed in Fig. 2 that there is a corresponding shift (dashed curves) in the resonant attenuation peaks. This transduction mechanism is commonly used for detecting the bulk analyte refractive index changes, as well as monitoring formation of the nanometer-thin biolayers on top of a metallized sensor surface (Hassani and Skorobogatiy 2006). In this particular design, there is considerably more field penetration into the analyte-filled channels for the first plasmonic peak as compared to that of the second. This makes it more sensitive than the second with regards to analyte refractive index changes. All subsequent analysis will be based on the first plasmonic peak as it is the most sensitive to refractive index change.

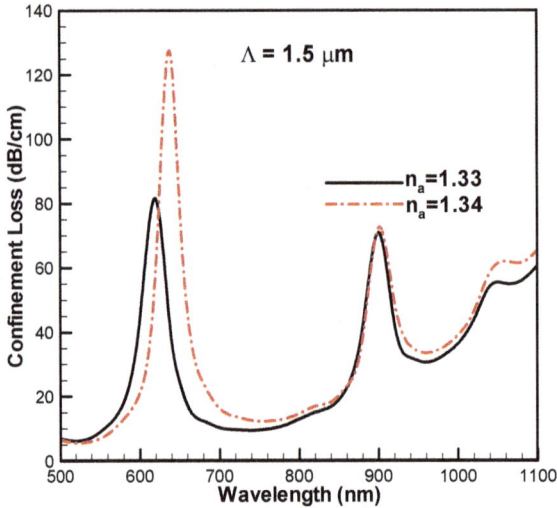

Fig. 2. Calculated loss spectra of the fundamental modes. Loss spectra (solid curves) feature several attenuation peaks corresponding to the excitation of plasmonic modes on the surface of metallized channels filled with aqueous analyte (n_a = 1.33). A change in analyte refractive index (dashed curves) leads to a corresponding shift in the points of phase matching between the core-guided and plasmon modes.

4.1 Optimisation of structural parameters

In order to optimise the several structural parameters of the proposed PCF SPR biosensor for high spectral sensitivity, it is important to understand the effects these parameters have on the sensor properties. Surface plasmon waves, being surface excitations, are very sensitive to the thickness of metallic layers. We therefore investigate the changes in the spectra for the first plasmonic peak when the thickness of gold t_{Au}, is varied from 30 nm to 50 nm. The analyte in this case is fixed at a refractive index of 1.33 and all other structural parameters are kept constant. The confinement loss for the fundamental mode is calculated for each case of t_{Au} and plotted to give the spectra in Fig. 3.

Figure 3 shows a general decrease in modal propagation loss at resonance when the thickness of the gold layer increases. In addition, there is a shift in the resonant wavelength towards longer wavelengths for every increase in t_{Au}. Specifically, the resonant wavelength shifts from about 580 nm to 640 nm for t_{Au} values of 30 nm and 50 nm respectively. This sensitivity of the resonant wavelength to the gold thickness can be used in the study of metal nanoparticle binding events on the metallic surface of the sensor (Hassani and

Skorobogatiy 2006). Practically, this can be used in the monitoring of concentration of metal nanoparticles attached to the photosensitive drugs in photodynamic cancer therapy (Hassani and Skorobogatiy 2006).

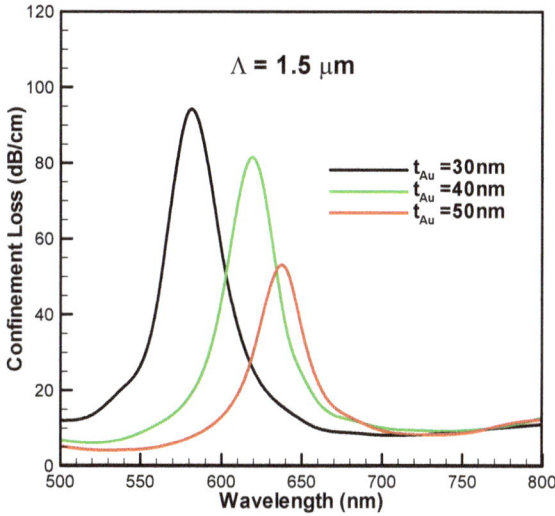

Fig. 3. Loss spectra of proposed PCF SPR biosensor in the vicinity of the first plasmonic peak for variation in gold layer thickness (t_{Au}). Analyte refractive index (na=1.33), $d_{01}/\Lambda = 0.15$, $d_1/\Lambda = 0.2$, $d_2/\Lambda = 0.35$.

In this particular design, a central air hole of diameter d_1, has been employed to tune the phase matching condition. In what follows, we investigate the effect of size variation of the central air hole, with a view of tuning and optimizing plasmon excitation by the core-guided mode of the proposed PCF SPR biosensor. In order to achieve this, d_1/Λ is varied from 0.15 to 0.25 whilst keeping all other structural parameters constant. The micro-fluidics slots are filled with analyte of refractive index, $n_a = 1.33$ for each calculation of the loss spectra for every change in the value of d_1/Λ. Figure 4 shows that there is an overall increase in the modal losses of the fundamental mode for the larger diameters of the central air hole. An increase in the size of the central air hole promotes expulsion of the modal field from the fiber core. This in turn, leads to the greater modal presence near the metallic interface, resulting in higher propagation losses.

Another consequence of the modal expansion from the fiber core into the air-filled microstructure is reduction of the modal refractive index (Fig. 5), leading to the shift of a plasmonic peak toward longer wavelengths (Fig. 4).

In particular, the resonant wavelength changes from about 615 nm to 620 nm for d_1/Λ values of 0.15 and 0.25 respectively.

The next step of the analysis focuses on the influence of the extra air holes, d_{01}, on the confinement loss and resonant wavelength of the PCF SPR sensor. To achieve this, d_{01}/Λ is varied from 0.10 to 0.20, whilst keeping all other structural parameters constant. The micro-fluidics slots are filled with analyte of refractive index, $n_a = 1.33$ for each computation of the

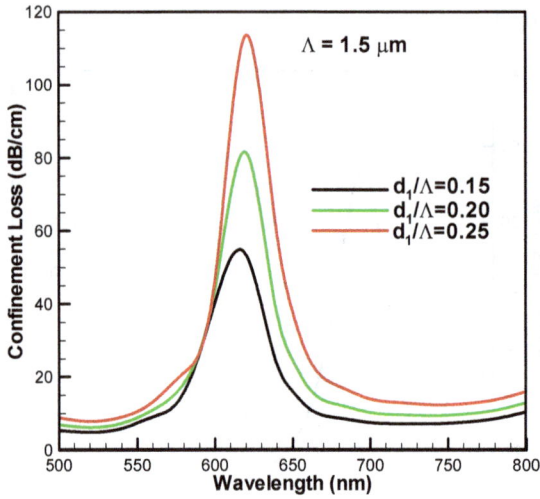

Fig. 4. Loss spectra of proposed PCF SPR biosensor in the vicinity of the first plasmonic peak for variation in d_1. Analyte refractive index (n_a=1.33), $d_{01}/\Lambda = 0.15$, $t_{Au} = 40$ nm, $d_2/\Lambda = 0.35$.

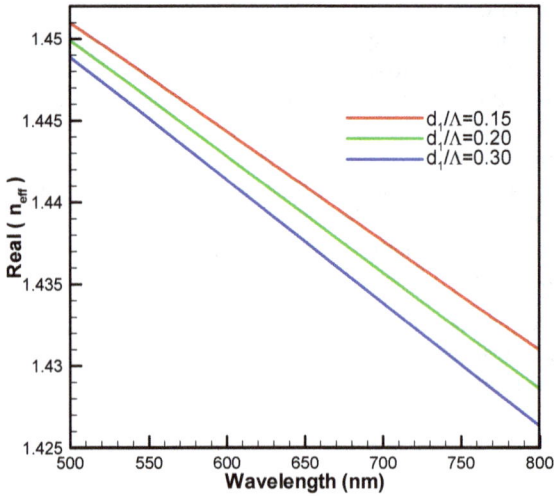

Fig. 5. Dispersion relation of the fundamental mode for variation in d_1. Analyte refractive index (n_a=1.33), $d_{01}/\Lambda = 0.15$, $t_{Au} = 40$ nm, $d_2/\Lambda = 0.35$.

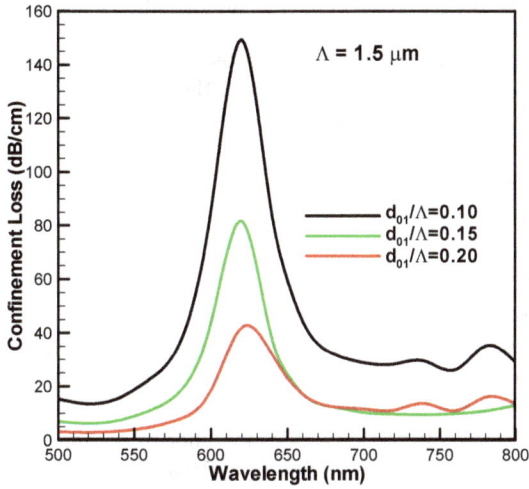

Fig. 6. Loss spectra of proposed PCF SPR biosensor in the vicinity of the first plasmonic peak for variation in d_{01}. Analyte refractive index (n_a=1.33), t_{Au} = 40 nm, d_1/Λ = 0.20, d_2/Λ = 0.35.

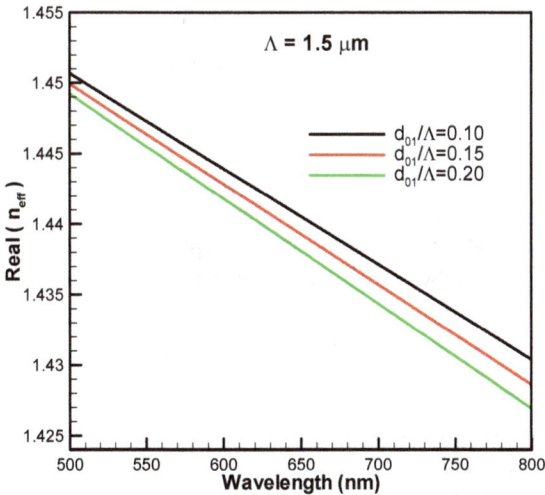

Fig. 7. Dispersion relation of the fundamental mode for variation in d_{01}. Analyte refractive index (n_a=1.33), t_{Au} = 40 nm, d_1/Λ = 0.20, d_2/Λ = 0.35.

loss spectra for a change in the value of d_{01}/Λ. Figure 6 shows that the extra air holes could be used to play a significant role in confinement loss reduction. Unlike d_1, increasing d_{01} reduces the confinement loss whilst shifting the resonant wavelength towards longer wavelength. However, d_{01} has less influence on the resonant wavelength as compared to d_1 and can be considered as a loss control parameter. An increase in d_{01} prevents expulsion of the modal field from the fibre core, thus ensuring better confinement to the core. This ultimately reduces the confinement loss at the resonant wavelength. The decrease in the modal effective index (Fig. 7) for an increase d_{01}/Λ is due to the fact that the "escaping" mode field from the core interacts with relatively larger air filled spaces.

4.1.1 Characterization of sensitivity of the proposed PCF SPR biosensor

The detection of changes in the bulk refractive index of an analyte is the simplest mode of operation of fibre - based SPR biosensors (Hassani and Skorobogatiy 2006; Piliarik, Párová, and Homola 2009; Homola 2003). There is a strong dependence of the real part of a plasmon refractive index on the analyte refractive index, which makes the wavelength of phase matching between the core-guided and plasmon modes sensitive to the changes in the analyte refractive index (Homola 2003; Piliarik, Párová, and Homola 2009). Amplitude (phase) and wavelength interrogation are two main detection methods (Homola 2003; Piliarik, Párová, and Homola 2009).

In the amplitude or phase based method, all measurements are done at a single wavelength (Homola 2003). This approach has the merit of its simplicity and low cost as there is no spectral manipulation needed (Homola 2003). The disadvantage however, is that a smaller operational range and lower sensitivity when compared with the wavelength interrogation methods, where the transmission spectra are taken and compared before and after a change in in analyte refractive index has occurred(Homola 2003).

The amplitude sensitivity is given by(Hassani and Skorobogatiy 2006):

$$S_A(\lambda) = -(\partial\alpha(\lambda,n_a) / \partial n_a) / \alpha(\lambda,n_a) \ [RIU^{-1}] \qquad (9)$$

where $\alpha(\lambda, n_a)$ represents the propagation loss of the core mode as a function of wavelength.

When the sensor operates in the wavelength interrogation mode, changes in the analyte refractive index are detected by measuring the displacement of a plasmonic peak. The sensitivity in this case is given by (Hassani et al. 2008; Homola 2003):

$$S_\lambda(\lambda) = \frac{\partial\lambda_{peak}}{\partial n_a} \ [nm \ / \ RIU] \qquad (10)$$

where λ_{peak} is the wavelength corresponding to the resonance (peak loss) condition. The proposed SPR sensor operates in wavelength interrogation mode. Thus all sensitivity analysis will be limited to spectral interrogation.

It can be observed from Fig. 8 that the thickness of the gold layer (t_{Au}) in the microfluidic slots has an influence on the shift in resonant wavelength. Specifically, the change in resonant wavelength (λ_{peak}) is inversely proportional to t_{Au}. This is due to the fact that when the thickness of the gold layer becomes significantly larger than that of its skin depth, (~20–30 nm), the fibre core mode becomes effectively screened from a plasmon, resulting in a low coupling efficiency, culminating in low sensitivity. Hence, the maximum shift of 20 nm occurs for t_{Au} of 30 nm (Fig. 8). This results in a maximum sensitivity of 2000 nm/RIU according to Eqn. (9).

The next step involves investigating the influence of d_1 on the spectral sensitivity of the PCF SPR sensor under consideration. It can be observed from Fig. 9 that the effect of d_1 on the spectral sensitivity follows the same trend as that of t_{Au}. Specifically, the sensitivity increases with d_1/Λ to a maximum of approximately 2100 nm / RIU for d_1/Λ value of 0.25. An increase in d_1/Λ ensures more leakage of the fundamental mode into the metal / analyte layer in the microfluidic slots, resulting in greater sensitivity of analyte refractive index change.

5. An optimised structure for higher spectral sensitivity

The analysis done so far gives some insight into the effects the structural parameters have on sensor performance. These results are summarised in Table 1.

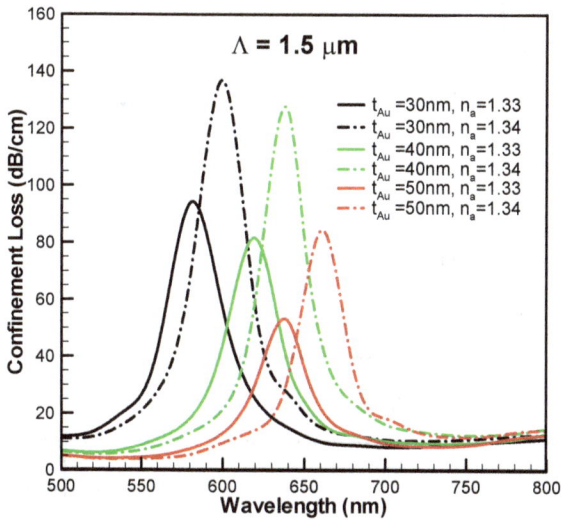

Fig. 8. Shift in resonant wavelength of the loss spectrum for a variation in tAu of the proposed SPR sensor. Analyte refractive index (n_a =1.33), d_{01}/Λ = 0.15, d_1/Λ = 0.2, d_2/Λ = 0.35.

According to Table 1, d_{01}, t_{Au} and d_1 can be considered as loss control parameters. With regards to spectral sensitivity, the main candidates to consider are t_{Au} and d_1. Of these two parameters, d_1 appears relatively easier to control as compared to t_{Au}. It will therefore be more convenient to fix t_{Au} to an appropriate value and optimise d_1 to achieve the desired sensitivity. It must also be noted that there is a limit to which d_1 can be increased due to the associated high confinement losses. To minimise the confinement losses, d_{01} can optimised to keep much of the field inside the core without compromising much sensitivity. By taking all these factors into consideration, the final set of device parameters to maximise sensitivity whilst maintaining an appreciable confinement loss are; $\Lambda = 1.5$ µm, $d_1/\Lambda = 0.50$, $d_{01}/\Lambda = 0.20$, $t_{Au} = 40$ nm.

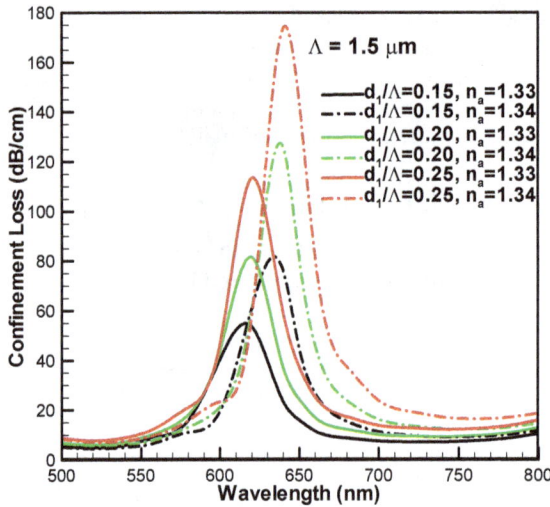

Fig. 9. Shift in resonant wavelength of the loss spectrum for a variation in d_1 of the proposed SPR sensor. Analyte refractive index ($n_a=1.33$), $d_{01}/\Lambda = 0.15$, $t_{Au} = 40$ nm, $d_2/\Lambda = 0.35$.

Parameter	Sensitivity $S_\lambda(\lambda)$	Resonant Wavelength (λ_{peak})	Confinement Loss
$d_1\uparrow$	↑	↑	↑
$d_{01}\uparrow$	↑ - But has less influence on it due to the fact that there is a relatively small change in λ_{peak} for a slight increase in d_{01}.	↑ - But it has less influence on it as compared to t_{Au} and d_1.	↓
$t_{Au}\uparrow$	↑	↑	↓

Table 1. Summary of influence of structural parameters on properties of the proposed PCF SPR biosensor. ↑ represents an increase in a parameter or property whilst ↓ represents a decrease.

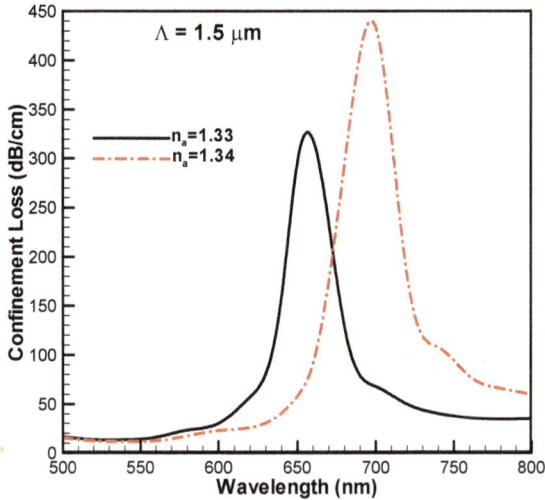

Fig. 10. Shift in resonant wavelength of the optimised PCF SPR biosensor. Structural parameters; $\Lambda = 1.5\ \mu m$, $d_1/\Lambda = 0.50$, $d_{01}/\Lambda = 0.20$, $t_{Au} = 40\ nm$.

Figure 10 shows the shift in resonant wavelength for a change in analyte refractive index from $n_a = 1.33$ to $n_a = 1.34$ for the optimised structure. It indicates an improvement in the spectral sensitivity, which is now approximately 4000 nm / RIU. If the assumption is made that 0.1nm change in the position of a resonance peak can be detected reliably, the resulting sensor resolution is 2.5×10^{-5} RIU, which is better than the 3×10^{-5} RIU and 1×10^{-4} RIU reported by (Hassani and Skorobogatiy 2006) and (Hautakorpi, Mattinen, and Ludvigsen 2008)respectively. The optimisation procedure presented so far can give an indication of the manufacturing tolerances acceptable, to maintain the estimated sensor sensitivity. Of particular interest is the fabrication tolerance of the air holes of the PCF SPR sensor structure. It can be concluded from Fig. 6, 9 and 10 that the maximum allowable change in both d_{01}/Λ and d_1/Λ to maintain the estimated sensitivities, is 50%. Hence, the proposed design can maintain its estimated sensitivity of 4000 nm / RIU provided the holes are fabricated within the 50% tolerance assuming all other conditions remain constant. The current state of advanced PCF fabrication technologies, make fabrication within this tolerance possible.

Since sensor length is inversely proportional to the modal loss, optimization of the PCF structural parameters allows design of PCF SPR sensors of widely different lengths (from millimetre to meter), while having comparable sensitivities. Due to the relatively high loss of our proposed PCF SPR sensor, its length is limited to the centimetre scale. Therefore, the proposed sensor should be rather considered as an integrated photonics element than a fibre.

6. Conclusion

The design and optimisation of a novel PCF SPR biosensor has been presented in this chapter. The loss spectra, phase matching conditions and sensitivity of the proposed biosensor have been presented using a full – vector FEM with PML.

It has been shown that the proposed PCF SPR sensor can be optimised to achieve a sensitivity of 4000 nm/RIU with regards to spectral interrogation, which is much higher than the 1000 nm/RIU and 3000 nm/RIU reported by (Hassani and Skorobogatiy 2006) and (Hautakorpi, Mattinen, and Ludvigsen 2008) respectively. In addition, the PCF SPR sensor incorporates the micro-fluidics setup, waveguide and metallic layers into a single structure. This makes the proposed design compact and more amenable to integration as compared to conventional fibre SPR biosensors.

With regards to fabrication, the proposed structure should be relatively easy to fabricate due to the notably large micro fluidic slots. Deposition of metal layers inside of the micro fluidic slots can be performed either with the high-pressure chemical vapor deposition technique (Sazio 2006) or electroless plating techniques used in fabrication of metalized hollow waveguides and microstructures (Harrington 2000; Takeyasu, Tanaka, and Kawata 2005).

7. References

Ademgil, H., S. Haxha, T. Gorman, and F. AbdelMalek. 2009. Bending Effects on Highly Birefringent Photonic Crystal Fibers With Low Chromatic Dispersion and Low Confinement Losses. *Journal of Lightwave Technology* 27 (5-8):559-567.

Akowuah, E. K., T. Gorman, S. Haxha, and J. V. Oliver. 2010. Dual channel planar waveguide surface plasmon resonance biosensor for an aqueous environment. *Optics Express* 18 (24):24412-24422.

Buksas, M. W. 2001. Implementing the perfectly matched layer absorbing boundary condition with mimetic differencing schemes - Abstract. *Journal of Electromagnetic Waves and Applications* 15 (2):201-202.

Dhawan, Anuj, Michael D. Gerhold, and John F. Muth. 2008. Plasmonic structures based on subwavelength apertures for chemical and biological sensing applications. *Ieee Sensors Journal* 8 (5-6):942-950.

Ferrando, A., E. Silvestre, J. J. Miret, P. Andres, and M. V. Andres. 2000. Vector description of higher-order modes in photonic crystal fibers. *Journal of the Optical Society of America a-Optics Image Science and Vision* 17 (7):1333-1340.

Gauvreau, Bertrand, Alireza Hassani, Majid Fassi Fehri, Andrei Kabashin, and Maksim Skorobogatiy. 2007. Photonic bandgap fiber-based surface plasmon resonance sensors. *Optics Express* 15 (18):11413-11426.

Harrington, J. A. 2000. A review of IR transmitting, hollow waveguides. *Fiber and Integrated Optics* 19 (3):211-227.

Hassani, A., B. Gauvreau, M. F. Fehri, A. Kabashin, and M. Skorobogatiy. 2008. Photonic crystal fiber and waveguide-based surface plasmon resonance sensors for application in the visible and Near-IR. *Electromagnetics* 28 (3):16.

Hassani, A., and M. Skorobogatiy. 2006. Design of the microstructured optical fiber-based surface plasmon resonance sensors with enhanced microfluidics. *Optics Express* 14 (24):11616-11621.

Hautakorpi, Markus, Maija Mattinen, and Hanne Ludvigsen. 2008. Surface-plasmon-resonance sensor based on three-hole microstructured optical fiber. *Optics Express* 16 (12):8427-8432.

Hoa, X. D., A. G. Kirk, and M. Tabrizian. 2007. Towards integrated and sensitive surface plasmon resonance biosensors: A review of recent progress. *Biosensors & Bioelectronics* 23 (2):151-160.

Homola, Jirí, Sinclair S. Yee, and Günter Gauglitz. 1999. Surface plasmon resonance sensors: review. *Sensors and Actuators B: Chemical* 54 (1-2):3-15.

Homola, Jiří. 2003. Present and future of surface plasmon resonance biosensors. *Analytical and Bioanalytical Chemistry* 377 (3):528-539.

Jha, R., and A. K. Sharma. 2009. High-performance sensor based on surface plasmon resonance with chalcogenide prism and aluminum for detection in infrared. *Optics Letters* 34 (6):749-751.

Johnson, P. B., and R. W. Christy. 1972. Optical - Constants of Noble-Metals. *Physical Review B* 6 (12):4370-4379.

Jorgenson, R.C., and S.S. Yee. 1993. A fiber-optics chemical sensor based on surface plasmon resonance. *Sens. Actuators B* 12:213-220.

Koshiba, M. 1992. *Optical Waveguide Theory by the Finite Element Method*: KTK Scientific Publishers.

Koshiba, M., and K. Saitoh. 2001. Numerical verification of degeneracy in hexagonal photonic crystal fibers. *Ieee Photonics Technology Letters* 13 (12):1313-1315.

Monzon-Hernandez, D., and J. Villatoro. 2006. High-resolution refractive index sensing by means of a multiple-peak surface plasmon resonance optical fiber sensor. *Sensors and Actuators B-Chemical* 115 (1):227-231.

Mouvet, C., R. D. Harris, C. Maciag, B. J. Luff, J. S. Wilkinson, J. Piehler, A. Brecht, G. Gauglitz, R. Abuknesha, and G. Ismail. 1997. Determination of simazine in water samples by waveguide surface plasmon resonance. *Analytica Chimica Acta* 338 (1-2):109-117.

Nooke, A., U. Beck, A. Hertwig, A. Krause, H. Krueger, V. Lohse, D. Negendank, and J. Steinbach. 2010. On the application of gold based SPR sensors for the detection of hazardous gases. *Sensors and Actuators B-Chemical* 149 (1):194-198.

Patskovsky, S., M. Meunier, P. N. Prasad, and A. V. Kabashin. 2010. Self-noise-filtering phase-sensitive surface plasmon resonance biosensing. *Optics Express* 18 (14):14353-14358.

Piliarik, Marek, Lucie Párová, and Jirí Homola. 2009. High-throughput SPR sensor for food safety. *Biosensors and Bioelectronics* 24 (5):1399-1404.

Piliarik, Marek, Milan Vala, Ivo Tichý, and Jirí Homola. 2009. Compact and low-cost biosensor based on novel approach to spectroscopy of surface plasmons. *Biosensors and Bioelectronics* 24 (12):3430-3435.

Saitoh, K., and M. Koshiba. 2001. Full-vectorial finite element beam propagation method with perfectly matched layers for anisotropic optical waveguides. *Journal of Lightwave Technology* 19 (3):405-413.

Sazio, P J A. 2006. Microstructured optical fibers as highpressure microfluidic reactors. *Science* 311:1583-1586.

Sellmeier, W. 1871. Zur Erklärung der abnormen Farbenfolge im Spectrum einiger Substanzen. *Ann. Phys. Chem* 6 (219):272-282.

Takeyasu, N., T. Tanaka, and S. Kawata. 2005. Metal deposition deep into microstructure by electroless plating. *Japanese Journal of Applied Physics Part 2-Letters & Express Letters* 44 (33-36):L1134-L1137.

Wang, K., Z. Zheng, Y. L. Su, Y. M. Wang, Z. Y. Wang, L. S. Song, J. Diamond, and J. S. Zhu. 2010. High-Sensitivity Electro-Optic-Modulated Surface Plasmon Resonance Measurement Using Multilayer Waveguide-Coupled Surface Plasmon Resonance Sensors. *Sensor Letters* 8 (2):370-374.

9

Thin Chalcogenide Films for Photonic Applications

Rossen Todorov, Jordanka Tasseva and Tsvetanka Babeva
Institute of Optical Materials and Technologies "Acad. J. Malinowski"
Bulgarian Academy of Sciences
Bulgaria

1. Introduction

In nature glass is formed as a result of rapid cooling and solidification of rock melts of volcanic origin. That hints at the existence of glass to have dated back since the dawn of creation. Man has quite soon borrowed the pattern to synthesize glass for their needs: in the production of glass beads and glazes for ceramic pots and vases (3500 BC), hollow glass items (1500 BC) - these developing further through the discovery of glassblowing (between 27 BC and AD 14), leading to fabrication of cast glass windows by Romans (around AD 100) and glass sheets (11th century) (http://glassonline.com/infoserv/history.html). Starting its applications as a non-transparent material for decoration purposes, glass has made a breakthrough to become used in a wide range of fields - from housing construction through optics (for lenses and protective coatings for mirrors), information processing (optical fibres and, in general, waveguides) and storage to photonics. The ingredients of the first accidentally man-synthesized glass were nitrate and sand, the latter being mostly composed of silica (SiO_2), usually in the form of quartz. On its part, quartz is the second most abundant mineral in the Earth's continental crust and is thus justifiable as a basic material in glass-production. Its high glass-transition temperature, however, places some limitations for its applications and other substances, such as sodium carbonate, aluminium, boron, calcium, cerium, magnesium, lead, thorium oxides are added to facilitate the processing and modulate the optical properties, e.g. refractive index. Alternatively, there are non-silica containing glasses as: borate, fluoride, phosphate and chalcogenide. Namely "chalcogenides" stay in the scope of the present chapter to be revealed as superb materials for photonics and in particular – photonic crystals (PhC). It would be traced in Section 3 what are the properties that we strive to derive out from chalcogenides in order to conform to the theoretical considerations. In the following lines, however we will see what chalcogenides originally offer.

2. Chalcogenide glasses and their properties

The term "chalcogen" – meaning "ore former" (from *"chalcos"* old Greek for "ore") – is a characteristic name that was proposed by Fischer in 1932 to refer to the group of elements O, S, Se, and Te (Fischer, 2001). "Chalcogenides" is used to address chalcogen's compounds with elements such as As (arsenic), Ag (silver), Bi (bismuth), Cu (copper), Cd (cadmium), Fe (iron), Ga (gallium), Ge (germanium), In (indium), Pb (lead), Na (sodium), Sb (antimony), Si (silicon), Sn (tin), Tl (thallium), Zn (zinc), etc. As suggested by the above variety of elements entering in the formation of the chalcogenide glassy alloys, chalcogenides offer wide range of variation in their properties – e.g., in their index of refraction and optical band gap, as seen in the figure below. It is seen that the refractive index could be freely varied with the composition in 2 – 3.6 range and the optical band-gap from 1.5 to nearly 2.8 eV.

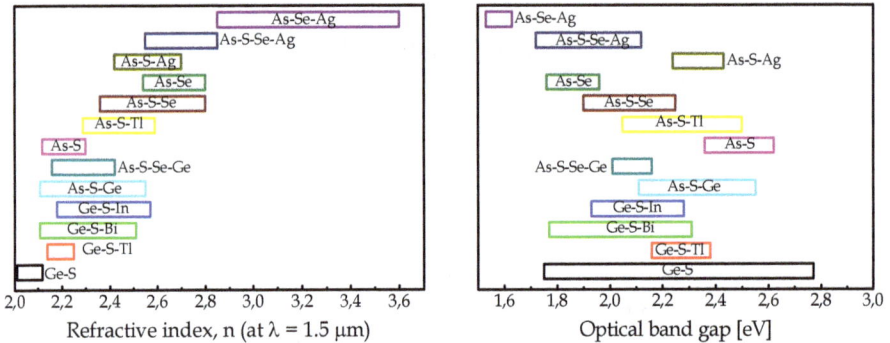

Fig. 1. Compositional variation of the refractive index and optical band-gap of chalcogenide glass thin films. The data depicted is taken from own studies and published elsewhere (Todorov et al., 2001, 2003; Tasseva et al., 2005, 2007, 2010; Kincl et al., 2009; Knotek et al., 2009; Petkov et al. , 2009).

Chalcogenide glasses, as implied by the latter term, are characterized by lack of a long-range order in the arrangement of the structural units within the glassy network, and are consequently in their nature – metastable. That fact is determining for the interesting effects resulting from the light-matter interactions in chalcogenides – electromagnetic radiation leads to modification of their structure and structural bonds. Chalcogens are two-fold coordinated with their atoms being in possession of lone-pair electrons, occupying states at the top of the valence band. Thus, exposure to light is associated with absorption of photons with certain energy, consumed for excitation of these anti-bonding electrons. That results to the formation of structural defects of one- or threefold coordinated chalcogen units (Kastner et al. 1976, Fritzsche, 1998, Liu et al., 2005). Considering all of the above and keeping in mind that the optical band gap for these materials is defined as the difference between the energies corresponding to the top of the valance and those, separating localized from delocalized states in the conductive band, we can interpret the above given values of the optical band gap as energies below which our materials are transparent. That provides transparency in the visible to IR region, where main telecommunication wavelengths are positioned, as well as the characteristic features of the so-called greenhouse gasses (water vapour, carbon dioxide, methane, nitrous oxide, and ozone) and biological molecules are located. That well-known fact can be utilized in fabrication of mid-IR guiding optical fibers for space and underground CO_2 storage monitoring (Houizot et al., 2007, Charpentier et al., 2009), optical biosensors (Anne et al., 2009), etc.

To follow the thread started above for the interesting interactions between light and chalcogenides, we should say that these interactions when low- or high-intensity electromagnetic field involved are manifested respectively as a linear or nonlinear response of the chalcogenide media. Photoinduced darkening or bleaching, expressed by the shift of the absorption edge towards longer or shorter wavelengths, accompanied by increase or decrease of the refractive index and sometimes by contraction or expansion of the films are among the linear effects. The listed phenomena have been of a continuous interest for the scientific groups, dealing with chalcogenides (Barik et al., 2011; Skordeva et al., 2001) and will not be in the focus of the considerations in this section, but further. We will just report here some characteristic values for the photoinduced refractive index change of our materials since, as it will be shown in the next sections, a target parameter in the fabrication of PhC is the high optical contrast, that can be post-tuned by simply exposing the structure to light, accounting namely on the induced modifications in the photosensitive medium.

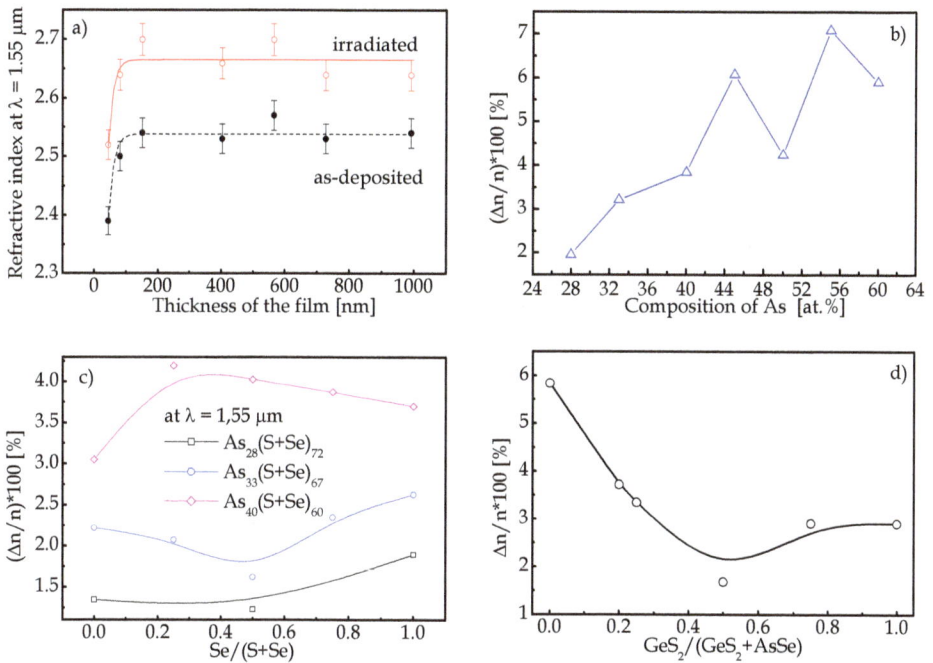

Fig. 2. a) Refractive index (at $\lambda = 1.55$ μm) of as-deposited and light irradiated films of AsSe, depending on the thickness and absolute relative (to the as-deposited) refractive index change after exposure of thin films from b) As-Se, c) As-S-Se and d) As-S-Se-Ge systems

As seen from Fig. 2, the photostimulated refractive index changes can reach up to 7 % in the system As-Se (fig. 2b), up to 4 % in the system As-S-Se (fig 2 c) and up to 5.8 % in As-S-Se-Ge system (fig. 2d). The data, presented in fig. 2 is a confirmation of the well-known fact that the sensitivity of As-containing chalcogenide films increases with the addition of arsenic. Though a tendency is observed for enhancing the photosensitivity of the coatings with the increase of the selenium in the As-S-Se system, the refractive index change is poorly influenced by the substitution of Se for sulphur. For the compositions within As-S-Se-Ge system (fig. 2d), obtained by co-evaporation of AsSe and GeS$_2$ - the details of which were published elsewhere (Tasseva et al., 2005) - it can be concluded that films enriched in AsSe undergo higher photoinduced changes. The poorest is the sensitivity of the coatings, containing equal fractions

AsSe and GeS$_2$ (50 %). The lower sensitivity of the GeS$_2$ films in respect to the AsSe ones and in general to the bi-component As-containing films is justified by the fact that the tetrahedral structural units forming the amorphous network of the Ge-containing glasses are with a greater volume and therefore are hardly reoriented in comparison with the structural units of As-containing glasses (Raptis et al., 1997). The microhardness measurements, presented in our previous work (Tasseva et al., 2005) come also to support that suggestion. At this point it should be emphasized the result from our investigation that the refractive index value is not affected by the thickness of the film in the range 80 – 1000 nm (fig. 2a). The deviation from the plateau observed for the film of 40 nm thickness is possibly attributed to voids due to the sooner interruption of the deposition process, depriving the initial formations of the possibility to grow as an integral coating. The possibility to produce films with invariable refractive index still varying their thickness has a special importance for the fabrication of multilayered structures, as would be seen in the next sections.

Further, the discussed possibilities to modify the refractive index either by exposing the coatings to light or simply changing their composition within a chalcogenide system are utilizable in the production of focusing elements – microlenses or diffraction gratings (Wágner & Ewen, 2000; Saithoh et. al., 2002; Kovalskiy et al., 2006; Eisenberg et al., 2005; Teteris & Reinfelde, 2004; Arsh et. al. 2004; Freeman et. al. 2005; Beev et. al. 2007; Vlaeva et. al. 2011). The phenomenon of total internal reflection staying at the basis of applications such as optical fibers or planar waveguides (Su et. al. 2008; Liao et. al. 2009; Conseil et. al. 2011; Savović & Djordjevich, 2011, Ung & Skorobogatiy 2011, Rowlands et. al. 2010, Ganjoo et al. 2006; Riley et al. 2008), could also benefit from the opportunity of retaining interface losses low when creating refractive index profile in compatible materials.

Let us now consider the non-linear response of the chalcogenide medium to intense light with photon energies lower than the optical band-gap. We know that two-photon absorption will be involved in the interband transitions in that case (Boyd, 2003). One of the associated effects is the induction of a non-linear refractive index, n_2 [esu] or γ [m^2/W] that gives the rate at which the intensity dependent refractive index n' changes with increasing light intensity, I:

$$n' = n + \gamma I = n + \frac{n_2}{2}|E|^2 \tag{1}$$

where n is the linear, weak-field refractive index, I - the intensity and E denotes the strength of the applied optical field. A detailed discussion on the origin of the nonlinear response in chalcogenides is published in (Bureau et al., 2004, Zakery & Elliott, 2007, etc). The electron shells of the chalcogens are such that favor the induction of polarization under strong electromagnetic field that is, in general, directly associated with the nonlinearity (Boyd, 2003). It is discussed in (Bureau et al., 2004) that the coordination of the chalcogen atoms is always pseudo-tetrahedral, consisting of two bonding and two anti-bonding electron pairs. One consequence of the presence of the unpaired electrons is that they occupy levels in the energy diagram, located between the bonding and non-bonding levels, significantly lowering in that way the optical band gap. The latter strongly influences the non-linear refractive index, according to the formula developed by Sheik-Bahae et al. (1990):

$$\gamma = K \frac{\hbar c \sqrt{E_p}}{2n^2 E_g^{opt4}} G_2(\hbar\omega / E_g^{opt}) \tag{2}$$

where E_p = 21 eV, K is found to be 3.1 x 10^{-8} in units such that E_p and E_g^{opt} are measured in eV, and γ is measured in m^2/W, $\hbar$ is the Dirac's constant, c the speed of light in vacuum and G_2 - a universal function:

$$G_2(x) = \frac{-2 + 6x - 3x^2 - x^3 - \frac{3}{4}x^4 - \frac{3}{4}x^5 + 2(1-2x)^{3/2}\Theta(1-2x)}{64x^6} \qquad (3)$$

where Θ is the Heaviside step function. n_2 and γ are related by:

$$n_2[esu] = \frac{cn}{40\pi}\gamma[SI] \qquad (4)$$

Originally developed for crystalline materials, Eq. 2 was shown to be applicable as a rough approximation as well for glasses (Tanaka, 2007), providing with a possibility for calculation of the dispersion of the nonlinear refractive index. In the same approximation, the two photon absorption, β, defined by $\alpha = \alpha_0 + \beta I$ (where α_0 is the linear absorption coefficient of the medium), can be expressed:

$$\beta = K\frac{\sqrt{E_p}}{n^2 E_g^{opt3}}F_2(2\hbar\omega / E_g^{opt}) \qquad (5)$$

where

$$F_2(2x) = \frac{(2x-1)^{3/2}}{2x^5} \text{ for } 2x > 1 \qquad (6)$$

and $F_2(2x)$ = 0 otherwise. That means that two photon absorption may occur for photon energies higher than at least half of the optical band-gap. It was shown (Mizrahi, 1989) that high values of β, that are likely to accompany the high values of the nonlinear refractive index, γ, would impose strong limitations upon the applicability of any third order nonlinear material. Qualitatively, that limitation would be expressed through a simple criterion set upon the so called figure-of-merit (FOM). For the effective operation of a given nonlinear device, the inequality should be observed:

$$FOM = \frac{2\beta\lambda}{\gamma} < 1 \qquad (7)$$

The above criterion implies that when optimizing chalcogenide compositions in order to enhance devices performance, we should carefully consider the trade-off between high nonlinearity and low optical losses due to two photon absorption. Following below, the results are depicted from the calculations of two photon absorption, nonlinear refractive index and FOM for thin films from As_2S_3, As_2Se_3 and GeS_2 (Fig. 3).

Chalcogenides, as suggested in table 1, offer a nonlinear refractive index up 700 times higher than that of fused silica, simultaneously detaining low two photon absorption, i.e. FOM. It is seen that small additions of silver (the thickness of the silver layer used for the process of photodiffusion, as reported elsewhere (Tasseva et al., 2010), was 25 nm) increase 3 times γ

still keeping low values of the two photon absorption. The values of the non-linear refractive index predicted for thin As-S-Se films are commensurable with those measured for bulk samples (Cardinal et al., 1999). The ultrafast response, i.e. the induction of nonlinear refractive index in chalcogenides exposed to the influence of strong electromagnetic field could be explained in the terms of the electronic structure of the glasses, considering ionization of the atom and distortion of the electron orbits (Liu et al., 2005).

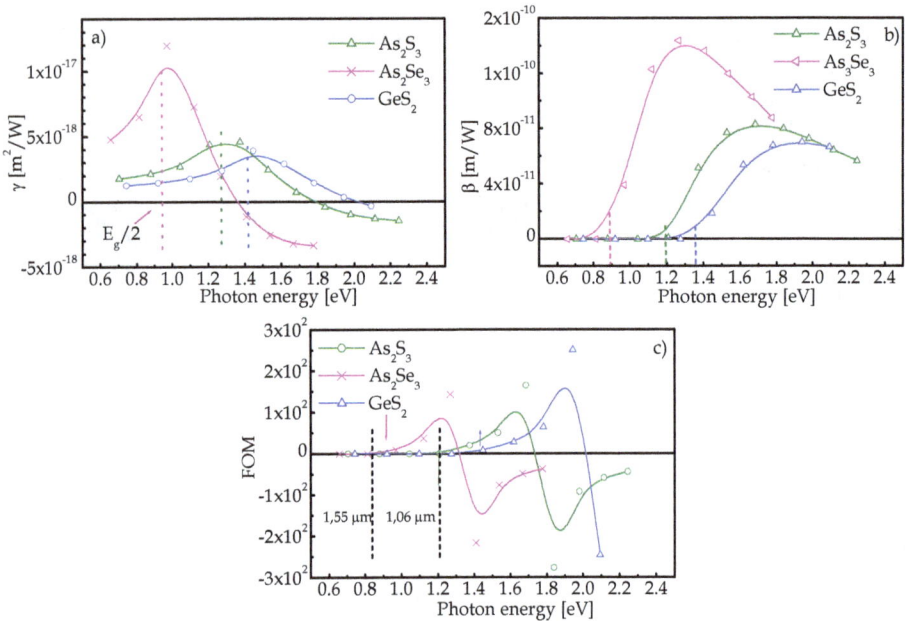

Fig. 3. Dispersion of the a) nonlinear refractive index, b) two photon absorption and c) FOM of thin chalcogenide films, calculated from eqs (2), (5) and (7)

Composition of the thin film	γ [m²/W] x10⁻¹⁸	γ/γ SiO₂	FOM
As₄₀S₆₀	1.96	71	0
As₄₀Se₆₀	6.32	231	0
As₄₀S₃₀Se₃₀	3.58	131	0
As₃₉.₂Se₅₅.₁Ag₅.₇	19.1	697	1
As₄₁Se₂₈S₂₆Ge₅	4.46	163	0
Ge₃₃S₆₇	1.47	54	0

Table 1. Nonlinear refractive index values, γ and FOM for low-loss thin chalcogenide films of various compositions at λ = 1.55 µm

In the process of selection of chalcogenide compositions for optimized photonic applications, as discussed in the lines above, one should be careful in tailoring the properties of the materials. It should be taken account for transparency of the films in the designation spectral region, high nonlinearities and low nonlinear absorption, high photosensitivity.

3. Theory

3.1 Short description of theory of photonic crystals

Photonic crystals are artificial structures usually comprising two media with different dielectric permittivity arranged in periodic manner with periodicity of the order of wavelength for the visible spectral range. Generally photonic crystals are divided into one-, two- or three dimensional PhC referred to as 1D, 2D and 3D PhC, depending on the dimensionality of the periodicity.

Photonic crystals occur in nature. Spectacular examples can be found in the natural opal, multilayered structures of pearls, flashing wings of some insects etc. A close inspection with an electron microscope shows that many species of butterflies and beetles have photonic crystal structures in some part of their bodies, resulting in a variety of optical effects such as structural colours, for example.

Although photonic crystals have been studied in one form or another since 1887, the term "photonic crystal" appeared about 100 years later, after Eli Yablonovitch and Sajeev John published two papers on photonic crystals (see Yablonovitch, 1987, John, 1987). It is very important to note that the periodicity is not a sufficient condition for a certain structure to be called photonic crystal. There is another requirement, namely the optical contrast, i.e. the difference between dielectric permittivity of the two constituent media, to be high enough (Yablonovitch, 2007).

One of the most striking features of photonic crystals is associated with the fact that if suitably engineered, they may exhibit a range of wavelengths over which the propagation of light is forbidden for all directions. The band of forbidden wavelengths is commonly referred to as "photonic band gap-PBG" and as "complete (or 3D) photonic band gap" if it is realized for all light propagation directions. These photonic bands enable various applications of PhC in linear, non-linear and quantum optics.

As mentioned above, the concept of 3D PBG materials was independently introduced by Yablonovitch (1987) and John (1987). Extensive numerical calculations conducted few years later (Ho et al. 1990) shown that 3D structures with a certain symmetry do indeed exhibit complete PBG. The "ideal" photonic crystal, defined as the one that could manipulate light most efficiently, would have the same crystal structure as the lattice of the carbon atoms in diamond. It is clear that diamonds cannot be used as photonic crystals because their atoms are packed too tightly together to manipulate visible light. However, a diamond-like structure made from appropriate material with suitable lattice constant would create a large "photonic bandgap". The first 3D photonic crystal was fabricated in 1991 in the group of Eli Yablonovitch (Yablonovitch et al., 1991) and is called Yablonovite. It had a complete photonic band gap in the microwave range. The structure of Yablonovite had cylindrical holes arranged in a diamond lattice. It is fabricated by drilling holes in high refractive index material.

Two-dimensional structures with a complete photonic band gap are neither known, nor likely to occur. Nevertheless, there is a growing scientific interest in 2D structures. The scientific efforts are focused on introduction of functional defects in 2D structures in order to realize waveguide structures (Brau et al., 2006).

The widely accepted concept for a one-dimensional photonic crystal is a quarter-wave stack of alternating low- and high-refractive index layers (Joanopoulus et al., 1997). For a wave propagating normally to the stack (zero angle of incidence), a one-dimensional photonic band gap exists that is shifted towards smaller wavelengths with increasing the incident angle. Considering all possible angles and polarizations, one can mistakenly conclude that a 1D structure has no three-dimensional photonic band gap. Fortunately, it has been shown that if the optical contrast (the difference in dielectric permittivity between stack constituents) and number of the layers are sufficiently high, an omnidirectional (OD) reflectance band could be open (reflectance of the stack is close to unity for all incident directions and polarization of light) (Fink et al., 1998). As it will be shown in the next section, there are special requirements for low and high refractive index values of the two media of the stack.

Here we will give some details of calculation of the photonic band gap in the case of 1D PhC. The calculation methods for 2D and 3D cases are not presented here because they are out of the scope of this chapter. Very comprehensive description can be found in (Sakoda K., 2005) for example.

As already mentioned, 1D PhC can be realized by deposition of alternating high and low refractive index layers with a quarter-wave optical thicknesses. If the refractive index and thicknesses of the materials are represented by n_i and d_i, where the subscripts i is H or L for high and low refractive index material, then the characteristic matrices M_P and M_S of each layer for p and s – polarization can be written in the form:

$$M_P = \begin{pmatrix} \cos\Delta_i & \dfrac{-in_i\sin\Delta_i}{\cos\theta_i} \\ \dfrac{-i\cos\theta_i\sin\Delta_i}{n_i} & \cos\Delta_i \end{pmatrix} \text{ and } M_S = \begin{pmatrix} \cos\Delta_i & \dfrac{-i\sin\Delta_i}{n_i\cos\theta_i} \\ -in_i\cos\theta_i\sin\Delta_i & \cos\Delta_i \end{pmatrix}, \quad (8)$$

In eq.8 $\Delta_i = 2\pi n_i d_i \cos\theta_i / \lambda$ is the optical phase thickness of the layer and θ_i is connected with the angle of incidence θ_0 by the Snell-Decarte's law $n_0 \sin\theta_0 = n_i \sin\theta_i$, where n_0 is the refractive index of incident medium.

The multilayered stack composed of q pairs of high (H) and low (L) refractive index layers, is presented by the matrix multiplication:

$$Q_P = (n_s/n_0)*I*H*(LH)^q*S \text{ , for p-polarization} \quad (9)$$

$$Q_S = 0.5*I*H*(LH)^q*S \text{ , for s-polarization,} \quad (10)$$

where the sign "*" denotes matrix multiplication, the matrices H and L are the characteristic matrices M_P and M_S for high and low refractive index materials; I and S are the characteristic matrices of the two surrounding media, usually air and substrate with refractive indices n_0 and n_s, respectively:

$$I_P = \begin{pmatrix} 1 & \dfrac{-n_0}{\cos\theta_0} \\ 1 & \dfrac{-n_0}{\cos\theta_0} \end{pmatrix}, I_S = \begin{pmatrix} 1 & \dfrac{-1}{n_0\cos\theta_0} \\ 1 & \dfrac{1}{n_0\cos\theta_0} \end{pmatrix}, S_P = \begin{pmatrix} 1 & 1 \\ \dfrac{-\cos\theta_s}{n_s} & \dfrac{\cos\theta_s}{n_s} \end{pmatrix} \text{ and } S_S = \begin{pmatrix} 1 & 1 \\ -n_s\cos\theta_s & n_s\cos\theta_s \end{pmatrix} \quad (11)$$

where the angle θ_s is associated with θ_0 by the Snell-Decarte's law $n_0 \sin\theta_0 = n_s \sin\theta_s$. Note in Eqs 9 and 10 that for high reflectance the stack should be terminated with high refractive index layer.

Transmittance and reflectance of the stack are finally obtained from the respective matrix elements $Q(i,j)$:

$$T_P = \frac{n_s \cos\theta_s}{n_0 \cos\theta_0}\left|\frac{1}{Q_P(1,1)}\right|^2 \text{ and } T_S = \frac{n_s \cos\theta_s}{n_0 \cos\theta_0}\left|\frac{1}{Q_s(1,1)}\right|^2 \tag{12}$$

$$R_P = \left|\frac{Q_P(2,1)}{Q_P(1,1)}\right|^2 \text{ and } R_S = \left|\frac{Q_S(2,1)}{Q_S(1,1)}\right|^2 \tag{13}$$

The short and the long wavelength edges of the band are function of n_H, n_L and the angle of incidence θ. They are calculated according to (Kim et. al, 2002) and are presented in Fig. 4a. Fig. 4 b,c and d presents the reflectance of the stack for p and s-polarized light incident at angles of 0, 45° and 85° as a function of the wavelength calculated from Eqs 12 and 13. It is seen that for higher incident angles the band shifts towards smaller wavelengths becoming wider for s-polarization and narrower for p-polarization. The shaded area in fig. 4 indicates the omnidirectional reflectance (ODR) band that is the spectral area of high reflectance for all incident angles and polarization types. It is seen that ODR band opens between short wavelength edge for 0° and long wavelength edge for p-polarization at incident angle of 90°.

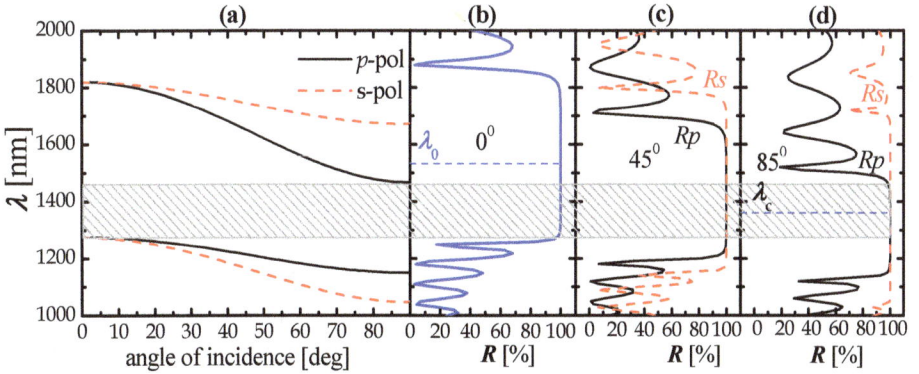

Fig. 4. a) Reflection band edges for both polarizations as a function of incident angle; Reflectance of the stack as a function of wavelength for incident angles of b) 0°, c) 45° and d) 85°. The shaded area represents the ODR band.

The short and long wavelength edges of ODR band can also be expressed explicitly as (Kim et. al., 2002):

$$\lambda_{short}^{ODR} = \frac{\pi}{2}\left[\cos^{-1}\left(-\frac{n_H - n_L}{n_H + n_L}\right)\right]^{-1} \tag{14}$$

$$\lambda_{long}^{ODR} = \frac{\pi}{4}\left(\frac{n_L\sqrt{n_H^2-n_0^2}+n_H\sqrt{n_L^2-n_0^2}}{n_H n_L}\right)\left[\cos^{-1}\left(\frac{n_H^2\sqrt{n_L^2-n_0^2}-n_L^2\sqrt{n_H^2-n_0^2}}{n_H^2\sqrt{n_L^2-n_0^2}+n_L^2\sqrt{n_H^2-n_0^2}}\right)\right]^{-1} \qquad (15)$$

Knowing n_H and n_L it is easy to calculate the width and position of ODR band using eqs. 14 and 15. It was shown (Kim et. al., 2002) that as higher is the optical contrast between the stack's constituents as wider is the ODR band. It is interesting to note that the smallest value of n_H for ODR to be opened is 2.264 (Kim et. al., 2002). It is clear from eqs 14 and 15 that the lowest refractive index for existence of ODR band can be calculated at a fixed n_H. For n_H = 2.264 the lowest n_L is 1.5132, i.e the smallest optical contrast Δn is around 0.75.

From material and technological viewpoint, there are various materials and deposition methods suitable for successful preparation of quarter-wave stacks with good optical contrast resulting in wide reflectance band with high reflectance value. But for achieving ODR band low and high refractive index materials should be carefully chosen to fulfil the additional requirements - n_H > 2.264 and Δn ~ 0.75.

One good opportunity is combining calcogenide glasses (n_H > 2.3 at 1550 nm, see Fig. 1) and polymer layer (n_L ~ 1.5-1.7). Quarter-wave stacks from $As_{33}Se_{67}$ (n_H = 2.64), $Ge_{20}Se_{80}$ (n_H = 2.58) and $Ge_{25}Se_{75}$ (n_H =2.35) as high refractive index materials and polyamide–imide films (PAI) (n_L = 1.67) and polystyrene (PS) (n_L = 1.53) as low refractive index materials are prepared (Kohoutek et. al., 2007a, Kohoutek et. al., 2007b) exhibiting high reflection band around 1550 nm. We prepared a quarter wave stack combining As_2S_3 with n_H = 2.27 at 1550 nm and Poly(methyl methacrylate) (PMMA) polymer with n_L = 1.49 thus achieving an optical contrast of about 0.78. Further it is shown that the addition of thin Au film with thickness of 50 nm as a layer close to the substrate in $Ge_{33}As_{12}Se_{55}$ / PAI stack increases the width of ODR band three times (Ponnampalam et. al., 2008)).

Another possibility for achieving ODR band is combining two suitably chosen chalcogenide glasses, i.e fabrication of all-chalcogenide reflectors. Combinations of Ge-S / Sb-Se (Kohoutek et al. 2009) with optical contrast of more than 1 and exposed As_2Se_3 / GeS_2 with optical contrast more than 0.8 have been already realized. (Todorov et al. 2010b).

3.2 Factors which influence the properties and quality of 1D photonic crystals

In section 3.1 it was shown that both the optical contrast between the two component of 1D PhC and n_H values are factors that influence the width of the omnidirectional reflectance band. Here we will show that the number of the layers and small optical losses due to absorption or scattering are additional factors that should be considered during the design of an OD reflector.

Fig. 5 presents the value of the reflectance band for normal light incidence as a function of the number of the layers in the stack and the ratio of their refractive indices. It is seen that the reflectance increases both with increasing the number of the layers and the ratio of their refractive indices. Besides, the lower optical contrast can be compensated to some extent by increasing the number of layers in the stack. For example if n_H/n_L = 1.8 then 20 layers will be required for 100% reflectance whereas for n_H/n_L =2.2 only 14 layers will be sufficient.

It is well known that during the fabrication of the stack deviations of both refractive indices and thicknesses of the layers from their target values may occur. Our calculations showed that the deviation of refractive index influences mostly the optical phase thickness of the L-layer, while the deviation in thickness from the target values is pronounced by changes in the phase thickness of the H-layer. If the error in n is $\Delta n = \pm 0.02$, than the phase thickness $\Delta_i = 2\pi n_i d_i / \lambda$ changes with 1.3 % for $i=L$ and 1 % for $i=H$. The deviation of 5 nm in the thickness leads to 1.9 % and 2.9 % changes of phase thicknesses of low and high refractive index layers. Fig. 6 presents the influences of the error in thickness and small losses in the layers on the reflectance band for the particular case of As_2S_3/PMMA stack with target refractive indices $n_H=2.27$, $n_L=1.49$ and thicknesses $d_H=170$ nm and $d_L=260$ nm. From Fig 6(a) it is seen that the alteration of thicknesses with 5 nm leads to the band shift of 44 nm. The small absorption of the layers leads to decrease in R of 0.5 % for $k=0.001$ and 2.1 % for $k=0.005$. (Fig. 6 (b)).

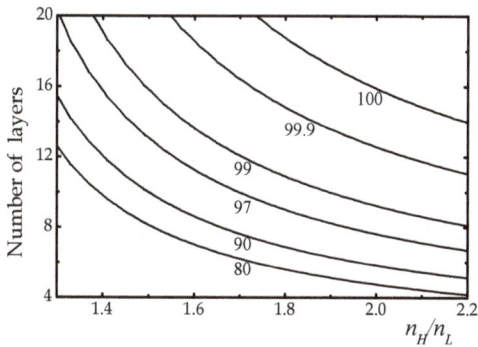

Fig. 5. Value of reflectance band for normal light incidence as function of number of the layers in the stack and ratio of their refractive indices

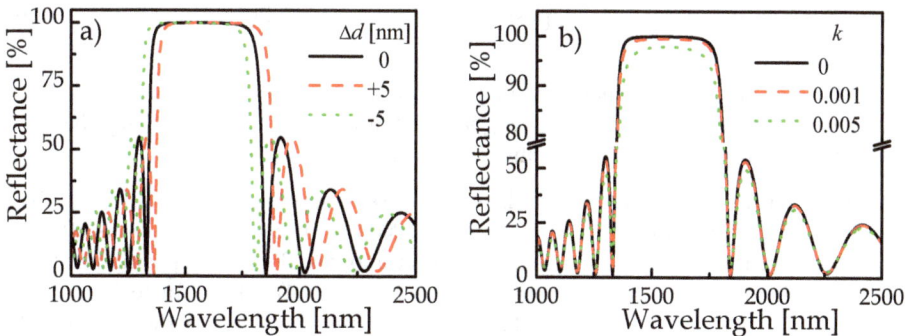

Fig. 6. a) Shift of reflectance band of As_2S_3 ($n_H=2.27$, $d_H=170$ nm) / PMMA ($n_L=1.49$, $d_L=270$ nm) stack due to error of 5 nm in thicknesses; b) influence of small absorption of layers on the reflectance value.

The comparison between the real and ideal 1D-PhC from As_2S_3 / PMMA is presented in Fig. 7. It is seen that there is an insignificant shift between the reflectance bands of both structures. The difference of 1.1 % between the measured and calculated reflectance can be due to slight absorption and scattering of the layers as well as to measurements errors. The most significant difference between the fabricated and simulated structures is in the side peaks. Note for example the first minima that are very high for the real structure and zero for the simulated one. Most probably this difference is due to random deviations of phase thicknesses of the constituent layers leading to violation of the conditions for destructive interference.

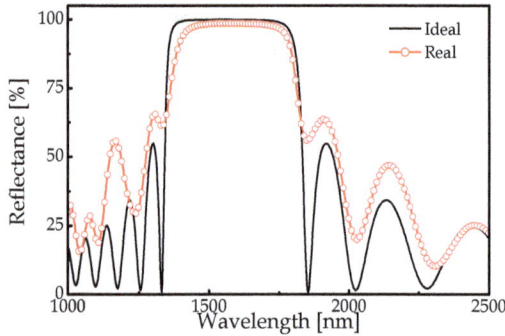

Fig. 7. Measured and simulated reflectance band of As_2S_3/PMMA stack

4. Device fabrication

4.1 Methods for fabrications of PhC's

In Section 3.1 it was mentioned that the widely accepted approach for production of 1D PhC is the alternating deposition of high and low refractive index materials with quarter-wave thicknesses. Depending on the materials chosen, different deposition methods as vacuum condensation, spin and dip coating, pulsed laser deposition, sol-gel etc. are used. The experimental conditions for deposition thin films of 1D PhC is presented below in section 4.2.

The most widely recognized methods for preparation of 2D and 3D photonic crystals are colloidal self-assembly, direct laser writing and lithography with deep etching, that can be E-beam, X-ray or holographic lithography (Lopez, 2003).

Typical colloidal PhC are face centred 3D arrays of self assembled monodispersed silica or polymer microspheres with diameter from 200 – 2000 nm. The sedimentation is often used for fabrication of thick samples, which are called bulk opals. Besides, deposition on vertical substrates with controlled moving meniscus (Egen et al., 2004) is used for thinner samples. Because the refractive index of silica and polymers spheres is around 1.5, the optical contrast of 0.5 is not sufficient for opening of a complete PBG. A post fabrication infiltration of colloidal crystals with high refractive index material is necessary. Following this strategy a colloidal crystals infiltrated with Si and Ge exhibiting complete PBG at wavelength of 1.5 μm are fabricated (A. Blanco et al., 2000).

Direct laser writing method through multi-photon absorption is another method for producing 3D PhC. The light from the laser is focused on a small spot of the dye-doped polymer used as a recording medium. The energy of the laser excites the dye molecules that initiate a local polymerization in the spot thus changing the refractive index of the polymer in the illuminated spot. A spatial resolution of 120 nm is reported in the literature (Kawata et al., 2001). The problem is that the optical contrast is very small and infiltration of the structure is needed. Another possibility is using the laser writing in high refractive index material. Promising candidates are chalcogenide glasses, particularly As_2S_3 that undergo changes in solubility upon exposure to light.

Conventional lithography and selective etching were used mainly for fabrication of 2D PhC. For producing 3D PhC a concept of layer-by-layer deposition has to be implemented that comprises repeated cycles of photolithography, wet and dry etching, planarization, and growth of layers (Blanco et. al., 2004).

The holographic lithography is very promising method enabling large-area defect-free 2D and 3D periodic structure to be produced in a single-step. In holographic lithography the sensitive medium (a photoresist) is exposed to a multiple-beam interference pattern (for N-dimensional structures at least N+1 beams are required) and subsequently developed, producing a porous structure. The interbeam angles and polarizations and the number of beams determine the type of symmetry of the recorded structures. It has been theoretically shown that all 14 Bravais lattices could be produced (Cai et. al., 2002). Usually SU-8 photoresist with refractive index of around 1.67 is used. A recognized drawback is the need for infiltration of the produced structure with high refractive index material. Otherwise the optical contrast is not sufficiently high for opening a complete photonic band. Difficulties such as optical alignment, vibrational instability, and reflection losses on the interface air/photoresist further complicate the recording processes. One possibility for overcoming the problem is rotating the sample between two consecutive exposures (Lai et. al., 2005). The second approach is implementation of specially design diffraction mask that provides the required number of beams with correct directions and polarizations reducing the alignment complexity and vibration instabilities in the optical setup (Divlianski et. al., 2002).

Two-dimensional structures have already been fabricated in chalcogenide glasses using holographic lithography (Feigel et. al., 2005; Su et. al., 2009). To the best of our knowledge holographic lithography has not been used yet for fabrication of 3D PhC from chalcogenide glasses. The three dimensional wood-pile photonic crystals made in chalcogenide glasses are fabricated by direct laser writing (Nicoletti et. al., 2008) or through layer-by-layer deposition (Feigel et. al., 2003).

4.2 Experimental procedure for deposition of 1D photonic crystals

In the present work we used the concept of layer-by-layer deposition of quarter-wave stacks of alternating suitably chosen films with low and high refractive indices for producing of one-dimensional photonic crystals.

The bulk chalcogenide glass was synthesized in a quartz ampoule by the method of melt quenching from elements of purity 99.999 %. The chalcogenide layers were deposited by thermal evaporation at deposition rate of 0.5 – 0.7 nm/s. The X-ray microanalysis showed that the film composition is close to that of the bulk samples (Todorov et al., 2010b).

The multilayer structures formed only from chalcogenide sublayers were produced in one cycle of thermal evaporation using two sources - As_2Se_3 and GeS_2. After deposition the samples were exposed in air to a mercury lamp (20 mW.cm^{-2}).

Multilayered stacks from chalcogenide glass and organic polymer comprising 19 layers are prepared by alternating vacuum evaporation of As_2S_3 with target thickness of 170 nm and spin coating of PMMA with target thickness of 260 nm. The stock solution of the polymer was prepared by dissolution at ambient temperature of one gram of PMMA (Poly (methyl methacrylate)) in 10 ml of Dichloroethane (Aldrich) using magnetic stirrer for accelerating the process. The polymer films with different thicknesses are obtained by the method of spin coating using the stock solution further diluted by adding dichloroethane. Polymer layers with thicknesses of 260 nm are obtained by dripping a drop of 0.5 ml of 2.2 wt % polymer solution on the preliminarily cleaned substrate. The speed and duration of spinning were 2000 rpm and 30 s. To remove the extra solvent the samples are annealed for 30 minutes at temperature of 60 °C.

4.3 Optical methods for control and characterization of thin films for photonic crystals fabrication

It is known that thin film's thickness can significantly affect their optical constants. In literature there is no unanimity about the dependence of the refractive index of thin chalcogenide films on their thickness (Abdel-Aziz et al. 2001). In photonic crystals it is necessary that the thin films are deposited with an exact optical thickness, (nd). For example, when a thin chalcogenide film is a part of a quarterwave stack and its refractive index has a value between 2.00 and 3.60 (see Fig.1), the thickness of the sublayers must be 110 – 193 nm for the working wavelength λ = 1550 nm.

The optical constants of thin films are usually determined by optical methods as spectrophotometry, prism-coupling technique and ellipsometry. We have demonstrated that the spectrophotometric and ellipsomeric methods offer a good accuracy for determination of the optical parameters of thin chalcogenide films from $\lambda/30$ to 2λ (λ is working wavelength) (Konstantinov et al. 1998, Babeva et al. 2001, Todorov et al. 2010a). Results on the reflectance response of photonic crystals from chalcogenide glass/polymer (DeCorby et al., 2005; Kohoutek et al., 2007a) or chalcogenide glass/chalcogenide glass, e.g. GeS_2/Sb_2Se_3 (Kohoutek et al., 2009b) by variable angle spectroscopic ellipsometry have been reported.

In the present work optical transmittance and reflectance measurements at normal incidence of light beam were carried out in the spectral range from 350 to 2500 nm using an UV–VIS–NIR spectrophotometer (Cary 05E, Australia). Reflectance measurements at oblique incidence of linearly polarized light were performed with VASRA (Variable Angle Specular Reflectance Accessory). For polarizing the incident radiation, a high quality Glan–Taylor polarizer is used that provides an extremely pure linear polarization with a ratio 100 000:1. The computer controlled stepper motor of the VASRA accessory ensures reproducible adjustment of the incident angle with an accuracy $\Delta\theta$ = ±0.25° (according to the Cary Operation Manual). A self-made reference Al-mirror, whose preparation and characterization are described in details elsewhere (Babeva et al., 2002), is used as a standard mirror.

5. Results and discussion

5.1 Optical properties of thin chalcogenide films

In this part the optical properties are presented of real multilayered structures consisting of alternating layers chalcogenide-chalcogenide glass and chalcogenide glass-organic polymer. The possibility of tuning their properties under external factors such as annealing or strong electric field is examined.

As it is mentioned in section 3.2, the knowledge of the optical properties of the single layers is important for the successful engineering of a multilayer structure. Firstly we have investigated the optical parameters of the single layers. The results of the investigation of the thickness dependence of the refractive indices of thin films from some basic chalcogenide glass formers As_2S_3, As_2Se_3 and GeS_2 are given in Fig. 8.

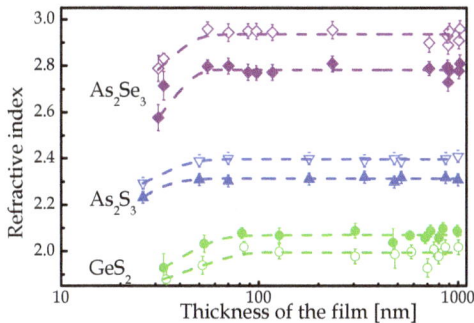

Fig. 8. Refractive index of GeS_2, As_2S_3 and As_2Se_3 thin films at $\lambda = 1550$ nm as a function of thickness before (solid symbols) and after irradiation (open symbols).

The Swanepoel's method (Swanepoel, 1983) was used for optical constants determination of the thin films with $d > 300$ nm and a combination of double (T, R) and triple (T, R, R) methods was applied in calculations for thinner layers (d < 300 nm) (Konstantinov et al., 1998; Babeva et al. 2001). The calculated values for optical parameters of thin films show that their refractive index is independent on the layer's thickness for $d > 50 - 70$ nm (Fig. 8). The calculated values for the refractive indices of thin films with composition As_2Se_3 and GeS_2 were 2.83 and 2.07 ($\lambda = 1550$ nm), respectively. Through the addition of metal in As_2Se_3 such as Ag and Cu thin films with a higher refractive index can be produced (Ogusu et al., 2004). The Ag-As_2Se_3 layer cannot be obtained by direct evaporation of the ternary Ag-As-Se glass since phase decomposition occurs. On the account of this, photodoping technique is used to produce As-As-Se film: evaporating Ag layer with a certain thickness on the As_2Se_3 layer, and exposing the stack to induce migration of the silver (Tasseva et al., 2010, Suzuki et al., 2005). We found a refractive index of 3.06 at $\lambda = 1550$ nm for thin As_2Se_3 film photodoped with 10 at % Ag (Tasseva et al., 2010). The photosensitivity of the chalcogenides arises from structural rearrangements induced by the absorption of photons at energies near the optical band gap of the material (Shimakawa et al., 1995). These structural rearrangements lead to changes in the optical properties. The magnitude and sign of these photoinduced changes can be dependant on the chemical composition of the glasses (Tanaka et al., 1979) and on the processing history of the sample (Okuda et al., 1979). It is seen from Fig.8 that the arsenic containing thin chalcogenide films with compositions As_2S_3 and As_2Se_3 demonstrate increasing of the refractive index up to 2.39 and 2.93 after exposure to light, respectively while for thin GeS_2 layer n decreases to 2.03.

Furthermore, these changes may be either reversible or irreversible. The irreversible changes result from light exposure or thermal annealing of as-deposited thin films in a non-equilibrium state (Biegelsen et al., 1980) while the reversible changes result from structural rearrangement leading from one quasi-stable state to another (Street, 1977, Biegelsen et al., 1980). In Fig. 9 the changes of the refractive index are shown of thin As_2Se_3 film after exposure and annealing. The as-deposited layer demonstrates an increase of the refractive index (Δn = +0.14 at λ = 1550 nm) that is an expression of the photo-darkening effect. A reversible change of Δn = -0.02 after annealing at 160°C in vacuum was observed.

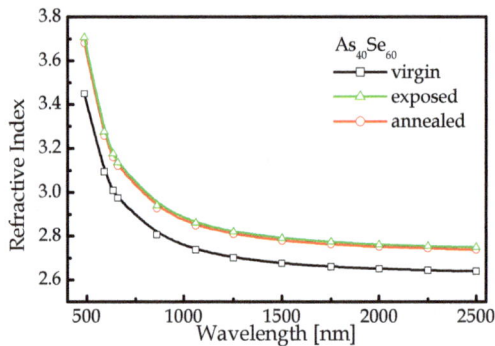

Fig. 9. Dispersion of the refractive index, n of as-deposited, exposed and subsequently annealed at 160°C in vacuum thin As_2Se_3 film.

5.2 Optical properties of a double layered structure and multilayered coatings

Further, we use the data plotted in Fig. 8 for modelling of double- and multilayered structures. The reflectance spectrum of a double layered coating from As_2Se_3 and GeS_2 (BK-7/As_2Se_3/GeS_2) was discussed in (Todorov et al. 2010b). Having determined the optical parameters and thickness of the single layers we calculated the theoretical reflectance spectrum of the complex structure using the equations from section 3.1. A good coincidence between the theoretical and measured spectrum of the double layered structure was obtained.

In Fig. 10 the measured transmittance and reflectance spectra at normal light incidence (θ = 0°) and angle θ = 70° of 19 layers quarter-wave structure from alternating As_2Se_3 and GeS_2 layers deposited on a glass and on an absorbing Si wafer substrate are presented. The shift of the photonic band gap to shorter wavelengths with the increase of the angle of light incidence, discussed in section 3, is clearly seen.

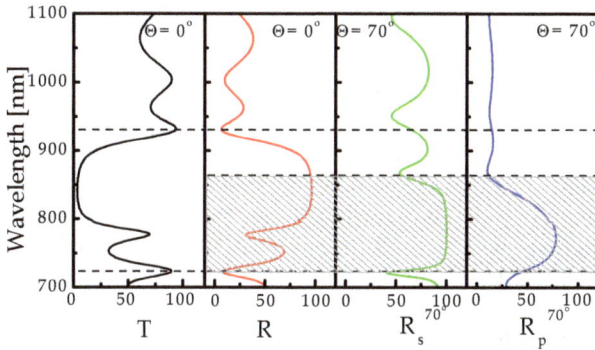

Fig. 10. Spectra of transmittance, T and reflectance, R of a multilayer coating on a glass substrate at normal incidence; and reflectance spectra at angle of light incidence 70° for p – and s- light polarization. The shaded area represents the ODR band.

5.3 Thermo-induced changes in chalcogenide glasses and tuning of the photonic band gap

The ability to tune or modulate the optical properties of photonic crystals would increase their functionality and open up new possibilities for a variety of applications for integrated optics (Lee et al., 2007). Photosensitivity has been successfully utilized for post-tuning in 2D photonic crystals (Lee et al., 2007). In (Todorov et al., 2010b) we demonstrated a possibility of shifting the fundamental reflectance band of as-deposited multilayer As_2Se_3/GeS_2 quarter-wave slab exposing it to light and thus inducing changes in both materials building the coating. It is seen from Fig. 1 that thin films from Ge - S and As - Se systems ensure high optical contrast $\Delta n \sim 0.8$. The exposure to light leads to opposite effects in the thin films – photodarkening in As_2Se_3 and photobleaching in GeS_2 layers and increase of Δn up to 1.0. The good knowledge of the photoinduced changes enabled a designed 50 nm expansion of the high reflectance band width. The presence of reversible changes in chalcogenide glasses allows production of photonic structures with dirigible optical properties.

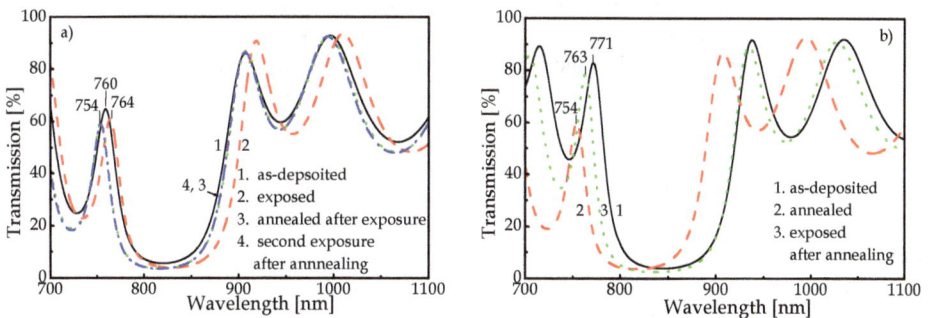

Fig. 11. Spectra of transmittance at normal light incidence of multilayer coating consisting of 19 alternating As_2Se_3 and GeS_2 layers in different consequences of treatment: as-deposited - exposed - annealed - exposed (a); and as-deposited - annealed – exposed (b).

In Fig. 11 the changes in the transmittance spectra are presented of the multilayer coatings referred to in Fig. 10 after exposure to light from a mercury lamp and after annealing at 160°C in vacuum. The annealing temperature was selected to be with 20°C lower then the glass forming temperature for As_2Se_3 glass (Li et al., 2002). The exposure to light leads to 17 nm or 30 nm red shift of the resonant band for as-deposited or annealed samples, respectively, at T = 10 %. It is seen that light illumination results in expansion of the resonant band to longer wavelengths and parallel shift of the pass band in both as-deposited and thermally treated samples. The annealing of the samples causes the stop band to move to shorter wavelengths - with 14 and 28 nm (T = 10 %) for exposed and as-deposited samples, respectively. Due to the effect of thermobleching, observed in thin As_2Se_3 and GeS_2 films (De Neufville et al. 1974, Tichy et al. 1993) the optical contrast between sublayers is changed after annealing. It is known that the reversible changes are increased with the increasing of the arsenic in the As_xSe_{100-x} system for x > 40 at % and allow making of many cycles between both states applying the light 'recording' and 'erasing' by annealing of the films (Lyubin, 1984). Unfortunately, the increase of the arsenic content for $40 \leq x \leq 60$ leads to a reduction of the refractive index (Petkov et al., 2009). Further investigations would involve the selection of suitable chalcogenide glasses with optimal reversible changes for creating of multilayered structure with variable width of the stop band.

5.4 Electrostatic tuning of the photonic band gap

The phenomena of electroabsorption due to the effect of Franz-Keldysh in crystals and glasses from As_2S_3 were observed by (Kolomiets et al., 1970). The absorption edge is shifted to longer wavelengths due to decrease of the optical gap, E_g following the formula:

$$\Delta E_g = e^2 h^2 S^2 F^2 / 24m^*$$
(16)

where e and m^* are electric charge and mass of the electron, respectively, h is Plank's constant, S is slope of the absorption edge in the absence of the field and F is the intensity of the electric field. The phenomena of the electroabsorption in chalcogenide glasses are explained through the barrier-cluster model (Banik, 2010). It assumes that an amorphous semiconductor consists of microscopic regions - perhaps closed clusters - separated from each other by potential barriers. The strong electric field increases the probability of tunnelling and optical absorption, α as well. The increase of α in non-crystalline semiconductors is proportional to the squire of the intensity of the electric field – F^2:

$$\frac{\Delta \alpha}{\alpha} = const.F^2$$
(17)

Samples from "sandwich" type were prepared for investigation of the effect of electroabsorption on the optical properties of the thin films or multilayered structure. Firstly, thin transparent electrode from chromium with 10 nm thickness was deposited on a glass substrate by electron beam evaporation. The transmission coefficient of the electrode in the spectral range 1200-2500 nm is between 70 and 80 % and the electrical resistivity, $\rho = 2$ mΩ.cm. Subsiquently the investigated samples - thin layer or multilayered stack, were deposited on the chromium electrode. The structure was terminated by deposition of the second chromium electrode on the top.

According to (Kolomiets et al., 1970) the change in the transmission, ΔT, induced by electric field, depends on the photon energy ($E = h\nu$). He observed that the maximal change ΔT occured at $h\nu = 2.30$ eV applying sinusoidal electric voltage of 100 V. In Fig. 12a the change in transmission spectrum, $\Delta T = T_{voltage} - T_{as\text{-}dep.}$ is presented of thin film from As_2S_3 with thickness 2 µm due to the application of dc electric voltage U = 100 V. The maximal value for ΔT was observed at 500 nm, i.e. for photon energy 2.48 eV. The changes of the spectrum at longer wavelengths are smaller and are possibly "shadowed" by the interference fringes. In Fig. 12b the variation is shown of the transmittance at $\lambda = 500$ nm depending on the applied electric voltage. Mathematical extrapolations showed that when the electric voltage is applied the transmittance decreases following an exponential low of the type - $T = T_0 + A_1.exp(-t/B_1)$, where T_0 is the initial value of the transmittance, $T_0 = 19.9\%$; $A_1 = 1.85$ and $B_1 = 2.58$ min are parameters. The switching off of the electric field restores the initial value of T. In this case the increase can be described through the exponential low of the type $T = T_0 + A_2.[1 - exp(-t/B_2)]$ where $A_2 = 136.45$ and $B_2 = 3.63$ min are the parameters.

Considering the relationship between optical band gap and refractive index (see Penn, 1962; Wemple & DiDomenico, 1971) it is expected the value of n to increase when E_g decreases due to electroabsorption. The changes in the refractive indices of the sublayers in 1D-photonic crystals would shift the position of their stop band. An example of the influence of the electric field on the stop band of 1D quaterwave structure from As_2S_3/PMMA is presented on Fig. 13. The shift of the transmission spectrum of such a multilayer system sandwiched between two thin transparent chromium electrodes is clearly seen (Fig. 13a).

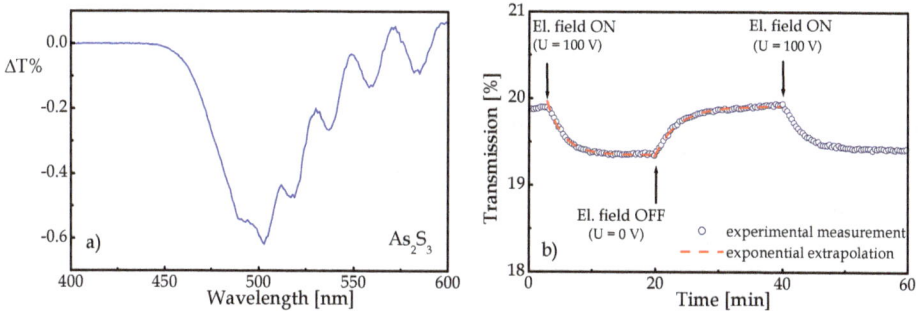

Fig. 12. Spectral dependence of the change of transmission spectrum, $\Delta T = T_{voltage} - T_{as\,dep.}$ of thin As_2S_3 film, under dc electric voltage 100 V (a); Time evolution of the transmittance at $\lambda = 500$ nm depending on the applied voltage (open symbols) and extrapolations with exponential functions (dashed line) (b).

The changes in the transmission at wavelength $\lambda = 1915$ nm are given in Fig. 13b. We observed cyclic reduction or enhancing of the transmission when switch *on* or *off* the electric field. It was observed that the changes of the transmission coefficient followed the same exponential laws as the thin film from As_2S_3 with parameters $A_1 = 3.59$ and $B_1 = 2.12$ min, when decreasing under the influence of the electric field, and $A_2 = 11120.68$ and $B_2 = 2.22$ min for the relaxation process after switching off the voltage.

The observations imply that the electric field induces reversible changes of the refractive index of the sublayers in the photonic structure. The applied electric voltage of 100 V is considerably high for applications in modern optoelectronic devices such as electrooptical modulators for high-speed time-domain-multiplexing (TDM) and wavelength-division-multiplexing systems (WDM). Further investigations would involve determination of the dependence of the electroabsorption effect on the composition of the chalcogenide glasses striving to reduce the voltage of the applied electric field.

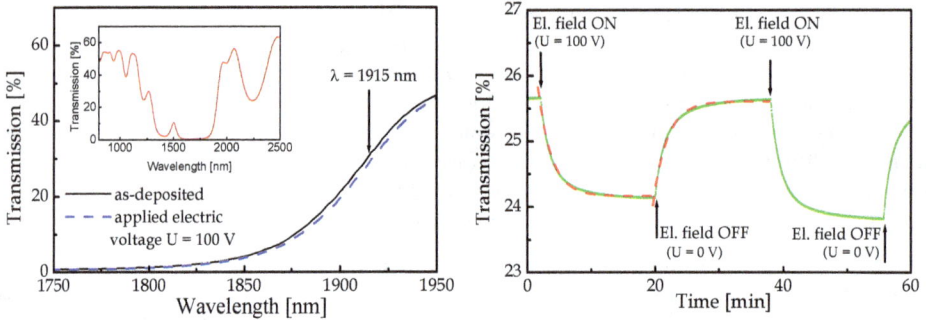

Fig. 13. Transmission spectrum of As_2S_3/PMMA multilayer slab between thin transparent chromium electrodes. In the inset the transmission spectrum of the same sample in wide spectral range - 800-2500 nm - is presented (a); Evolution of the transmission coefficient at $\lambda = 500$ nm in cyclic switching on and off of dc electric voltage U = 100 V. The exponential extrapolations are given with a dashed line (b).

6. Conclusion

In this chapter we present our results on fabrication and optimization of photonic structures exploiting the photo-, thermo- and electro-induced changes in thin chalcogenide films. It is demonstrated that the irreversible and reversible changes of the optical properties of chalcogenide glasses can be used for modification of the optical contrast of materials in 1D photon photonic crystal and shift of the reflectance stop band. It was observed that the exposure to light leads to significant red shift of the resonant band (17 nm or 30 nm for as-deposited or annealed samples, respectively), while the annealing of the samples causes the stop band to move to shorter wavelengths - with 14 and 28 nm for exposed and as-deposited samples, respectively.

High electric voltage induces reversible changes of the refractive index of the sublayers in the photonic structure. It was observed a 3 nm red shift of the photonic stop band and decrease of the transmittance, $\Delta T = 2$ % during the applying of the dc electric voltage. Although the electric voltage of 100 V is considerably high for applications in optoelectronics, our initial results indicate for a possible path for the implementation of reversible tuning of the stop band in 1 D photonic crystals.

Chalcogenide glasses possess peculiar optical properties such as high linear and non-linear refractive index and transmittance in wide range of the infrared region. These unique properties make them irreplaceable materials for mid-infrared sensing, integrated optics and ultrahigh-bandwidth signal processing; recently, a new term –"chalcogenide photonics' (Eggleton et al., 2011), has been introduced. Good knowledge of the properties of the chalcogenide glasses and their changes under influence of external factors such as exposure to light, annealing or electric field has a key role in understanding the processes in these materials and would support manufacturing of chalcogenide photonic crystals.

7. References

Abdel-Aziz, M.M.; El-Metwally, E.G.; Fadel, M.; Labib, H.H. & Afifi, M.A. (2001). Optical properties of amorphous Ge-Se-Tl system films. *Thin Solid Films*, Vol.386, No.1, (May 2001), pp.99-104, ISSN 0040-6090

Anne M.-L., Keirsse J., Nazabal V., Hyodo K., Inoue S., Boussard-Pledel C., Lhermite H., Charrier J., Yanakata K., Loreal O., Le Person J., Colas F., Compère C. & Bureau B. (2009). Chalcogenide Glass Optical Waveguides for Infrared Biosensing. *Sensors*, Vol.9, (September 2009), pp. 7398-7411, ISSN 1424-8220

Arsh, A., Klebanov, M., Lyubin, V., Shapro, L., Feigel, A., Veigner, M. & Sfez, B. (2004). Glassy $mAs_2S_3 \cdot nAs_2Se_3$ photoresist films for interference laser lithography. *Opt. Mater.* Vol.26, No.3, (August 2004), pp. 301-304, ISSN 0925-3467

Babeva, Tz.; Kitova, S. & Konstantinov, I. (2001). Photometric methods for determining the optical constants and the thicknesses of thin absorbing films: Selection of a combination of photometric quantities on the basis of error analysis. *Appl. Opt.* Vol.40, No.16, (June 2001), pp.2675-2681, ISSN: 0003-6935

Babeva, Tz., Kitova S., Mednikarov, B. & Konstantinov, I. (2002). Preparation and characterization of a reference aluminium mirror. *Appl. Opt.* Vol. 41, pp. 3840-3846

Banik, I. (2010). Photoconductivity In Chalcogenide Glasses In Non-Stationary Regime And The Barrier-Cluster Model. *Acta Electrotechnica et Informatica*, Vol. 10, No. 3, (July 2010), pp. 52–58, ISSN 1335-8243

Biegelsen, D.K. & Street, R.A. (1980). Photoinduced defects in chalcogenide glasses, *Phys. Rev. Lett.* Vol.44 (12), (March 1980), pp. 803-806, ISSN: 0031-9007

Bowden, B.F. & Harrington J.A, (2009), Fabrication and characterization of chalcogenide glass for hollow Bragg fibers. *Appl. Opt.* Vol.48, No. 16, (June 2009), pp.3050-3054

Barik, A.R., Adarsh, K.V., Naik, R., Ganesan, R., Yang, G., Zhao, D., Jain, H. & Shimakawa, K. (2011). Role of rigidity and temperature in the kinetics of photodarkening in $Ge_xAs_{(45-x)}Se_{55}$ thin films. *Opt. Express*, Vol.19, No.14, (July 2011), pp. 13158-13163, ISSN: 1094-4087

Beev, K., Sainov, S., Stoycheva-Topalova R. (2007). Total internal reflection holographic recording in chalcogenide glass films. *J. Optoelectron. Adv. M.*, Vol.9, No.2, (February 2007), pp. 341-343, ISSN 1454-4164

Blanco, A., Chomski, E., Grabtchak, S., Ibisate, M., John, S., Leonard, S.W., Lopez, C., Meseguer, F., Miguez, H., Mondla, J.P., Ozin, G.A., Toader, O., Van Driel, H.M. (2000). Large-scale synthesis of a silicon photonic crystal with a complete three-dimensional bandgap near 1.5 micrometres. *Nature*, Vol. 405, No. 6785, (May 2000), pp.437-440 ISSN: 0028-0836

Blanco, A., Busch, K., Deubel, M., Enkrich Ch., Von Freyman, G., Hermatschweiler, M., Koch, W-P., Linden, S., Meisel, D.C., Wegener, M. (2004) Three-dimensional lithography of photonic crystals, In: Photonic crystals – advances in design, fabrication and characterization", K. Busch, S. Lolkes, R.B. Wehrspohn, and H. Foll, (Ed.'s), Wiley – VCH, Weinheim, 153-173, Darmstadt, Germany

Boyd, R.W. (2003). *Nonlinear Optics*, second edition, Academic Press, Elsevier Science, USA

Braun, P.V., Rinne, S.A. & Garcia-Santamaria, F. (2006). Introducing defects in 3D photonic crystals: State of the art. *Adv. Mat.* Vol.18, No. 20, pp.2665-2678, ISSN: 09359648

Bureau, B., Zhang, X.H., Smektala, F., Adam, J.-L., Lucas, J., Troles, J., Ma, H.-L., Boussard-Pledel, C., Lucas, P., Le Coq, D., Riley, M.R. & Simmons, J.H. (2004). Recent advances in chalcogenide glasses. *J. Non-Cryst. Solids*, Vol.345&346, (October 2004), pp. 276-283, ISSN 0022-3093

Cai, L.Z., Yang, X.L. & Y.R. Wang, Y.R. (2002). All fourteen Bravais lattices can be formed by interference of four noncoplanar beams. *Opt. Lett.* Vol. 27, No.11, (June 2002), pp.900-902, ISSN: 0146-9592

Cardinal, T., Richardson, K.A., Shim, H., Shulte, A., Beatty, R., Le Foulgoc, K., Viens, J. F., Villeneuve, A. (1999). Non-linear optical properties of chalcogenide glasses in the system As-S-Se. *J. Non-Cryst. Solids*, Vol.256&257, pp.353-360, ISSN 0022-3093

Charpentier, F., Bureau., B., Troles, J., Boussard-Plédel, C., Michel-Le Pierrès, K., Smektala, F., Adam, J.-L. (2009). Infrared monitoring of underground CO_2 storage using chalcogenide glass fibers. *Opt. Mater.* Vol.31, No.3, pp.496–500, ISSN 0925-3467

Conseil, C., Coulombier, Q., Boussard-Plédel, C., Troles, J., Brilland, L., Renversez, G., Mechin, D., Bureau, B., Adam, J.L. & Lucas, J. (2011). Chalcogenide step index and microstructured single mode fibers. *J. Non-Cryst. Solids*, Vol.357, No.11-13, (June 2011), pp. 2480-2483, ISSN 0022-3093

DeCorby, R.G., Nguyen, H.T., Dwivedi, P.K. & Clement, T.J. (2005). Planar omnidirectional reflectors in chalcogenide glass and polymer. *Opt. Express* Vol.13, No.16 pp.6228-6233, ISSN: 1094-4087

De Neufville, J.P., Moss, S.C. & Ovshinsky, S.R. (1974), Photostructural transformations in amorphous As_2Se_3 and As_2S_3 films, *J. Non-Cryst. Solids*, Vol.13, No.2, (January 1974), pp. 191-223, ISSN: 0022-3093

Divlianski, I., Mayer, T.S., Holliday, K.S. & Crespi, V.H. (2003). Fabrication of three-dimensional polymer photonic crystal structures using single diffraction element interference lithography. *Appl. Phys. Lett.* Vol.82, No.11, (March 2003), pp.1667-1169, ISSN: 0003-6951

Eggleton, B.J., Luther-Davies, B. & Richardson, K. (2011). Chalcogenide photonics. Nature Photonics, Vol.5, No.3, (March 2011), pp.141-148, ISSN: 17494885

Egen, M., Zentel, R., Ferrand, P., Eiden, S, Maret, G., & Caruso, F. (2004), Preparation of 3D photonic crystal from opals, In: *Photonic crystals – advances in design, fabrication and characterization*, K. Busch, S. Lolkes, R.B. Wehrspohn, & H. Foll, (Ed.'s), 109-128, Wiley – VCH, Weinheim, ISBN: 978-3-527-40432-2, Darmstadt, Germany

Eisenberg, N.P., Manevich, M., Arsh, A., Klebanov, M. & Lyubin, V., (2005), Arrays of micro-prisms and micro-mirrors for infrared light based on $As_2S_3-As_2Se_3$ photoresists., *J. Optoelectron. Adv. M.* Vol.7, No.5, pp. 2275-2280, ISSN 1454-4164

Feigel, A., Veinger, M., Sfez, B., Arsh, A., Klebanov, M., & Lyubin, V. (2005). Two dimensional photonic band gap pattering in thin chalcogenide glassy films, Thin Solid Films Vol.488, No.1-2, (September 2005), pp.185-188, ISSN: 0040-6090

Feigel, A., Veinger, M., Sfez, B., Arsh, A., Klebanov, M. & Lyubin, V. (2003). Three-dimensional simple cubic woodpile photonic crystals made from chalcogenide glasses. *Appl. Phys. Lett.* Vol.83, No.22, (Dec. 2003), pp.4480-4482, ISSN: 0003-6951

Fink, Y., Winn, J.N., Fan, S., Chen, C., Michel, J, Joannopoulos, J.D. & Thomas, E.L. (1998). A dielectric omnidirectional reflector, *Science*, Vol.282, No. 20, (November 1998), pp. 1679-1682, ISSN: 00368075

Fischer, W. (2001). A Second Note on the Term "Chalcogen". *J. Chem. Educ.* Vol.7, No.10, pp. 1333, ISSN 0021-9584

Freeman, D., Madden, S. & Luther-Davies, B. (2005). Fabrication of planar photonic crystals in a chalcogenide glass using a focused ion beam. *Opt. Express*, Vol.13, No.8, pp.3079-3086, ISSN 1094-4087

Fritzsche. H. (1998). Toward understanding the photoinduced changes in chalcogenide glasses. *Semiconductors*, Vol.32, No.8, (August 1998), pp. 850-854, ISSN 1063-7826

Ganjoo, A., Jain, H., Yu, C., Song, R., Ryan, J.V., Irudayaraj, J., Ding, Y.J. & Pantano, C.G. (2006). Planar chalcogenide glass waveguides for IR evanescent wave sensors. *J. Non-Cryst. Solids*, Vol.352, (May 2006), pp. 584-588, ISSN 0022-3093

Ho, K., Chan, C. & Soukoulis, C. (1990). Existence of a photonic gap in periodic dielectric structures. *Phys. Rew. Lett.* Vol.65, No.23, pp. 3152-3155, ISSN 1079-7114

Houizot, P., Boussard-Plédel, C., Faber, A. J., Cheng, L. K., Bureau, B., Van Nijnatten, P. A., Gielesen, W. L. M., Pereira do Carmo, J., Lucas, J. (2007). Infrared single mode chalcogenide glass fiber for space. *Opt. Express* Vol.15, (Sep. 2007), pp. 12529-12538.

Joanopoulus, J.D., Meade, R.D. & Winn, J.N. (1995). *Photonic crystals: Molding the Flow of Light*, Princeton University Press, ISBN: 9781400828241, Princeton, USA

John, S. (1987). Strong localization of photons in certain disordered dielectric superlattices, *Phys. Rew. Lett.* Vol.58, No.23, pp. 2486–2489, ISSN 1079-7114

Kastner, M., Adler, D., Fritzsche, H. (1976), Valence-Alternation Model for Localized Gap States in Lone-Pair Semiconductors. *Phys. Rev. Lett.* Vol.37, No.22, pp. 1504-1507, ISSN 0031-9007

Kawata, S., Sun, H.B., Tanaka, T., & Tanaka, K. (2001). Finer features for functional microdevices, *Nature*, 412, No. 6848, (August 2001), pp. 697-698, ISSN: 0028-0836

Kim, S.H & C K. Hwangbo, C.K. (2002). Design of omnidirectional high reflectors with quarter-wave dielectric stacks for optical telecommunication bands. *Appl. Opt.* Vol. 41, No. 16, (June 2002), pp. 3187-3192, ISSN: 0003-6935

Kincl, M., Tasseva, J., Petkov, K., Knotek, P., Tichy, L., (2009) On the photo-induced shift of the optical gap in amorphous $Ge_6As_{43}S_{35}Se_{16}$ film., *J. Optoelectron. Adv. M.*, Vol.11 No.4, (April 2009), pp. 395-398, ISSN 1454-4164

Kohoutek, T., Wagner, T., Orava, J., Krbal, M., Ilavsky, J., Vesely, D. & Frumar, M. (2007a). Multilayer systems of alternating chalcogenide As-Se and polymer thin films prepared using thermal evaporation and spin-coating techniques. *J. Phys. Chem. Solids*, Vol.68, No.5, (May 2007), pp. 1268-1271, ISSN: 0022-3697

Kohoutek, T., Orava, J., Hrdlicka, M., Wagner, T., Vlcek, Mil., Frumar, M, (2007b). Planar quarter wave stacks prepared from chalcogenide Ge-Se and polymer polystyrene thin films. *J. Phys. Chem. Solids* Vol.68 No.12, (December 2007), pp. 2376-2380

Kohoutek, T., Orava, J., Prikryl, J., Wagner, T., Vlcek, Mil., Knotek, P. & Frumar, M. (2009b). Planar chalcogenide quarter wave stack filter for near-infrared. *J. Non-Cryst. Solids*, Vol.355, No.28-30, (August 2009), pp.1521-1525, ISSN: 0022-3093

Kolomiets, B.T., Mazets, T.F. & Efendief, Sh.M. (1970), On The Energy Spectrum Of Vitreous Arsenic Sulphide, *J. Non-Cryst. Solids*, Vol.4, (April 1970), pp.45-56, ISSN: 0022-3093

Konstantinov, I.; Babeva, Tz.; & Kitova, S. (1998), Analysis of errors in thin-film optical parameters derived from spectrophotometric measurements at normal light incidence, *Appl. Opt.* Vol.37, No.19, pp.4260 -4267, ISSN: 0003-6935

Kovalskiy, A., Vlček, M., Jain, H., Fiserova, A., Waits, C.M. & Dubey, M. (2006). Development of chalcogenide glass photoresists for gray scale lithography. *J. Non-Cryst. Solids*, Vol.352, No.6-7, (May 2006), pp. 589 – 594, ISSN 0022-3093

Knotek, P., Tasseva, J., Petkov, K., Kincl, M. & Tichy, L. (2009). Optical properties and scanning probe microscopy study of some Ag-As-S-Se amorphous films. *Thin Solid Films*, Vol.517, No.20, (August 2009), pp. 5943-5947, ISSN 0040-6090

Lai, N.D., Liang, W.P., Lin, J.H., Hsu, C.C. & Lin, C.H. (2005). Fabrication of two- and three-dimensional periodic structures by multi-exposure of two beam interference technique. *Opt. Express*, Vol.13, No.23, (Nov. 2005), pp. 9605-9611, ISSN: 094-4087

Lee, M.W.; Grillet, Ch; Smith, C.L.C.; Moss, D.J., Eggleton, B.J.; Freeman, D.; Luther-Davies, B., Madden, S., Rode, A.; Ruan, Y. & Lee, Y-h. (2007). Photosensitive post tuning of chalcogenide photonic crystal waveguides, *Opt. Express* Vol. 15, No. 3, pp. 1277-1285, ISSN: 094-4087

Li, W., Seal, S., Rivero, C., Lopez, C., Richardson, K., Pope, A., Schulte, A., Myneni, S., Jain, H., Antoine, K. & Miller, A.C. (2002). X-ray photoelectron spectroscopic investigation of surface chemistry of ternary As-S-Se chalcogenide glasses. *J. Appl. Phys.* Vol.92, No.12, (December 2002), pp. 7102-7108, ISSN: 00218979

Liao, M., Chaudhari, C., Qin, G., Yan, X., Kito, C., Suzuki, T., Ohishi, Y., Matsumoto, M. & Misumi, T. (2009). Fabrication and characterization of a chalcogenide-tellurite composite microstructure fiber with high nonlinearity. *Opt. Express*, Vol.17, No.24, (November 2009), pp. 21608-21614, ISSN 1094-4087

Liu, Q., Zhao, X., Gan, F., Mi, J., Qian, S. (2005). Ultrafast optical kerr effect in amorphous $Ge_{10}As_{40}S_{30}Se_{20}$ films induced by ultrashort laser pulses. *J. Optoelectron. Adv. M.*, Vol.7, No.3, (March 2007), pp. 1323-1328, ISSN 1454-4164

Lopez, C. (2003). Materials Aspects of Photonic Crystals. *Adv. Mater.* Vol.15, No.20, (October 2003), pp. 1679-1704, ISSN: 09359648

Lyubin, V. (1984). *Photographic processes on the base of the chalcogenide glassy semiconductors*, In: Non-silver photographic processes, A. Kartuzhanskovo (Ed.), pp. 188-208, Khimiyia, Leningrad (St. Petersburg), Russia (in Russian)

Marquez, E., Gonzalez-Leal, J.M., Bernal-Oliva, A.M., Wágner, T., & Jimenez-Garay, R. (2007), Preparation and optical dispersion and absorption of Ag-photodoped $Ge_xSb_{40-x}S_{60}$ (x ≤ 10, 20 and 30) chalcogenide glass thin films, *J. Phys. D: Appl. Phys.* Vol.40, No.17, (September 2007), pp. 5351-5357, ISSN: 0022-3727

Milam, D. (1998) Review and Assessment of Measured Values of the Nonlinear Refractive-Index Coefficient of Fused Silica. *Appl. Opt.* Vol.37, No.3, (January 1998), pp. 546-550, ISSN 1559-128X

Mizrahi, V., DeLong, K.W., Stegeman, G.I., Saifi, M. A., Andrejco, M. J., (1989), Two-photon absorption as a limitation to all-optical switching. *Opt. Lett.* Vol.14, No.20, pp.1140-1142, ISSN 0146-9592

Nicoletti, E., Zhou, G., Jia, B., Ventura, M.J., Bulla, D., Luther-Davies, B. & Gu, M. (2008). Observation of multiple higher-order stopgaps from three-dimensional chalcogenide glass photonic crystals. *Opt. Lett.* Vol.33, No.20, (October 2008), pp.2311-2313, ISSN: 0146-9592

Okuda, M., Tri Nang, T. & Matsushita, T. (1979). Photo-induced absorption changes in selenium-based chalcogenide glass films. *Thin Solid Films*, Vol.58, No.2, (April 1979), pp. 403-406, ISSN 0040-6090

Ogusu, K., Maeda, S., Kitao, M., Li, H. & Minakata, M. (2004). Optical and structural properties of Ag(Cu)–As_2Se_3 chalcogenide films prepared by a photodoping. *J. Non-Cryst. Solids* Vol.347, No.1-3, (November 2004), pp. 159-165, ISSN: 00223093

Penn, D.R. (1962). Wave-Number-Dependent Dielectric Function of Semiconductors, *Phys. Rev.* Vol.128, pp. 2093-2097, ISSN: 0031-899X

Petkov, K., Todorov, R., Tasseva, J. & Tsankov, D. (2009). Structure, linear and non-linear optical properties of thin As_xSe_{1-x} films. *J. Optoelectron. Adv. M.*, Vol.11, No.12, (December 2009), pp.2083-2091, ISSN 1454-4164

Ponnampalam, N. & DeCorby R. (2008). Analysis and fabrication of hybrid metal-dielectric omnidirectional Bragg reflectors. Appl. Opt. Vol. 47, pp. 30-37, ISSN 1464-4258

Raptis, C., Kotsalas, I.P., Papadimitriou, D., Vlcek, M., Frumar, M. (1997). *Physics and Applications of Non-crystalline Semiconductors in Optoelectronics, NATO ASI Series, 3. High Technology*, 36, p. 291, M. Bertolotti, A. Andriesh (Eds.)

Riley, B.J., Sundaram, S.K., Johnson, B.R., Saraf, L.V. (2008). Differential etching of chalcogenides for infrared photonic waveguide structures. *J. Non-Cryst. Solids*, Vol.354, No.10-11, (February 2008), pp. 813-816, ISSN 0022-3093

Rowlands, C.J., Su, L. & Elliott, S.R. (2010). Rapid prototyping of low-loss IR chalcogenide-glass waveguides by controlled remelting. *Chem.Phys.Chem.*, Vol.11, No.11, (August 2010), pp. 2393-2398, ISSN 1439-4235

Saithoh A., Gotoh, T. & Tanaka, K. (2002). Chalcogenide-glass microlenses for optical fibers. *J Non-Cryst. Solids*, Vol.299-302, (April 2002), pp. 983-987, ISSN 0022-3093

Sakoda K. (2005) Optical Properties of Photonic Crystals, W. T. Rhodes (Ed), Springer-Verlag Berlin, Heidelberg, Germany, ISSN 0342-4111

Savović, S. & Djordjevich, A. (2011). Mode coupling in chalcogenide-glass optical fibers. *Opt. Laser. Eng.* Vol.49, No.7, (July 2011), pp. 855-858, ISSN 0143-8166

Sheik-Bahae, M., Hagan, D. J., Van Stryland, E. W. (1990). Dispersion and band-gap scaling of the electronic Kerr effect in solids associated with two-photon absorption. *Phys. Rev. Lett.*, Vol.65, No.1, pp. 96-99, ISSN 0031-9007

Shimakawa, K., Kolobov, A. & Elliott, S.R. (1995). Photoinduced effects and metastability in amorphous semiconductors and insulators, *Adv. Phys.* Vol.44, No.6, (November 1995), pp.475-588, ISSN: 00018732

Skordeva, E., Arsova, D., Aneva, Z., Vuchkov, N., Astadjov, D. (2001). Laser induced photodarkening and photobleaching in Ge-As-S thin films. *Proceedings of SPIE - The International Society for Optical Engineering*, Vol.4397, pp. 348-352, ISSN 0277786X

Street, R.A. (1977). Non-radiative recombination in chalcogenide glasses. *Solid State Communications*, Vol.24, No.5, (November 1979), pp. 363-365, ISSN: 00381098

Su, L., Rowlands, C.J., Lee, T.H., Elliott, S.R. (2008). Fabrication of photonic waveguides in sulfide chalcogenide glasses by selective wet-etching, *Electron. Lett.*, Vol.44, No.7, pp. 472-474, ISSN 0013-5194

Su, L., Rowlands, C.J. & Elliott, S.R. (2009). Nanostructures fabricated in chalcogenide glass for use as surface-enhanced Raman scattering substrates. *Opt. Lett.* Vol.34, No.11, (June 2009), pp.1645-1647, ISSN: 0146-9592

Swanepoel, R. (1983). Determination of the thickness and optical constants of amorphous silicon, *J.Phys. E: Sci. Instrum.* Vol.16, No.12, pp.1214-1222, ISSN: 0022-3735

Tanaka, K. & Ohtsuka, Y. (1979). Composition dependence of photo-induced refractive index changes in amorphous As-S films, *Thin Solid Films*, Vol.57 No. 1, (February 1979), pp. 59-64, ISSN 0040-6090

Tanaka, K. (2007). Nonlinear optics in glasses: How can we analyze? *J. Phys. Chem. Solids*, Vol.68, No.5-6, (May 2007), pp. 896-900, ISSN 0022-3697

Tasseva, J., Petkov, K., Kozhuharova, D. & Iliev, Tz. (2005). Light-induced changes in the physico-chemical and optical properties of thin Ge-S– Se-As films. *J. Optoelectron. Adv. M.*, Vol.7, No.3, (March 2005), pp. 1287-1292, ISSN 1454-4164

Tasseva, J., Lozanova, V., Todorov, R., Petkov, K. (2007). Optical Characterization of Ag/As-S-Se thin films. *J. Optoelectron. Adv. M.*, Vol.9, No.10, pp. 3119-3124, ISSN 1454-4164

Tasseva, J., Todorov, R., Babeva, Tz. & Petkov, K. (2010). Structural and optical characterization of Ag photo-doped thin $As_{40}S_{60-x}Se_x$ films for nonlinear applications., *J. Opt.*, Vol.12, (June 2010), pp. 065601 (9 pp), ISSN 1464-4258

Teteris, J., Reinfelde, M. (2004). Subwavelength-period gratings in amorphous chalcogenide thin films. *J. Opt. A: Pure Appl. Opt.* Vol.6, No.3, pp. S151 -S154, ISSN 1464-4258

Tichy, L., Ticha, H., Blecha, J. & Vlcek, M. (1993). Compositional trend of the blue shift of the gap in Ge_xS_{100-x} thin amorphous films induced by annealing and illumination. Mater. Lett. Vol.17, No.5, pp. 268-273

Tikhomirov, V.K., Furniss, D., Seddon, A.B., Savage, J.A., Mason, P.D., Orchard D.A. & Lewis K.L. (2004). Glass formation in the Te-enriched part of the quaternary Ge-As-Se-Te system and its implication for mid-infrared optical fibres. *Infrared Phys. Technol.* Vol.45, No.2, (March 2004), pp. 115-123, ISSN: 13504495

Todorov, R. & Petkov, K, (2001). Light induced changes in the optical properties of thin As - S - Ge(Bi, Tl) films. *J. Optoelectron. Adv. M.*, Vol.3, No.2, pp. 311-317, ISSN 1454-4164

Todorov, R., Iliev, Tz., Petkov, K. (2003). Light-induced changes in the optical properties of thin films of Ge-S-Bi(Tl, In) chalcogenides. *J. Non-Cryst. Solids*, Vol. 326-327, (October 2003), pp. 263-267, ISSN 0022-3093

Todorov, R., Paneva, A., & Petkov K. (2010a). Optical characterization of thin chalcogenide films by multiple-angle-of-incidence ellipsometry, *Thin Solid Films*, Vol.518, No.12, (April 2010), pp.3280-3288, ISSN: 00406090

Todorov, R., Tasseva, J., Babeva, Tz. & Petkov K. (2010b). Multilayered As_2Se_3/GeS_2 quarterwave structures for photonic applications, *J. Phys. D: Appl. Phys.* Vol.43 No.50, (December 2010) art. num. 505103 (8pp), ISSN: 0022-3727

Ung, B. & Skorobogatiy, M. (2011). Extreme nonlinear optical enhancement in chalcogenide glass fibers with deep-subwavelength metallic nanowires. *Opt. Lett.* Vol.36, No.12, (June 2011), pp. 2527-2529, ISSN 0146-9592

Vlaeva, I., Petkov, K., Tasseva, J., Todorov, R., Yovcheva, T. & Sainov, S. (2010). Electric charging influence on the diffraction efficiency of total internal reflection holograms, recorded in very thin chalcogenide films, *J. Opt.*, Vol.12, No.12, (December 2010), pp. 124008 (6pp), ISSN 1464-4258

Wágner, T. & Ewen, P.J.S., (2000), Photo-induced dissolution effect in $Ag/As_{33}S_{67}$ multilayer structures and its potential application. *J. Non-Cryst. Solids*, Vol.266-269, (May 2000), pp. 979-984, ISSN 0022-3093

Wemple S. & DiDomenico, Jr., M. (1971). Behavior of the Electronic Dielectric Constants in Covalent and Ionic Materials, *Physical Review B*, Vol.3, No.14, (February 1971), pp.1338-1351, ISSN: 0163-1829

Yablonovitch, E. (1987). Inhibited Spontaneous Emission in Solid-State Physics and Electronics, *Phys. Rew. Lett.* Vol.58, No. 20, pp.2059–2062, ISSN 1079-7114

Yablonovitch, E., Gmitter, T.J. & Leung, K. M. (1991). Photonic band gap structure: The face-centered-cubic case employing nonspherical atoms. *Phys. Rev. Lett.* Vol. 67, No.17, (October 1991), pp. 2295-2301, ISSN: 0031-9007

Yablonovitch, E. (2007). Photonic crystals: What's in a name? *Optics and Photonics News*, Vol.18, No.3, pp.12-13, ISSN: 10476938

Zakery A. & Elliott, S.R. (2007). *Optical Nonlinearities in Chalcogenide Glasses and their Applications*, Springer Series in optical sciences, ISBN-10 3-540-71066-3, Springer Berlin, Heidelberg, New York

http://glassonline.com/infoserv/history.html

10

Ultra-Broadband Time-Resolved Coherent Anti-Stokes Raman Scattering Spectroscopy and Microscopy with Photonic Crystal Fiber Generated Supercontinuum

Hanben Niu[1,2] and Jun Yin[1,2]

[1]*College of Optoelectronic Engineering, Shenzhen University, Shenzhen*
[2]*Key Laboratory of Optoelectronic Devices and Systems of Ministry of Education and Guangdong Province, Shenzhen University, Shenzhen*
China

1. Introduction

Optics, one of the oldest natural sciences, has been promoting the developments of sciences and technologies, especially the life science. The invention of the optical microscope eventually led to the discovery and study of cells and thus the birth of cellular biology, which now plays a more and more important role in the biology, medicine and life science. The greatest advantage of optical microscopy in the cellular biology is its ability to identify the distribution of different cellular components and further to map the cellular structure and monitor the dynamic process in cells with high specific contrast. In 1960's, with the invention of laser, an ideal coherent light source with high intensity was provided. Since then the combination of optical microscopy with laser has been expanded. Many novel optical microscopic methods and techniques were developed, such as various kinds of fluorescence microscopy. In fluorescence microscopy, the necessary chemical specificity is provided by the labeling samples with extrinsic fluorescent probes [1, 2]. With the developments of ultra-short laser, fluorescent labeling technique and modern microscopic imaging technique, the fluorescence spectroscopy and microscopy with high spatial resolution, sensitivity and chemical specificity has become the key promotion of the life science and unveiled many of the secrets of living cells and biological tissues [3, 4]. In particular, the confocal fluorescent microscope (CFM), with the confocal detection [5] and multi-photon excitation [6, 7], can obtain the 3D sectioning images of cells and tissues with high spatial resolution. Today, fluorescence microscopy has become a powerful research tool in life science and has achieved the great triumph. Nevertheless, the disadvantages of fluorescence microscopy, such as the photo-toxicity and photo-bleaching, can not be ignored [8]. Furthermore, some molecules in cells, such as water molecule and other small biomolecules, can not be labeled until now. Finally, for the biological species that do not fluoresce, the extrinsic fluorescent labels will unavoidablely disturb the original characteristics and functions of biological molecules, which will limit the applicability of fluorescence microscopy. Therefore, it is very necessary to develop some complementary methods with molecular specificity, high spatial and temporal resolution, but without any extrinsic labels.

Now some imaging techniques without any extrinsic labels have been developed, such as the harmonic generation microscopy (second harmonic generation, SHG [9-13] and third harmonic generation, THG [14, 15]), infrared microscopy [16], Raman microscopy [17-19], Stimulated Raman Scattering (SRS) microscopy [20] and coherent anti-Stokes Raman scattering (CARS) microscopy [21, 22]. The specificity and contrast of SHG and THG imaging arise from the unique non-centrosymmetric structure of specific biological tissues, such as the collagen fibrils, microtubules and myosin, but the structural information can not be obtained at molecular level. Because of longer wavelength involved and vibrational absorption of the common solvent, the spatial resolution of infrared imaging technique is too low. With the Raman scattering spectrum, we can distinguish different molecules [17-19]. However, higher average power laser is necessary due to small Raman scattering cross-section and autofluorescence background that limit its applicability in life science. Both SRS and CARS are essentially Raman scattering, the molecules of a sample can be distinguished by their active vibrational modes of specific molecular bonds, which provide the necessary imaging contrasts and specificities [20-26]. As a coherent resonance-enhance process, the CARS signal is not only many orders higher than spontaneous Raman scattering signal but also well oriented in emission direction, which significantly reduce the integration time. The wavelength of CARS signals is blue-shift relative the excitation, the interference of strong one-photon fluorescence background can be avoided. Furthermore, as a nonlinear optical process, the CARS signals are generated in a small focal volume under the tight-focusing condition. Thus, the CARS microscopy can provide high temporal and spatial resolution, and strong 3D sectioning imaging ability. The information of molecular distributions of a biological sample can be obtained without any disturbance on sample.

In 1965, CARS was first reported by P. D. Maker and R. W. Terhune [27]. They found that a very strong signal could be obtained when two coherent light beams at frequencies ω_1 and ω_2 were used to drive an active vibrational Raman mode at frequency $\Omega_R = \omega_1 - \omega_2$. It was named as the CARS process, whose signals are 10^5 stronger than spontaneous Raman scattering process [28]. As a tool for spectral analysis, the CARS spectroscopy was extensively studied and used in physics and chemistry [29-32]. The first CARS microscopy with noncollinear beams geometry was demonstrated by M. D. Duncan et al in 1982 [21]. In 1999, A. Zumbusch and his colleagues revived the CARS microscopy by using an objective with high numerical aperture, collinear beams geometry and detection of CARS signals in forward direction that boosted wide research in the improvements of CARS microscopy [22]. The distinguished research works of Prof. S. Xie and his colleagues proved that the CARS microscopy was a very effective noninvasive optical microscopic approach with high spatial resolution and imaging speeds matching those of multi-photon fluorescence microscopy and had a great prospect in biology, medicine and life science [22, 33].

In a traditional CARS microscopy, the contrast and specificity are based on single or few chemical bonds of molecule due to the limitation of laser line-width. It is not adequate to distinguish various biological molecules with just single chemical bond. In order to effectively distinguish different molecules, a method for simultaneously obtaining the complete molecular vibrational spectra is required. For this purpose, many methods for extending the simultaneously detectable spectral range of CARS spectroscopy and microscopy are presented [34-38]. With the advent and the progress of supercontinuum (SC) generated by photonic crystal Fibre (PCF) pumping with ultra-short laser pulses [39], the broadband CARS microscopy based on SC has been developed that provided better feasibilities [40-44].

As well known, CARS microscopy is not background-free [29]. Strong nonresonant background signals (NRB), from the electronic contributions to the third-order susceptibility of sample and the solvent, always company with the CARS signals in a broad spectral range. It often interferences or overwhelms CARS signals from small targets that limits the spectral accuracy, imaging contrast and systemic sensitivity. In the broadband CARS with a broadband pump source, it is not beneficial to simultaneously distinguish various biological molecules because of the influence of strong NRB.

In this chapter we briefly introduce the theoretical basics of the CARS process and characteristics of CARS spectroscopy and microscopy. The classical and quantum mechanical descriptions of Raman scattering and CARS processes are qualitatively reviewed in order to be helpful for understanding the physical mechanisms of these two light scattering processes. The main characteristics and applications of a CARS spectroscopy and microscopy are specifically emphasized, such as the condition of momentum conservation; the generation and suppression of NRB noise. In order to simultaneously obtain the complete molecular vibrational spectra, SC laser is a proper pump source. We will briefly review the history and characteristics of PCF. A summarization of the theoretical analysis of SC generation with PCF pumping by ultra-short (picosecond or femtosecond) laser pulses will be outlined. The necessities, developments and characteristics of the broadband CARS spectroscopy and microscopy will be briefly described. An ultra-broadband time-resolved CARS technique with SC generated by PCF will be especially emphasized. By this method, the complete molecular vibrational spectra without NRB noise can be obtained. During recent years, the ways to improve spatial resolution of CARS microscopy have become one of attractive questions all over the world. We will briefly review and outlook the feasible ways.

2. Theories of CARS process

As a coherent Raman scattering process, CARS is a typical three-order nonlinear optical process. In order to well understand the mechanism of the CARS process, the brief classical and quantum mechanical discussions of Raman scattering are necessary. Based on the theoretical analysis of Raman scattering, a theoretical description of the CARS process and the conditions for the generation of CARS signals will be outlined. The general classical description will give an intuitive picture. And the quantum mechanical description will enable the quantitative analysis of the CARS process. Here, we just carry out the qualitatively theoretical analysis, and the descriptions of physical mechanism will be highlighted. The emphases are the main characteristics of a CARS microscopy, such as the conditions for generation of CARS signals, and generation and suppression of NRB noise.

2.1 Raman scattering process

When a beam of light passes through a media, one can observe light scattering phenomenon besides light transmission and absorption. Most of elastically scattered photons (so called

Rayleigh scattering) from atom or molecule have the same energy (frequency) as the incident photons, as shown in figure 1 (a). However, a small part of the photons (approximately 1 in 10 million photons) are inelastically scattered with the frequencies different from the incident photons [45]. This inelastic scattering of light was theoretical predicted by A. Smekal in 1923 [46]. In 1928, Indian physicist Sir C. V. Raman first discovered this phenomenon when a monochromatic light with frequency of ω_P was incident into a medium. He found that the scattered light components contained not only Rayleigh scattering with frequency of ω_P, but also some weaker scattering components with frequencies of $\omega_P \pm \Omega_R$, which come from the inelastic scattering phenomenon of light named as Raman scattering or Raman effect [47, 48], as shown in figure 1 (b) and (c) in energy level diagram. Raman scattering originates from the inherent features of molecular vibration and rotation of individual or groups of chemical bonds. The obtained Raman spectra contain the inherent molecular structural information of the medium and can be used to identify molecules. Because of this significant feature, it has been widely used as a tool for analyzing the composition of liquids, gases, and solids.

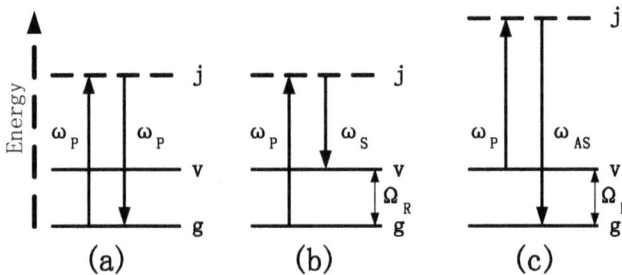

Fig. 1. Energy level diagram of the elastic Raleigh scattering (a), the inelastic Stokes Raman scattering (b) and anti-Stoke Raman scattering (c), where ω_P, ω_S, ω_{AS} and Ω_R represents the frequency of incident light, Stokes scattering light, anti-Stoke scattering light and resonance respectively. The ground state level, the vibrational level and the virtual intermediate level is labeled with g, v, and j respectively.

2.1.1 Classical description of raman scattering

The Raman scattering phenomenon arises from the interactions between the incident photon and the electric dipole of molecules. In classical terms, the interaction can be viewed as fluctuation of the molecule under the influence of the applied optical field. With the used optical field of frequency ω_P expressed as $E(t) = E(r)e^{-i\omega_P t}$, the induced dipole moment of a molecule is:

$$\mu(t) = \alpha(t)E(t), \tag{2.1}$$

where $\alpha(t)$ is the polarizibility of the material. When the incident optical field interacts with the molecules, the polarizibility can be expressed as a function of the nuclear coordinate Q, and expanded to the first order in a Taylor series [49]:

$$\alpha(t) = \alpha_0 + \frac{\partial \alpha}{\partial Q}Q(t), \tag{2.2}$$

where Q(t) is a simple harmonic oscillator, $Q(t)=2Q_0\cos(\omega_R t)$. So the induced dipole moment can be rewritten as:

$$\mu(t) = \alpha_0 E(r) e^{-i\omega_p t} + \frac{\partial \alpha}{\partial Q} E(r) Q_0 e^{\pm i(\omega_p \pm \Omega_R)t}.$$ (2.3)

On the right-hand side of equation (2.3), the first term corresponds to Rayleigh scattering with the same frequency of incident light. The second term describes the Raman frequency shift of $\omega_p \pm \Omega_R$. Because the Raman frequency shifting term depends on $\partial \alpha / \partial Q$, Raman scattering occurs only when the incident optical field induces a polarizibility change along the specific molecular vibrational mode. This specific mode is an active Raman mode, which is the basis of the selection rule of Raman spectroscopy.

The differential scattering cross-section is one of key parameters to express the intensity of Raman scattering signal. In a solid angle $\Delta\Omega$, it can be defined as the amount of scattered intensity divided by the incident intensity:

$$\sigma_{diff} = \frac{\partial \sigma}{\partial \Omega} = \frac{I_{Raman}}{VNI_p \Delta \Omega},$$ (2.4)

where I_{Raman} is the intensity of Raman scattering light in $\Delta\Omega$, V is the volume of scattering medium, N is the molecular density, I_p is the intensity of incident optical field. Therefore, the total intensity of the Raman scattering light in whole solid angle can be described as the summation of contributions from all N molecules:

$$I_{Raman} = NI_p \int \sigma_{diff} d\Omega.$$ (2.5)

Obviously, the spontaneous Raman is a linear optical process, because the intensity of scattering signal has the linear relationship with the intensity of incident optical field and number of scattering molecules respectively. The classical description of Raman process only provides a qualitative relationship between the Raman scattering cross-section and intensities of the Raman signals. In order to achieve the quantitative study for the Raman process, a quantum mechanical description is necessary.

2.1.2 Quantum mechanical description of raman scattering

When the interaction between incident optical field and medium is studied with quantum mechanical method, the molecular system of medium should be quantized. The Raman scattering is a second-order process in which two interaction processes between incident optical field and medium are involved. The quantum mechanical explanation of the Raman scattering process is based on the estimation of the transition rate between the different molecular states. In quantum physics, the Fermi's golden rule is a common way to calculate the first-order transition rate between an initial state $|g\rangle$ and a final state $|v\rangle$ that is proportional to the square modulus of the transition dipole μ_{vg}. But in order to describe the Raman process, we need to calculate the second-order transition rate. The second-order transition rate τ^{-1} can be presented as [50, 51]:

$$\frac{1}{\tau} = \sum_v \sum_R \frac{\pi e^4 \omega_p \omega_R n_R}{2\varepsilon_0^2 \hbar^2 V^2} \left| \sum_j \left\{ \frac{\mu_{vj}\mu_{jg}}{\omega_j - \omega_p} + \frac{\mu_{vj}\mu_{jg}}{\omega_j + \Omega_R} \right\} \right|^2 \delta(\omega_v + \omega_R - \omega_p), \tag{2.6}$$

where e is the electron charge, ε_0 is the vacuum permittivity, $\hbar$ is Planck's constant, n_R is the refractive index at Raman frequency, and δ is the Dirac delta function. ω_P is the frequency of incident optical field and ω_R is the frequency of Raman scattering light. The frequencies ω_v and ω_j are the transition frequencies from the ground state to the final state $|v\rangle$ and intermediate state $|j\rangle$, respectively.

In Raman scattering process, an incident optical field first converts the material system from the ground state $|g\rangle$ to an intermediate state $|j\rangle$, which is an artificial virtual state. Then, the transition from the intermediate state $|j\rangle$ to the final state $|v\rangle$ happens that is considered as an instantaneous process. A full description of Raman scattering thus incorporates a quantized field theory [52]. From the quantized field theory, we can find the number of photon modes at frequency of ω_R in volume of medium V [52], and perform the summation over R in equation (2.6). From the argument of the Dirac delta function, the only nonzero contributions are the ones for which the emission frequency $\omega_R=\omega_P-\omega_v$, the red-shifted Stokes frequencies. With $\sigma_{diff}=d/d\Omega(V/cn_R\tau)$, the expression for the transition rate can be directly presented with the differential scattering cross-section [51]:

$$\sigma_{diff} = \sum_v \frac{e^4 \omega_p \omega_R^3}{16\pi^2 \varepsilon_0^2 \hbar^2 c^4} \left| \sum_j \left\{ \frac{\mu_{vj}\mu_{jg}}{\omega_j - \omega_p} + \frac{\mu_{vj}\mu_{jg}}{\omega_j + \omega_R} \right\} \right|^2. \tag{2.7}$$

This is the Kramers-Heisenberg formula, which is very important in quantum mechanical description of light scattering [50]. For Raman scattering process, the differential scattering cross-section for a particular vibrational state $|v\rangle$ can be simplified to:

$$\sigma_{diff} = \frac{\omega_p \omega_R^3}{16\pi \varepsilon_0^2 \hbar^2 c^4} |\alpha_R|^2, \tag{2.8}$$

where the Raman transition polarizibility α_R can be written as:

$$\alpha_R = \sum_j \left\{ \frac{\mu_{vj}\mu_{jg}}{\omega_j - \omega_p} + \frac{\mu_{vj}\mu_{jg}}{\omega_j + \omega_R} \right\}. \tag{2.9}$$

From the above discussion, we can know that Spontaneous Raman scattering is a weak effect because the spontaneous interaction through the vacuum field occurs only rarely. Although Raman scattering is a second-order process, the intensity of Raman signal depends linearly on the intensity of the incident optical field. When the frequency approaches the frequency of a real electronic state of molecule, the Raman scattering is very strong, which is known as the resonant Raman and is one of effective methods to improve the efficiency of Raman scattering. If the spontaneous nature of the j→v transition can be eliminated by using a second field of frequency ω_R, the weak Raman scattering can also be enhanced, such as CARS process.

2.2 CARS process

The disadvantage of the Raman scattering is the low conversion efficiency due to the small scattering cross-section. Only 1 part out of 10^6 of the incident photons will be scattered into the Stokes frequency when propagating through 1cm of a typical Raman active medium. It makes Raman spectroscopy and microscopy more complex and costly that limits its broad applications. As one of nonlinear techniques with coherent nature, intensity of CARS signal is about 10^5 stronger than spontaneous Raman. Therefore, the CARS spectroscopy and microscopy have been widely used in physics, chemistry, biology and many other related domains [22-26].

In the CARS process, three laser beams with frequencies of ω_P, $\omega_{P'}$ and ω_S are used as pump, probe and Stokes, the energy level diagram of CARS is shown in figure 2. The primary difference between the CARS and Raman process is that the Stokes frequency stems from an applied laser field in the former. We can simply consider the joint action of the pump and Stokes fields as a source for driving the active Raman mode with the difference frequency ω_P-ω_S. Here, we will first describe CARS process with the classical model, after that a quantum mechanical explanation will be applied for finding the correct expression of the third-order nonlinear susceptibility.

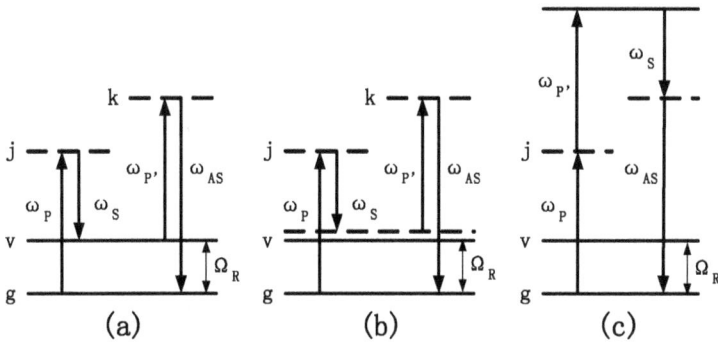

Fig. 2. Energy level diagram of CARS. (a) resonant CARS, (b) nonresonant electronic contribution and (c) electronically enhanced nonresonant contribution. Solid lines indicate real states (g and v); dashed lines denote virtual states (j and k).

2.2.1 Classical description of CARS

The classical description of an active vibrational mode driven by the incident optical field is a model of damping harmonic oscillator. The equation of motion for the molecular vibration along Q is [53]:

$$\frac{d^2Q(t)}{dt^2} + 2\gamma\frac{dQ(t)}{dt} + \omega_v^2 Q(t) = \frac{F(t)}{m}, \tag{2.10}$$

where γ is the damping constant, m is the reduced nuclear mass, and F(t) is the external driving force of the oscillation from the incident optical fields. In the CARS process, F(t) is provided by the incident pump and Stokes fields:

$$F(t) = \left(\frac{\partial \alpha}{\partial Q}\right)_0 E_P E_S^* e^{-i(\omega_P - \omega_S)t} , \qquad (2.11)$$

where the time-varying driven force oscillates at the beat frequency of the incident optical fields. A solution to equation (2.10) can be written as:

$$Q = Q(\omega_P - \omega_S)\left\{e^{-i(\omega_P - \omega_S)t} + e^{i(\omega_P + \omega_S)t}\right\} ,$$

where $Q(\omega_P-\omega_S)$ is the amplitude of the molecular vibration. Then, from the equation (2.10), it can be worked out:

$$Q(\omega_P - \omega_S) = \frac{(1/m)[\partial \alpha/\partial Q]_0 E_P E_S^*}{\omega_v^2 - (\omega_P - \omega_S)^2 - 2i(\omega_P - \omega_S)\gamma} . \qquad (2.12)$$

From the equation (2.12), we know that the amplitude of molecular vibration is proportional to the product of the amplitudes of the driving fields and the polarizability change. When the frequency difference of the pump and the Stokes fields equals to the resonant frequency ω_v, the molecular vibration of active Raman mode will be resonantly enhanced. When a probe field with frequency of ω_{Pr} passes through the medium, it will be modulated by the resonant enhanced molecular vibrational mode, resulting in a component at the anti-Stokes frequency, $\omega_{Pr}+\omega_P-\omega_S$.

The total nonlinear polarization is the summation of all N dipoles:

$$P(t) = N\mu(t) = N\left[\left(\frac{\partial \alpha}{\partial Q}\right)_0 Q\right]E_{Pr}(t) . \qquad (2.13)$$

In order to simplify the experimental system, the pump field provides the probe field. The frequency of generated anti-Stokes signal is $\omega_{AS}=2\omega_P-\omega_S$. The total nonlinear polarization can be written as:

$$P(t) = P(\omega_{AS})e^{-i(\omega_{AS})t} . \qquad (2.14)$$

With (2.12), (2.13), and (2.14), we can deduce the amplitude of total nonlinear polarization:

$$P(\omega_{AS}) = \frac{(N/m)[\partial \alpha/\partial Q]_0^2}{\omega_v^2 - (\omega_P - \omega_S)^2 - 2i(\omega_P - \omega_S)\gamma}E_P^2 E_S^* = 3\chi_r^{(3)}(\omega_{AS})E_P^2 E_S^* . \qquad (2.15)$$

From above discussion, we know that the amplitude of the total nonlinear polarization is proportional to the product of three incident optical fields. Here, we define the vibrational resonant third-order susceptibility $\chi_r^{(3)}(\omega_{AS})$:

$$\chi_r^{(3)}(\omega_{AS}) = \frac{(N/3m)[\partial \alpha/\partial Q]_0^2}{\omega_v^2 - (\omega_P - \omega_S)^2 - 2i(\omega_P - \omega_S)\gamma} . \qquad (2.16)$$

It is an inherent property of medium that describes the medium's response to the incident optical fields. When the frequency difference of incident optical fields matches with frequency of a vibrational mode, $\omega_P-\omega_S=\omega_v$, $\chi_r^{(3)}(\omega_{AS})$ will maximize.

The intensity of CARS signal is proportional to the square modulus of the total polarization:

$$I_{CARS} \propto \left|P^{(3)}(\omega_{AS})\right|^2 = 9\left|\chi_r^{(3)}(\omega_{AS})\right|^2 I_P^2 I_S. \tag{2.17}$$

It scales quadratically with the pump intensity, linearly with the Stokes intensity, and quadratically with the third-order susceptibility of the medium.

Although the classical description of CARS can provide a picture of CARS process and a simplified relationships among the medium, intensities of incident optical fields and CARS signals, it is unable to account for the interaction of the fields with the quantized states of the molecule. More accurate numerical estimates can only be achieved with a quantum mechanical description of the CARS process.

2.2.2 Quantum mechanical description of CARS

The quantum mechanism of CARS process can be effectively described by the time-dependent third-order perturbation theory. In the quantum mechanical description, the system is usually expressed in terms of the density operator:

$$\rho(t) \equiv \left|\psi(t)\right\rangle\left\langle\psi(t)\right| = \sum_{nm} \rho_{nm}(t)\left|n\right\rangle\left\langle m\right|, \tag{2.18}$$

where the wave functions are expanded in a basis set $\{|n\rangle\}$ with time-dependent coefficients $c_n(t)$, and $\rho_{nm}(t) \equiv c_n(t)c_m^*(t)$.

The expectation value for the electric dipole moment is then given by:

$$\left\langle\mu(t)\right\rangle = \sum_{n,m} \mu_{mn}\rho_{nm}(t). \tag{2.19}$$

The third-order nonlinear susceptibility for the CARS process is found by calculating the third-order correction to the density operator through time-dependent perturbation theory and with the relation $\chi_r^{(3)} = N\left\langle\mu(t)\right\rangle/3E_P^2 E_S^*$:

$$\chi_r^{(3)} = \frac{N}{V}\sum_v \frac{A_v}{\omega_v - (\omega_P - \omega_S) - i\Gamma_v}, \tag{2.20}$$

where Γ_v is the vibrational decay rate that is associated with the line width of the Raman mode R. The amplitude A_v can be related to the differential scattering cross-section:

$$A_v \propto \frac{(\pi\varepsilon_0)^2 c^4 n_P}{\hbar\omega_P\omega_S^3 n_S}\left(\rho_{gg} - \rho_{vv}\right)\sigma_{diff}, R, \tag{2.21}$$

where n_P and n_S are the refractive index at the pump and Stokes frequency, ρ_{gg} and ρ_{vv} is the element of the density matrix of the ground state and vibrationally excited state, respectively. The CARS signal intensity is again estimated by substituting equation (2.20) into (2.7).

The quantum mechanical description of CARS process can be qualitatively presented by considering the time-ordered action of each laser field on the density matrix $\rho_{nm}(t)$. Each electric field interaction establishes a coupling between two quantum mechanical states of the molecule, changing the state of the system as described by the density matrix. Before interaction with the laser fields, the system resides in the ground state ρ_{gg}. An interaction with the pump field changes the system to ρ_{jg}. Then the system is converted into ρ_{vg} by the following Stokes field. The density matrix now oscillates at frequency $\omega_{vg}=\omega_{jg}-\omega_{vj}$ that is a coherent vibration. When the third incident optical field interact with medium, the coherent vibration can be converted into a radiating polarization ρ_{kg}, which propagates at $\omega_{kg}=\omega_{jg}+\omega_{vg}$. After emission of the radiation, the system is brought back to the ground state.

As a coherent Raman process, the intensity of CARS signal is more than five orders of magnitude greater than that of spontaneous Raman scattering process. Because the radiating polarization is a coherent summation, the intensity of CARS signal is quadratic in the number of Raman scattering. Because of the coherence, the CARS signal is in certain direction that allows a much more efficient signal collection than Raman scattering. CARS signal is blue-shifted from incident beams, which avoids the influence from any one-photon excited fluorescence.

2.3 Resonant and nonresonant signals in CARS

2.3.1 Source of nonresonant background signals

From the theory of the CARS process, we can know that CARS signal comes from the third-order nonlinear susceptibility. The total CARS signal is proportional to the square modulus of the nonlinear susceptibility [46]:

$$I(\omega_{AS}) \propto \left|\chi^{(3)}(\omega_{AS})\right|^2 = \left|\chi_r^{(3)}(\omega_{AS})\right|^2 + \left|\chi_{nr}^{(3)}\right|^2 + 2\chi_{nr}^{(3)}\,\mathrm{Re}\left\{\chi_r^{(3)}(\omega_{AS})\right\}. \qquad (2.22)$$

The total third-order nonlinear susceptibility is composed of a resonant ($\chi_r^{(3)}$) and a nonresonant ($\chi_{nr}^{(3)}$) part:

$$\chi^{(3)} = \chi_r^{(3)} + \chi_{nr}^{(3)}, \qquad (2.23)$$

where resonance $\chi_r^{(3)}$ is a complex quantity, $\chi_r^{(3)} = \chi_r^{(3)\,'} + i\chi_r^{(3)\,''}$ and represents the Raman response of the molecules. When the frequency difference between the pump and Stokes fields equals to the vibrational frequency of an active Raman mode, a strong CARS signal is induced. It provides the inherent vibrational contrast mechanism of CARS microscopy. However, it is not the only components in the total anti-Stokes radiations. In the absence of active Raman modes, the electron cloud still has oscillating components, at the anti-Stokes frequency $\omega_{AS} = 2\omega_P - \omega_S$, coupling with the radiation field. It is the purely electronic nonresonant contribution from $\chi_{nr}^{(3)}$ that is frequency-independent and a real quality. Two energy level diagrams of nonresonant contribution are depicted in figure 2 (b) and (c), when all three incident optical fields overlap in time. As shown in figure 2 (b), a radiating polarization at $2\omega_P-\omega_S$ is established via field interactions with virtual levels. When $2\omega_P$ is close to the frequency of a real electronic state, it is a nonresonant two-photon enhanced electronic contribution, as shown in figure 2 (c). The nonresonant contribution is a source of background that limits the sensitivity of CARS microscopy. For weak vibrational resonances, the nonresonant background may overwhelm the resonant information. In biological samples, the concentration of interested molecules usually is low, while the nonresonant background from the aqueous surrounding is generally ubiquitous. The mixing of the nonresonant field with the resonant field gives rise to the broadened and distorted spectral line shapes. Therefore, the suppression of the nonresonant contribution is essential for practical applications.

2.3.2 Suppression of NRB noise

Several effective methods have been developed in order to suppress the NRB noise. Here, we briefly discuss several widely used techniques for suppressing the NRB noise.

Epi-detection [54, 55]

In samples, every object will be the source of NRB noise. The aqueous environment produces an extensive NRB noise that may be stronger than the resonant CARS signal from a small object in focus. Because the epi-CARS (E-CARS) has the size-selective mechanism, the NRB noise from the aqueous surrounding can be suppressed while the signal from small objects will be retained. It should be noted that the NRB noise can not be directly reduced in E-CARS. When samples have comparative sizes or in a highly scattering media, such as tissues, this method will not work.

Polarization-sensitive detection

The polarization-sensitive detection CARS (P-CARS) is based on the different polarization properties of the resonant CARS and nonresonant signals to effectively suppress the NRB noise [56-58]. According to the Kleinman's symmetry, the depolarization ratio of the nonresonant field is $\rho_{nr} = \chi_{1221}^{(3)nr}/\chi_{1111}^{(3)nr} = 1/3$ [59]. However, the depolarization ratio of resonant field is $\rho_r = \chi_{1221}^{(3)r}/\chi_{1111}^{(3)r}$, which depends on the symmetry of the molecule and may vary from $1/3$. The nonlinear polarization, polarized at an angle θ, can be written as a function of the angle between the pump and Stokes fields:

$$P^i(\theta) = \frac{3}{4}\chi_{1111}^i \cos\phi\left\{\overline{e}_x + \rho^i \tan\phi\overline{e}_x\right\} E_P^2 E_S^* , \qquad (2.24)$$

where i is either the resonant or nonresonant component. The nonresonant field is linearly polarized along an angle $\theta_{nr} = \tan^{-1}(\tan(\varphi)/3)$. When detecting the signal at an angle orthogonal to the linearly polarized nonresonant background, the resonant signal is:

$$P^r(\theta) = \frac{3}{4}\chi_{1111}^r \cos\phi\sin\theta_{nr}\left(1-3\rho^r\right)E_P^2 E_S^* . \qquad (2.25)$$

When $\varphi=71.6°$ and $\theta_{nr}=45°$, the ratio of resonant and nonresonant signals reaches the maximum. Under this condition, the nonresonant background can be negligible. The P-

CARS has been successfully applied in spectroscopy and microscopy [60]. A schematic of a typical P-CARS system is shown in figure 3.

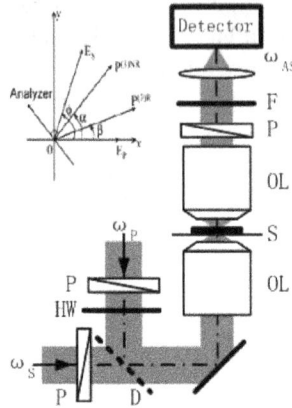

Fig. 3. Schematic of P-CARS. P, polarizer; HW, half-wave plate; D, dichroic beam splitter; OL, objective lens; S, specimen; F, filter.

In a P-CARS system, an analyzer in front of the detector is used to block the nonresonant signal, while the portion of the differently polarized resonant signal passes through the analyzer. Although the P-CARS can effectively suppress the NRB noise, the acquisition time is longer because of the loss of resonant signals

Time-resolved CARS detection

In the time-resolved CARS detection (T-CARS), the ultra-short laser pulse is used as excitation laser. The resonant and nonresonant contributions are separated in the time domain due to the different temporal response characteristics [61, 62]. Because of the instantaneous dephasing characteristics of nonresonant signal, it exists only when three laser pulses temporally overlap. In T-CARS, a pair of temporally overlapped laser pulses is used as the pump and Stokes pulse to resonantly enhance the molecular vibration. A laser pulse with time delay is used as the probe pulse. The resonant CARS signal decays in the finite dephasing time of the vibrational mode. The dephasing time is related to the width of spectral line of the corresponding Raman band and is typically several hundred femtoseconds (in solid) to a few picoseconds (in gas or liquid) [63]. Therefore, the NRB noise can be eliminated by introducing a suitable time delay between the pump/Stokes and probe pulses [64]. The detail discussions will be given in the next section.

Phase control

In the phase control method, a phase-mismatched coherent addition of nonresonant spectral components is introduced with phase shaping of the femtosecond laser pulses to suppress the nonresonant signal [65-67]. For CARS imaging with picosecond pulses, the phase control can be achieved by heterodyning the signal with a reference beam at the anti-Stokes wavelength [68, 69]. With the heterodyne CARS interferometry, the imaginary part of the third-order nonlinear susceptibility $(Im\{\chi_r^{(3)}\})$ can be separated to suppress NRB noise [70-72].

2.4 Condition of momentum conservation: Phase-matching

Unlike fluorescence or spontaneous Raman microscopy, the CARS process is a parametric process, in which the conditions of energy and momentum must conserve. The generation of CARS signals thus relies on not only the intensity of focused incident optical fields, but the phase conditions of focused fields.

$$l << l_C = \pi / |\Delta k|, \tag{2.26}$$

where l is the effective interaction length, l_C is the coherent length, the wave-vector mismatch $\Delta k = k_{AS} - (k_P + k_{P'} - k_S)$, k_P, $k_{P'}$, k_S and k_{AS} is the wave-vector of pump, probe, Stokes and anti-Stokes field respectively. Under the tight focusing condition, in the CARS microscopy with collinear geometry, a small excitation volume and a large cone angle of wave-vectors compensate the wave-vector mismatch induced by the spectral dispersion of the refractive index of the sample, and the phase matching condition can easily be fulfilled [73, 74]. Therefore, the collinear geometry is the best configuration choice of CARS microscopy.

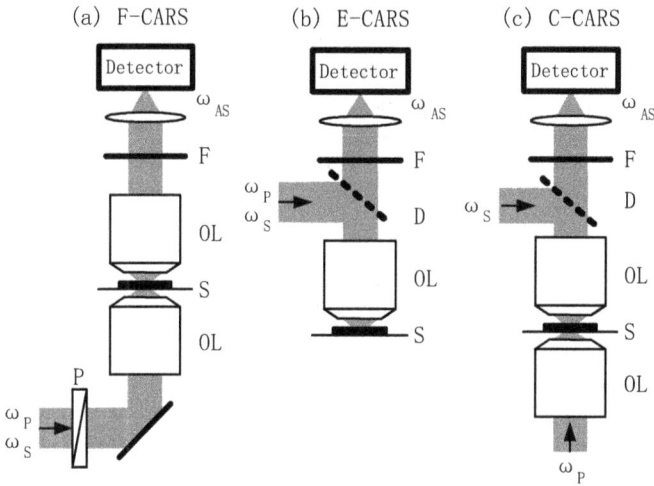

Fig. 4. Schematics of three typical CARS microscopy. (a) forward-detection CARS (F-CARS), (b) epi-CARS (E-CARS), and (c) counterpropagating CARS (C-CARS) microscopy. P, polarizer; D, dichroic beam splitter; OL, objective lens; S, specimen; F, filter.

Three typical geometries of CARS microscopes are shown in figure 4. In figure 4(a) and (b), both the pump and Stokes beams collinearly propagate, and the anti-Stokes signal is detected in the forward direction (a) and backward direction (b). For the forward-detection CARS (F-CARS), the phase matching condition can be easily fulfilled by using an objective with high numerical aperture (NA). For the epi-detection CARS (E-CARS) and counterpropagating CARS (C-CARS) (c) with collinearly propagating geometry, large wave-vector mismatch is introduced and is $|\Delta k| = 2|k_{AS}| = 4n\pi/\lambda_{AS}$, and $|\Delta k| = 2|k_S| = 4n\pi/\lambda_S$. Here, n is the refractive index of medium assumed to be independent of frequency. Therefore, the latter two CARS microscopes have higher sensitivity for object of much smaller than the interaction length.

2.5 Applications of CARS spectroscopy and microscopy

As one of noninvasive research tools with high sensitivity, specificity and resolution, CARS microscopy has attracted more and more attention and been widely used in physics, chemistry, biology, medicine and life science et al. The capabilities and availability of CARS microscopy has been further improved with the recent technique's advances. Many exciting results have been presented in many literatures.

Because of the label-free characteristic of CARS microscopy, it has been regarded in the biological research, especially in the unstained cells. The first CARS microscopy was used to obtain the structural image of epidermal cells of onion immersed in D_2O [21]. The water diffusion in live dictyostelium cells was researched with a broad vibrational resonance centered at 3300 cm-1, which could not be observed with fluorescence microscopy [75]. These early experimental results have proved that the CARS microscopy is an effective complementary method of fluorescence microscopy. Since many cellular processes take place on a subsecond timescale, high temporal resolution is required. By improving the temporal resolution, it is possible to image the chromosome distribution during mitosis using the symmetric stretching vibration of the DNA phosphate backbone [76]. Because of the good detectability of lipids, the structural and functional images of various living cells were obtained with CH bond of lipid [75, 77-79]. The sensitivity of CARS microscopy is high enough to detect lipid vesicles with sizes smaller than 300 nm in diameter [79]. Compared with fluorescence microscopy, CARS microscopy allows long-term investigations of cell without photobleaching. Therefore it can be used to long-term track biological molecules, such as lipid droplets, in living cells [80]. Nan and associates used the CARS microscopy to study the growth and transport of lipid droplets in live cells [79]. By tuning to the CH_2 lipid vibration, Cheng and his colleagues observed the apoptosis, and identified different stages in the apoptotic process [76]. Potma and his associates visualized the intracellular hydrodynamics with the CARS signal of the O-H stretching vibration of water [81].

On the basis of cell imaging, the CARS microscopy is used in the living animal's tissue imaging, in which the tissue's optical properties, such as absorbability and scattering, are of obvious concern. The method of epi-detection is a good solution in tissue imaging with CARS. CARS microscopy has been successfully used for imaging of nonstained axonal myelin in spinal tissues in vitro [82]. Both the forward and backward CARS signals from the tissue slab were detected. The lipid distributions in skin tissue of live animals have been observed [83]. These all preliminary experimental results show us a vast potential of CARS microscopy in biomedical imaging and early diagnosis of diseases.

3. Supercontinuum with photonic crystal fiber

As we have discussed in previous sections, in a CARS spectroscopy or microscopy, it is necessary that two ultra-short laser pulses with high peak power and different frequencies reach focus at the same time. In order to quickly distinguish different molecules in a complex system with the complete CARS spectra, such as various biological molecules in cells, it is required that the output of source must have not only a wide enough spectral range, but the spectral continuity and simultaneity of various spectral components [84]. Spectral broadening and the generation of new frequency components are inherent features of nonlinear optics. When ultra-short laser pulses propagate through a nonlinear medium, a dramatic spectral broadening will happen. This particular physical phenomenon, known as supercontinuum (SC) generation, was first demonstrated in the early 1970s [85-87]. With the advent of a new kind of optical waveguides in the late 1990s, photonic crystal fiber (PCF) has led to a great revolution in the generation of SC with ultra-broad spectral range and high brightness [39, 88, 89]. In this section, we will introduce the SC generation with PCFs by theoretical analysis and modeling. Based on the requirements of CARS, the method and conditions for realizing an ideal SC source are discussed.

3.1 Photonic crystal fiber used for supercontinuum generation

SC generation involves many nonlinear optical effects, such as self- and cross-phase modulation, four-wave mixing (FWM), stimulated Raman scattering (SRS) and solitonic phenomena, which add up to produce a output with an ultra-broadband spectra, sometimes spanning over a couple of octaves. With the developments of theories and techniques of modern nonlinear optics, various optical materials are realized and widely used in various fields. A photonic crystal fiber (PCF) (also called holey fiber (HF) or microstructure fiber (MF)) [90-92], based on the properties of two-dimension photonic crystal, is a kind of special optical fiber, which can confine the incident light passing through the entire length of fiber with its tiny and closely spaced air holes. Different arrangements of the air holes make PCFs with various optical characters, such as the single-mode propagation, high nonlinearity, and controllable dispersion.

According to different guiding mechanism, there are mainly two categories of PCF, photonic bandgap (PBG) PCF [93] and the total internal reflection (TIR) PCF, as shown in figure 5. The PBG PCF is usually used for transmission of high-energy laser pulses and optical signals. The major energy propagates through the hollow core of a PBG PCF with low loss, dispersion and nonlinear effects. The TIR PCF is used for the SC generating with wide spectral range. When high-intensity laser pulses with narrow line-width propagate in a TIR PCFs, the SC, sometimes spanning over a couple of octaves, could be generalized because of its high nonlinearity and group velocity dispersion effects.

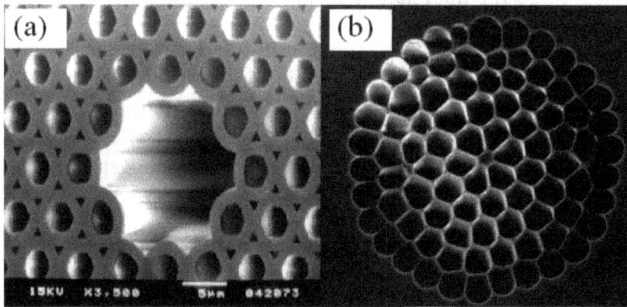

Fig. 5. Structures of a typical PBG PCF (a) and TIR PCF (b) obtained with scanning electron microscope.

3.2 Numerical modeling

The process of SC generation is a synthesis result of a variety of the nonlinear optical effects, when ultra-short laser pulses with high intensity propagate in a PCF [94, 95]. Used as the source for CARS, we have mostly concerned about, however, the single-mode propagation and temporal distributions of various spectral components of SC, called temporal-spectral distribution [96]. For a PCF with a given structure, a number of numerical modeling and computational methods have been constructed and reported to obtain the entire properties of a PCF. Here, we carry out an entire analysis on the SC generation with a common method that is mainly divided into three steps.

Firstly, as one of most effective methods, the finite element method (FEM) can be used to obtain the coefficients of the chromatic dispersion (the effective propagation constant β_{eff}) based on the structural parameters of PCFs (the diameter of air-hole, and the pitch between two holes). The dispersion coefficients β_k ($k \geq 2$) can be derived by the Taylor series expansion at the central frequency ω_0, and the nonlinear coefficient γ can be approximately calculated by $\gamma = n_2\omega_0/cA_{eff}$, with n_2 the nonlinear-index coefficient for silica, c the speed of light in vacuum, and A_{eff} the effective core area.

Secondly, a propagation equation is used to calculate the SC generation during the propagation of ultra-short laser pulses, although the generalized nonlinear Schrödinger Equation (GNLSE) is not the only way to realize it. The process of the pulses propagation was simulated with the split-step Fourier method (SSFM) to solve GNLSE [94].

$$\frac{\partial A}{\partial z} + \frac{\alpha}{2}A - \sum_{k \geq 2} \frac{i^{k+1}}{k!}\beta_k \frac{\partial^k A}{\partial T^k} = \\ i\gamma\left(1 + \frac{i}{\omega_0}\frac{\partial}{\partial T}\right)\left(A(z,t)\int_{-\infty}^{+\infty} R(T') \times |A(z,T-T')|^2 dT' + i\Gamma_R(z,T)\right) \qquad (3.1)$$

In equation (3.1), the linear propagation effects on the left-hand side and nonlinear effects on the right-hand side are given, where α and A are the loss coefficient and the spectral envelope with the new time frame $T=t-\beta_1 z$ at the group velocity β_1^{-1}. $R(T)$ presents the Raman response function. The noise Γ_R, which affects the spontaneous Raman noise, is neglected, $\Gamma_R=0$. It has more detailed explanation in the paper [94].

For a CARS spectroscopy or microscopy, the temporal-spectral distribution of SC is also an important factor. Therefore, thirdly, we have to figure it out in order to fully understand the temporal distribution of various spectral components in SC, although the spectral envelope of SC can be obtained in the second step. To obtain the temporal-spectral distribution of SC, cross-correlation frequency resolved optical gating method (XFROG) was applied for characteristic of SC and could be proved by an experimental instrument of XFROG [97]. The two-dimensional XFROG spectrogram can be plotted by using two electromagnetic fields and the following equation:

$$I_{XFROG}(\omega,\tau) = \left|\int_{-\infty}^{+\infty} E(t)E_{gate}(t-\tau)\exp(-i\omega t)dt\right|^2, \qquad (3.2)$$

where $E(t)$ is the calculated envelope of the SC with the variable t, and $E_{gate}(t-\tau)$ is the gating pulses with the delay time τ between the seed laser pulses and the SC. It can be concluded that XFROG measurement is a good way to characterize the temporal and spectral evolution of the SC generation and interpret the particular time and frequency domain information of the optical effects. With the above introduced method, we carried out simulation analysis in order to find out a way to achieve an ideal SC source for CARS spectroscopy and microscopy. Some representative results will be shown in the next section.

3.3 Supercontinuum generation with photonic crystal fiber

Some of our computational results are shown here in order to account for the whole processing course clearly. We have simulated the SC generation by using a PCF with two zero dispersion wavelengths (ZWD) [98]. The calculated group velocity dispersion (GVD) curve is shown in figure 6. By solving GNLSE with SSFM, the temporal and spectral distributions of the SC generation along the whole length of PCF are shown in figure 7. With the XFROG trace, the results of temporal-spectral distributions of PCFs with different lengths are described in figure 8.

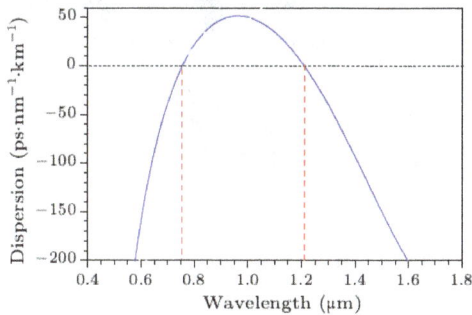

Fig. 6. Group velocity dispersion curve of the PCF with two ZWDs [98].

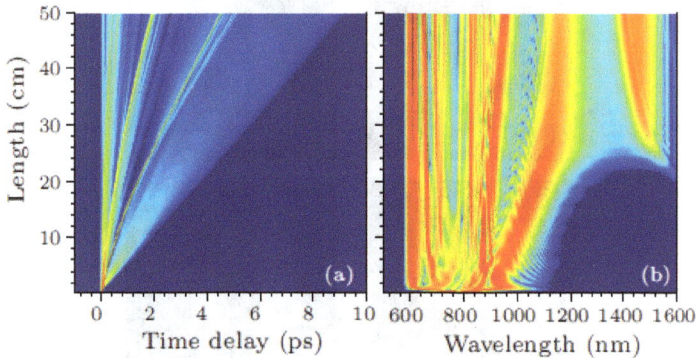

Fig. 7. Time (a) and spectrum (b) evolution of SC along the entire length of the PCF with the input pulse width 30 fs and peak power 10 kW [98].

In figure 8(c), the spectral range of generated SC is 500nm by using a PCF with two ZDW under proper pumping conditions. In SC, the spectral continuity, simultaneity and intensity of red-shifted SC components are all good enough for a source of CARS. But for this purpose, an ultra-short pulse laser system with pulse width of 30fs is needed, which is not easily sustainable during practically experimental operations. Therefore, we have tried to seek a simpler way to generate favorable SC for CARS applications. The simulation results are shown in figure 9, where we can see that the SC generated by a PCF with two ZDW is quite good for CARS applications when the laser pulse width is 300fs, as shown in figure 9 (c).

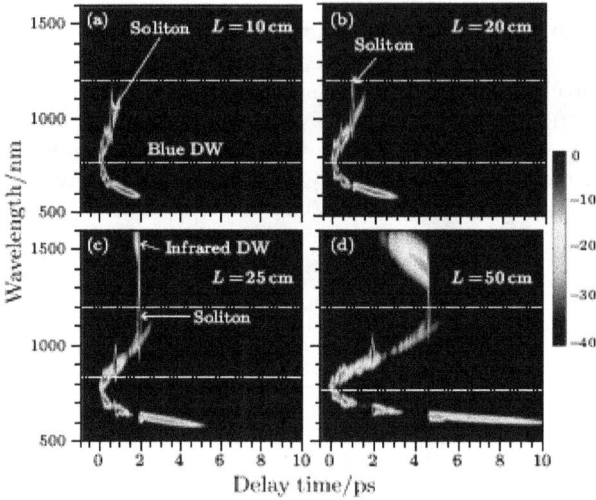

Fig. 8. Temporal-spectral distribution of SC when PCFs with lengthes of 10 cm (a), 20 cm (b), 25 cm (c), and 50 cm (d) pumped by laser pulse with pulse-width 30 fs, wavelength 780 nm, and peak power 10 kW [98].

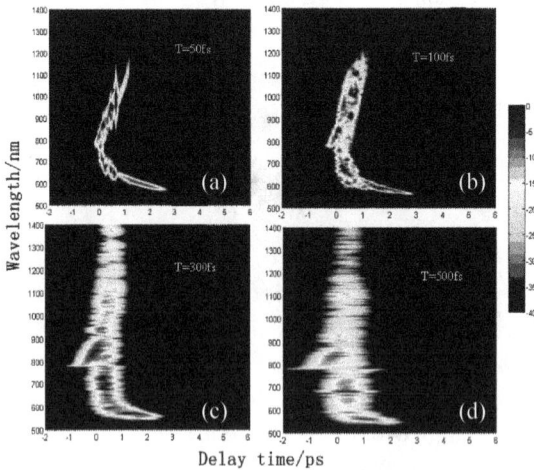

Fig. 9. Temporal-spectral distribution of the SC when using the femtosecond laser pulse with a central wavelength of 780 nm, peak power of 10 kW and pulse-width of 50 fs (a), 100 fs (b), 300 fs (c) and 500 fs (b) as seed pulse to pump a PCF with length of 10cm.

By numerical simulations, we clearly understood the effects of the parameters of PCF and pumping laser pulse on the generation of SC. All simulation results provide us an intuitive description of SC generation, and theoretical guides for experimental instruments. Dispersion and nonlinearity of a PCF can be modified and optimized by adjusting the air-hole structure of a PCF. Under specifically experimental conditions, a perfect SC source for CARS spectroscopy and microscopy can be achieved with an optimized PCF.

4. Broadband CARS spectroscopy and microscopy

In a traditional CARS microscopy, two or three ultra-short laser pulses with narrow line-width and different frequencies are used as excitation beams. It permits high-sensitivity imaging based on a particular molecular bond, called single-frequency CARS. But for a mixture with various or unknown components, it is not adequate to distinguish the interested molecules from a complex based on the signal of a single active Raman bond. The broadband even complete molecular vibrational spectra will be beneficial for obtaining the accurate information of various chemical compositions. Although it can be achieved by sequentially tuning the frequency of Stokes beam, it is time-consuming and unpractical for some applications. This problem can be circumvented by using the multiplex CARS (M-CARS) or broadband CARS spectroscopy with simultaneously detected wider band.

4.1 Introduction to broadband CARS

The M-CARS spectroscopy was first demonstrated by Akhamnov et al., a part of CARS spectra of a sample can be simultaneously obtained [34]. In M-CARS, a broadband laser beam is used as the Stokes beam for providing a required spectral range. A narrow line-width laser beam is used as the pump and probe beam that determines the spectral resolution of the system. The multiplex molecular vibrational modes of a sample can be resonantly enhanced, the corresponding CARS signals can be detected simultaneously, the energy diagram of M-CARS shown in figure 10.

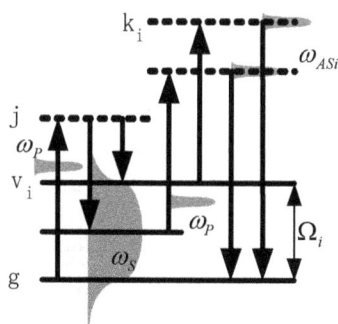

Fig. 10. Energy level diagram of M-CARS.

In the previous works, a narrowband and a broadband dye laser was used for the pump/probe and the Stokes beams respectively [36-38]. The recent progress in wavelength-tunable ultra-short pulse laser has been giving a powerful momentum to the development of M-CARS. The M-CARS micro-spectroscopy has been developed for fast spectral characterization of microscopic samples [35, 99, 100]. But because of the used laser limitation to the line-width, M-CARS is still unable to simultaneously obtain wider molecular vibrational spectra as required. With the progress of SC generation technique [85-87], especially with the advent of PCF [39], much wider spectra of M-CARS can be simultaneously obtained by broadening the spectral range of Stokes pulses with nonlinear optical fiber, such as the tapped optical fiber [101] or PCF [43, 44, 102]. By using a specially designed and achieved SC source, the simultaneously detectable spectral range of M-CARS spectroscopy and microscopy is greatly widened, which can be called the broadband CARS. Wider simultaneously detectable spectral range makes it possible to quickly distinguishing various components and real-time monitoring slight variations in a mixture [103-105]. At the same time, the system of the broadband CARS with SC is simplified and cost is reduced.

4.2 Suppression of NRB noise in broadband CARS with SC

In the M-CARS, NRB noise can not be avoided and many methods for suppressing it in a single-frequency CARS can also be used, but they can not be easily applied in the broadband CARS with SC, because of the complex polarization and phase of various spectral components in SC, as shown in section 3. The NRB noise can be eliminated with numerical fitting method by regarding it as a reference signal, but the Raman spectra of the samples are needed in advance [106].

As presented in above section, the time-resolved detection method can effectively eliminate the NRB noise by introducing a temporal delay between pump/Stokes pulses and probe pulse in order to temporally separate the resonant and nonresonant signals. In a T-CARS, three laser pulses, with frequencies at ω_P, $\omega_{P'}$, and ω_S, are used as the pump, probe and Stokes pulses respectively. The generation process of CARS signals can be described in three phases [107]. In the first phase, the inherent molecular vibration of active Raman mode is driven by simultaneous pump and Stokes pulses and is resonantly enhanced when $\Omega_R = \omega_P - \omega_S$. The amplitude of resonantly enhanced molecular vibration is [108]:

$$\frac{\partial Q_v}{\partial t} + \frac{Q_v}{T_2} = \frac{i}{4m\Omega_r}\left(\frac{\partial \alpha}{\partial Q}\right)E_P E_S^*\left(1 - 2n_a\right),$$ (4.1)

where Q_v is the amplitude of molecular vibration driven by the incident optical fields, T_2 is the dephasing time of the resonant enhanced molecular vibrational state. When the simultaneous ultrashort laser pulses are used as the pump and Stokes pulses, the intensity changes of Q_v with time is shown in figure 11. Q_v increases during the period of incident laser pulses and reaches its maximum when the pump and Stokes pulses just disappear.

In the second phase, with the disappearance of incident laser pulses, the resonantly enhanced molecular vibration will rapidly return to its original state that can be regarded as a free relaxation process. Equation (4.1) can be rewritten as [107]:

$$\frac{\partial Q_v}{\partial t} = -\frac{Q_v}{T_2}.$$ (4.2)

The solution of equation (4.2) is [107]:

$$|Q_v|^2 = A\exp\left(-2t/T_2\right),$$ (4.3)

where A is an integration constant. We can see that Q_v decays exponentially with time immediately after disappearance of the pump and Stoke pulses. Assuming the dephasing time is 10ps, the relaxation process of Q_v is shown in figure 12.

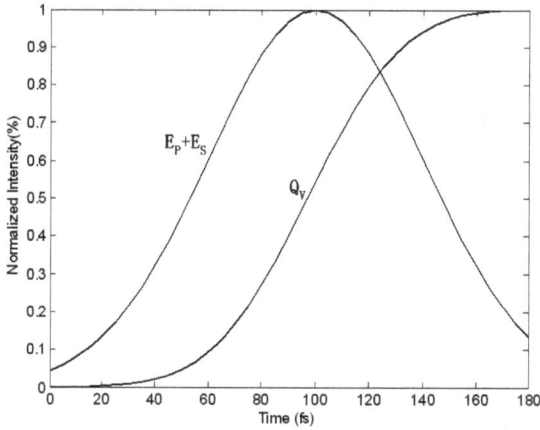

Fig. 11. Intensity of Q_v amplitude versus time, when the vibration of active Raman mode is resonantly enhanced by incident pump and Stokes pulses [107].

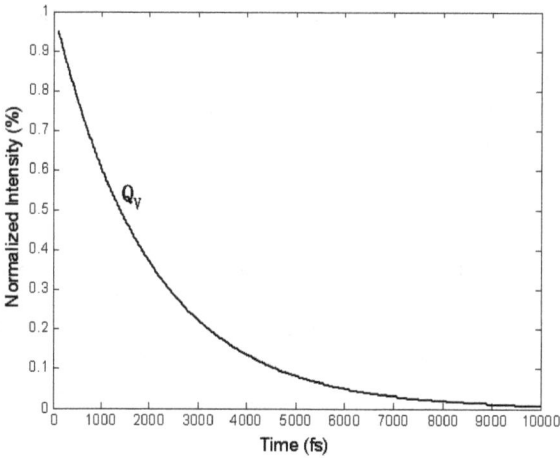

Fig. 12. The free relaxation process of a molecular vibration mode immediately after disappearance of the pump and Stokes laser pulses [107].

In the third phase, when the probe pulse reaches the focus with the delay time t_D, it will be modulated by the resonantly enhanced molecular vibration. The signal field at anti-Stokes frequency is generated and can be expressed as [107]:

$$\frac{\partial E_{AS}}{\partial z} + \frac{n_{AS}}{c}\frac{\partial E_{AS}}{\partial t} + \frac{\alpha_{AS}}{2}E_{AS} = \frac{i\omega_{AS}}{2n_{AS}c\varepsilon_0}\left[\frac{1}{2}N\left(\frac{\partial \alpha}{\partial Q}\right)E_{p'}Q_v e^{-i\Delta kz}\right]. \tag{4.4}$$

From the equation (4.4), we can find that the CARS signal field does not include the nonresonant component $\chi_{nr}^{(3)}$ that disappears with the end of the pump and Stokes pulses simultaneously. When the phase-matching condition is satisfied, $\Delta k=0$, the intensity changes of the CARS signal with time is shown in figure 13.

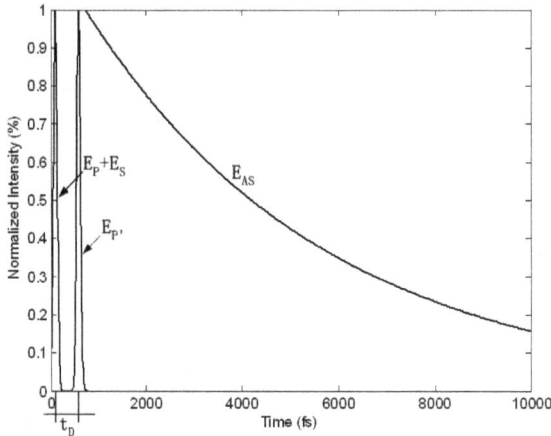

Fig. 13. Intensity of CARS signal changes with time [107].

In the T-CARS method, the resonant and nonresonant components have different temporal response characteristics. In order to effectively separate the resonant and nonresonant contributions and avoid the intensity loss of the CARS signal, the pulse-width of simultaneous pump ans Stokes pulse should be as short as possible and the rising edge of the probe pulse should be as steep as possible.

Recently the broadband T-CARS spectroscopy and microscopy with SC has been rapidly developed [43, 44], whose energy level diagram is shown in figure 14. In the broadband T-CARS, a well-designed SC is used as the pump and Stokes pulses and a temporally delayed laser pulse is used as the probe pulse. The simultaneously detectable spectral range and the spectral resolution are determined by the temporally overlapped spectral range of the SC and by the line-width of the probe pulse respectively. With the improvement of temporal-spectral distribution of the SC, the simultaneously detectable spectral range of system can be further extended, which is called the ultra-broadband T-CARS and will be discussed in detail in the next section.

4.3 Ultra-broadband T-CARS spectroscopy with SC generated by PCF[84, 96]

With a broadband T-CARS spectroscopy, we can obtain more specificities of the sample, not only the vibrational spectra reflecting the molecular structure and compositions, but also the dephasing time of various molecular vibrational modes reflecting the molecular responses to the external micro-environment, which are especially favorable for the study of the complicated interaction processes between molecules and their micro-environment such as solute-solvent interactions [108, 109], molecular dynamics[110-114], supramolecular structures[115] and excess energy dissipations in the fields of biology, chemistry and material science [64, 116-118].

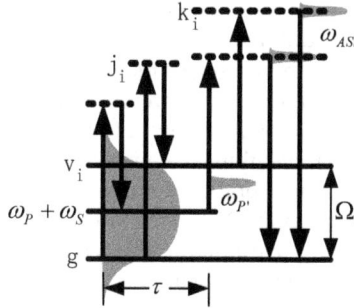

Fig. 14. Energy level diagram of the broadband T-CARS.

The principle of broadband T-CARS spectroscopy has been presented in section 4.2. As discussed in section 3 and 4, the simultaneously detectable spectral range of a broadband T-CARS spectroscopy is limited by the simultaneously generated spectral range and its continuity in the SC. An ultra-broadband T-CARS spectroscopy based on optimized SC has been developed to simultaneously obtain CARS signals corresponding to various molecular vibrational modes and Raman free induction decays (RFID) of these molecular vibrational modes in a single measurement [43, 84]. The schematic of the broadband T-CARS spectroscopy is shown in figure 15. A femtosecond laser pulse of a mode-locked Ti:sapphire laser oscillator (Mira900, Coherent) is split into two parts by a beam splitter. One beam of the laser pulse, used as the seed pulse, is introduced into a PCF with geometric length of 180 mm and ZDW of 850 nm respectively. After passing through a long-pass filter, the residual spectral components of SC are used as the pump and the Stokes. Another beam is used as the probe pulse after passing through a narrow-band-pass filter. Two beams of the laser pulses are collinearly introduced into a microscope and tightly focused into a sample with an achromatic microscopy objective. The generated CARS signals in the forward direction, passing through a short-pass filter, are collected with the same microscope objective and detected by a fibre spectrometer. The delay time between SC pulse and probe pulse can be accurately adjusted by a kit of time delay line.

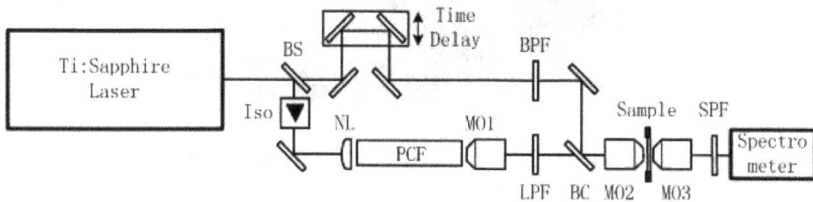

Fig. 15. The schematic of the ultra-broadband T-CARS system. BS, beam splitter; Iso, optical isolator; NL, non-spherical lens; PCF, photonic crystal fibre; BC, beam combiner; MO1-3, microscopy objective; BPF, narrow-band-pass filter; LPF, long-pass filter; SPF, short-pass filter [84].

With the ultra-broadband T-CARS spectroscopy, the time-resolved measurement is achieved by adjusting the delay time between the SC pulse and the probe pulse step by step [84]. The obtained time-resolved CARS spectral signals and CARS signals at specific delay time of pure benzonitrile and mixture solution are shown in figure 16. The molecular vibrational spectra for pure liquid benzonitrile in the range of 380-4000 cm^{-1} can be simultaneously obtained without any tuning of the system and its characteristics. The NRB noise can be effectively suppressed through tuning the delay time. For the pure benzonitrile, the obvious peaks at wavenumbers of 1016 cm^{-1}, 1190 cm^{-1}, 1608 cm^{-1}, 2248 cm^{-1} and 3090 cm^{-1} correspond to C-C-C trigonal breathing, C-H in plane bending, C-C in plane stretching, C≡N stretching, and C-H stretching vibrational modes respectively [119]. In the mixture, the peaks at wavenumbers of 1016 cm^{-1}, 2248 cm^{-1} and 3090 cm^{-1} of benzonitrile can be apparently seen [120]. Other peaks correspond to the typical molecular vibrational modes of methanol and ethanol. We can easily and accurately distinguish the various components in the mixture. The spectral resolution, depending on the line-width of the probe pulse and spectral resolution of the spectrometer, is 14 cm^{-1} in this case.

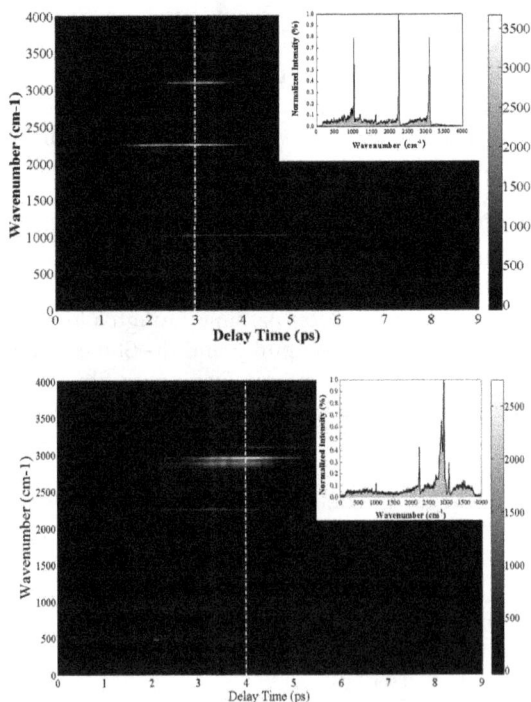

Fig. 16. Intensities of time-resolved CARS signals of pure benzonitrile and at the delay time of 3 ps (a). Intensities of time-resolved CARS signals of benzonitrile-methanol-ethanol mixture solution and at the delay time of 4 ps (b) detected with the ultra-broadband T-CARS spectroscopy [84].

By extracting the time evolutions of CARS signals corresponding to the molecular vibrational modes for a pure liquid benzonitrile and mixture, the RFID processes of various molecular vibrational modes can be measured at the same time. The dephasing times of various molecular vibrational modes can be obtained by fitting the data to a single exponential function as:

$$I(\tau) = A_0 \exp(-2\tau/T),\qquad(4.5)$$

where T is the vibrational dephasing time responding to each molecular vibrational mode; A_0 is a constant; τ is the delay time. The normalized intensities of five typical peaks corresponding to typical molecular vibrational modes for pure benzonitrile, at the wavenumbers of 1016 cm-1, 1190 cm-1, 1608 cm-1, 2248 cm-1 and 3090 cm-1, are plotted one by one as a function of τ and fitted to equation (4.5) in figure 17 (a)-(e) respectively [84].

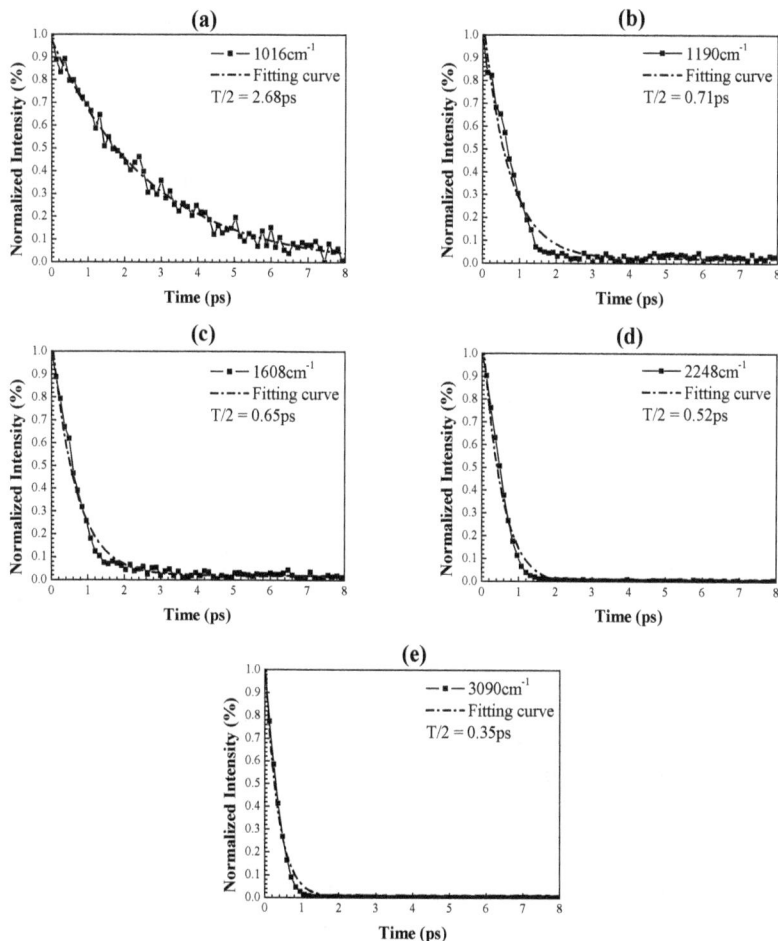

Fig. 17. Vibrational dephasing processes and their relevant dephasing times of various molecular vibrational modes for pure liquid benzonitrile shown in (a)-(e) respectively, with the solid line corresponding to the change of CARS signal with delay time and the dash dot line representing the fitting curve to a single exponential function [84].

In a benzonitrile-methanol-ethanol mixture solution, the benzonitrile molecule is regarded as the target molecule. The normalized intensities of three typical peaks corresponding to typical molecular vibrational modes for benzonitrile, at the wavenumbers of 1016 cm^{-1}, 2248 cm^{-1} and 3090 cm^{-1}, are plotted one by one as a function of τ and fitted to equation (4.5) in figure 18 (a)-(c), respectively.

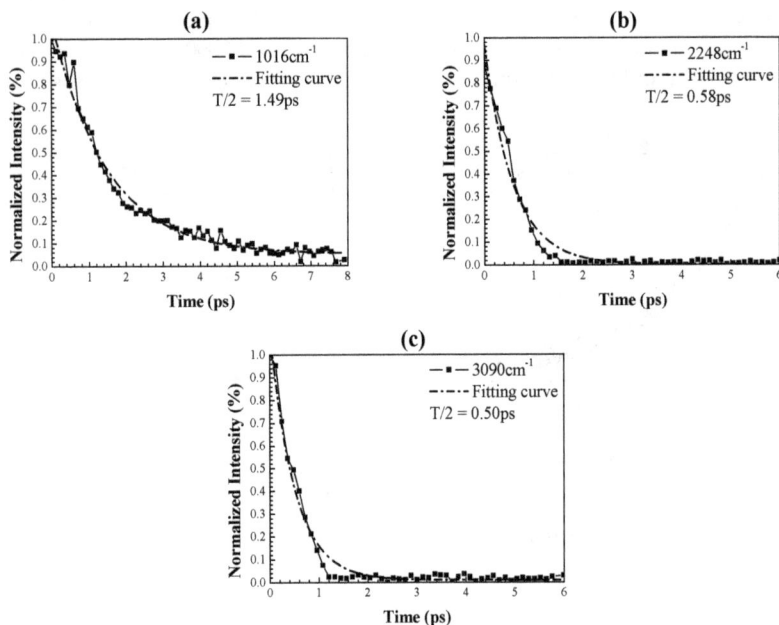

Fig. 18. Vibrational dephasing processes and their corresponding dephasing times of various molecular vibrational modes for benzonitrile in benzonitrile-methanol-ethanol mixture solution shown in (a)-(c) respectively, with the solid line corresponding to the variation of CARS signal with delay time and the dash dot line representing the single exponential fitting curve [84].

From experimental results, the intensities of the CARS signals corresponding to different molecular vibrational modes attenuate exponentially against the delay time in a large dynamic range. By fitting the intensity data of the CARS signal to a single exponential function for the molecular vibrational modes at different wave-numbers, half of vibrational dephasing time T/2 can be worked out as shown in figure 17, which are consistent with the previously published data [43, 121-123]. But in a benzonitrile-methanol-ethanol mixture solution, the experimental results show that the influence of solvent on the property of solute is reflected not by the Raman peak position but by the variations of the vibrational dephasing times for different molecular vibrational modes.

4.4 Simultaneously obtaining the complete molecular vibrational spectra

As discussed above, the simultaneously detectable spectral range of the ultra-broadband T-CARS with SC depends on the quality of the SC. It is of importance to simultaneously obtain the complete molecular vibrational spectra and the dephasing times of various molecular vibrational modes of the sample. The former is very useful for effectively and accurately distinguishing various kinds of components and understanding the mechanisms of chemical reactions in a dynamic process. The latter is very helpful for explanation of both solvent dynamics and solute-solvent interactions in the fields of biology, chemistry and material science. The question is whether we can reach above goal in the near future by optimizing the temporal-spectral distribution of the SC? Our answer is positive. As what we have known the existing molecules have the Raman wave-numbers in the range of about tens to 5000 cm[-1], which means that the simultaneously generated Stokes wavelength bandwidth should be not less than 350nm. As we have given in section 3 that the bandwidth of the simultaneously generated SC can be greater than 400nm, therefore it is very promising to achieve label-free microscopic imaging technique with high contrast and chemical specificity based on the simultaneously obtained complete molecular vibrational spectra.

5. Sub-diffraction-limited CARS microscopy

5.1 Methods of improving spatial resolution of CARS microscopy

As well known, there is a theoretical limitation of the spatial resolution for any far-field optical microscopes because of the existence of light diffraction. Ernst Abbe defined the diffraction limit as [124]:

$$d = \frac{0.61\lambda}{n\sin\phi} \cong \frac{\lambda}{2NA} ,\qquad (5.1)$$

where d is the resolvable minimum size, λ is the wavelength of incident light, n is the refraction index of the medium being imaged in, φ is the aperture angle of the lens, and NA is the numerical aperture of the optical lens. It is obvious that for an optical microscope, d is the theoretical limit of spatial resolution. The samples' spatial features, smaller than approximately half the wavelength of the used light, would never be able to be resolved.

In recent years, in order to meet the requirements on the study of life science and material science, ones have found several ways to overcome the optical diffraction limit and obtained sub-diffraction limited spatial resolution theoretically. In fluorescence microscopy, the success of the resolution enhancement techniques relies on the ability to control the emissive properties of fluorophores with a proper optical beam. The most important developments for breaking through the diffraction barrier are sub-diffraction-limited resolution fluorescent imaging techniques, such as photo activated localization microscopy (PALM) [125], stochastic optical reconstruction microscopy (STORM) [126], and stimulated emission depletion (STED) microscopy [127, 128], which have opened up notable prospect for sub-cellular structure and bio-molecular movement and interaction imaging.

As one of label-free nonlinear imaging techniques, the spatial resolution of CARS microscopy is higher (about 300nm lateral resolution) than the one of traditional linear optical microscopy, but it is still a diffraction-limited imaging technique. Today, how to achieve a sub-diffraction-limited CARS microscopy has become one of attractive topics all over the world. Compared with developments of the fluorescence nanoscopy, the method for breaking through the diffraction limitation in CARS microscopy is still under theoretical research.

In 2009, Beeker et al. firstly presented a way to obtain a sub-diffraction-limited CARS microscopy in theory [129]. With the density matrix theoretical calculations, they found that the molecular vibrational coherence in CARS can be strongly suppressed by using an annular mid-infrared laser to control the pre-population of the corresponding vibrational state. The energy level diagram is shown in figure 19. Thereby the emission of generated CARS signals in the annular area of point spread function could be significantly suppressed and the spatial resolution can be improved considerably.

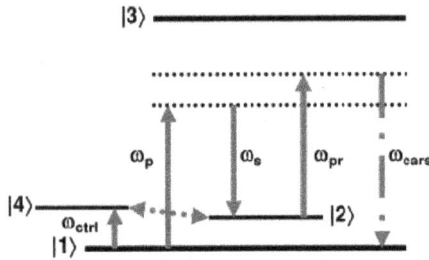

Fig. 19. Energy level diagram for CARS extended with an additional level $|4\rangle$. Energy level $|1\rangle$ - $|4\rangle$ are ground state, vibrational state, excited vibrational state and control state, respectively. ω_P, ω_{pr}, ω_S, and ω_{ctrl} are the frequencies of pump, probe, Stokes and control laser [129].

Alexei Nikolaenko et al provided their CARS interferometric theory in the same year [130]. In this theoretical research, a stabilized, phase-adjustable interferometer was used to achieve nearly complete interference between the local oscillator and the pump- and Stokes-induced CARS radiation. The schematic of the CARS interferometry setup is shown in figure 20. Their theoretical analysis showed that the energy loss in the anti-Stokes channel is accompanied by an energy gain in the pump and Stokes channels. This implied that the CARS interferometry provided a controllable switching mechanism of anti-Stokes radiation from the focal volume, which might be a possible technique for improving the spatial resolution of the CARS microscopy.

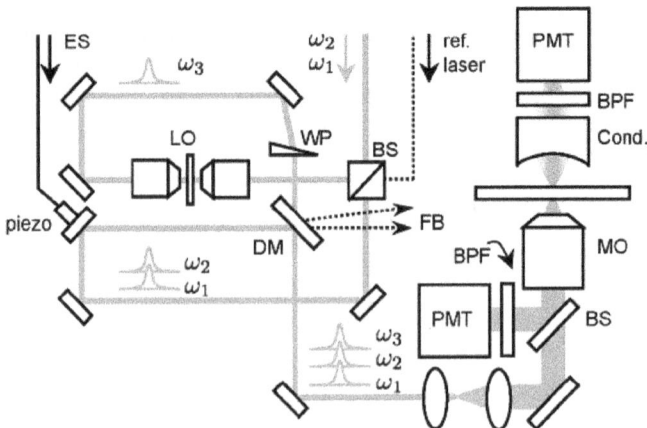

Fig. 20. Schematic of the CARS interferometry. ES, error signal; LO, unit for generating local oscillator; WP, wedged plate; BS, 50-50 beam splitter; DM, dichroic mirror; FB, optical feedback signal; MO, microscope objective; BPF, bandpass filter; Cond., condenser; PMT, photomultiplier [130].

Kim M. Hajek et al presented a theoretical analysis and simulation of a wide-field CARS microscopy with sub-diffraction-limited resolution in 2010 [131]. The configuration and a simulation result are shown in figure 21. In this method, two coherent pump beams were used and interfered in the sample plane, forming a standing wave with variable phase. The numerical simulation showed that a super-resolved image with three times better lateral resolution could be obtained by image processing method of standing-wave frequency theory.

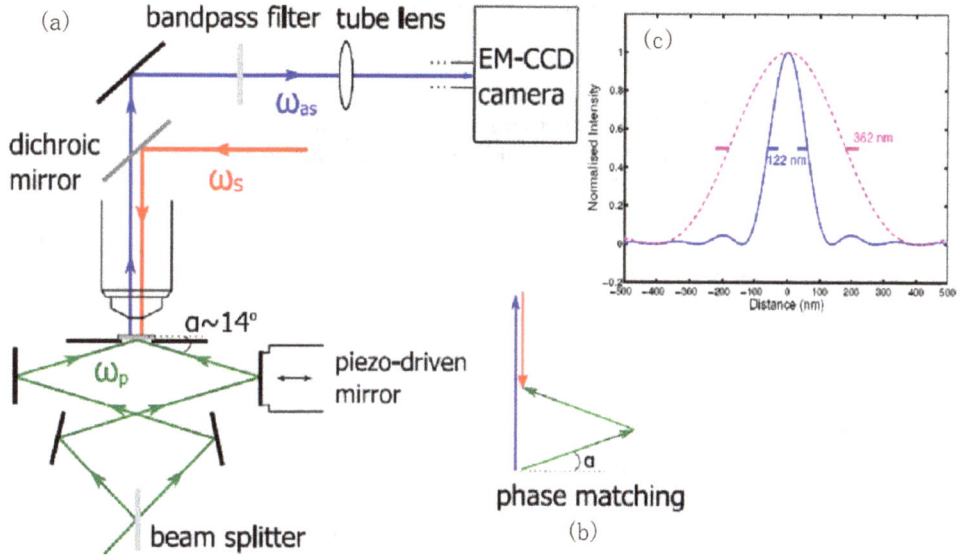

Fig. 21. (a) Schematic of stand-wave wide-filed CARS microscopy, (b) phase-matching condition, (c) simulation result of the point spread function (PSF) of the proposed method [131].

All above discussed sub-diffraction-limited CARS microscopy open up the possibility of achieving sub-diffraction-limited CARS microscopy. Unfortunately, these approaches can only be used in the single frequency CARS microscopy based on the signal of a single molecular vibrational mode.

5.2 Phonon depletion CARS microscopy[132]

As discussed above, successful methods of breaking through the theoretical diffraction limitation of CARS microscopy depend on the controllable emissive properties of the useful signals in the focus. But the above suggested methods for breaking through the diffraction limited resolution can only be used for dealing with the single bond signal. By researching the CARS process with quantum optics theory, we presented our method for breaking through the diffraction limitation, unlike the above methods, which is effective for ultra-broadband T-CARS microscopy.

In our theoretical model, all incident laser fields, generated signal field and the material system are all described with quantum mechanics theory. In the CARS process, the first light-matter interaction process involves resonant enhancements of all active molecular vibrational modes, in which the frequency differences of pump and Stokes fields equal the inherent vibrational frequencies of the molecular bonds respectively. The resonantly enhanced molecular vibrations exist in quantized forms which are called the phonons. Their numbers are equal to the numbers of generated Stokes photons respectively. When a probe field propagates through the matter, the photons of the probe interact with the generated phonons. The photons with anti-Stokes frequencies are generated, and phonons are annihilated at the same time.

Based on the whole quantized picture of the CARS process, we presented a phonon depletion CARS (PD-CARS) technique by introducing an additional probe beam with the frequency different from the one of the probe beam in the center of the focus. When the pump and Stokes simultaneously reaches the focus, the phonons are generated. The additional probe beam, which is shaped into a doughnut profile at the focus with a phase mask, reaches the focus a little bit earlier than the probe beam in the center of the focus. Therefore the wavelengths of the generated anti-Stokes signals at the peripheral region differ from the ones at the center of the focus and can be easily separated with a proper interference filter. By this way, the spatial resolution of the ultra-broadband T-CARS microscopy can be improved greatly. The simulation result of PSF is defined as [132]:

$$\Delta r = \sqrt{2}\frac{\lambda}{\pi n \sin\alpha \sqrt{3+\dfrac{I_{P1}^{max}}{I_{dep}}}} \approx \frac{0.9}{\sqrt{3+\dfrac{I_{P1}^{max}}{I_{dep}}}} \cdot \frac{\lambda}{2n\sin\alpha}, \tag{5.2}$$

where I_{dep} and I_{P1}^{max} are intensities of phonon field at the center of the focus and additional probe field for phonon depletion in the annual region respectively. From equation (5.2), we can know that the spatial resolution of CARS microscopy will be improved by increasing the intensity of additional probe beam. The simulation result of PSF is shown in figure 22. When I_{P1}^{max} is fiftyfold of I_{det}, the spatial resolution of the ultra-broadband T-CARS microscopy reaches 41nm.

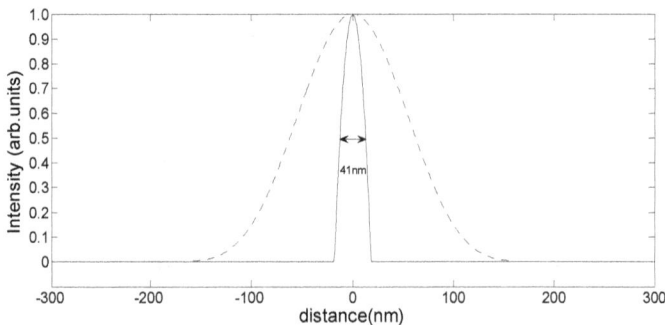

Fig. 22. Simulation result of the PSF in the PD-CARS microscopy [132].

6. Conclusions and prospects

In this chapter, we mainly introduce a kind of noninvasive label-free imaging technique – the ultra-broadband T-CARS spectroscopy and microscopy with SC generated by PCF. We describe the mechanisms of Raman scattering and T-CARS process with the classical and quantum mechanical theory. The CARS signals with much stronger strength and well-oriented direction originate from the coherent resonant enhancement between incident lights and molecular vibrations. In order to quickly and accurately distinguish different kinds of molecules in a complex, such as in a live cell, a method for simultaneously detecting the ultra-broadband CARS signals without the NRB noise has to be developed. On the basis of theoretical analysis and simulation of the SC generation with a PCF, a satisfied SC source can be achieved for obtaining the ultra-broadband even complete CARS spectra of the specimen by optimizing the parameters and the other experimental conditions of the PCF and ultra-short laser. At the same time, the NRB noise can be effectively suppressed in a broad spectral range with the time-resolved method. The method study for obtaining sub-diffraction limited spatial resolution is still on the stage of theoretical research. Some original techniques are presented in this chapter. The PD-CARS technique provides a possible route to the realization of ultra-broadband T-CARS microscopy with sub-diffraction limited spatial resolution, which will probably become an attractive imaging method in biology, medicine and life science in the near future.

7. References

[1] R P Haugland, Handbook of Fluorescent Probes and Research Chemicals, Eugene, OR: Molecular Probes, 1996.

[2] R Y Tsien and A Miyawaki, Biochemical imaging: Seeing the Machinery of Live Cells, Science, 1998, 280(5371): 1954-1955.

[3] E Kohen and J G Hirschberg, Cell structure and function by microspectrofluorometry, San Diego, CA: Academic, 1989.

[4] J B Pawley, Handbook of Biological Confocal Microscopy, New York: Plenum, 1995.

[5] T Wilson and C Sheppard, Theory and Practice of Scanning Optical Microscopy, Orlando, FL: Academic, 1984.

[6] W Denk, J H Strickler and W W Webb, Two-photon laser scanning fluorescence microscopy, Science, 1990, 248(4951): 73-76.

[7] Xu C, W Zipfel, J B Shear, et al, Multiphoton fluorescence excitation: new spectral windows for biological nonlinear microscopy, Proc. Natl. Acad. Sci. U.S.A., 1996, 93: 10763-10768.

[8] J Hoyland, Fluorescent and Luminescent Probes for Biological Activity, ed. W T Mason. San Diego, CA: Academic, 1999.

[9] G T Boyd, Y R Shen and T Hansch, Continuous-wave second-harmonic generation as a surface microprobe, Opt. Lett., 1986, 11(2): 97-99.

[10] U Gauderon, P B Lukins, and C J R Sheppard, Three-dimensional second-harmonic generation imaging with femtosecond laser pulses, Opt. Lett., 1998, 23(15): 1209-1211.

[11] J C Paul, C M Andrew, T Mark, et al, Three-dimensional high-resolution second-harmonic generation imaging of endogenous structural proteins in biological tissue, Biophys. J, 2002, 82(1): 493-508.

[12] I Freund and M Deutsch, Second-harmonic microscopy of biological tissue, Opt. Lett., 1986, 11(2): 94-96.

[13] R M Williams, W R Zipfel and W W Webb. Interpreting second-harmonic generation imaes of collagen I fibers, Biophys. J., 2005, 88:1377-1386.

[14] Y Barad, H Eisenberg, M Horowitz and Y Silberberg, Nonlinear scanning laser microscopy by third harmonic generation, Appl. Phys. Lett., 1997, 70(8): 922-924.

[15] D D Vbarre, W Supatto, A M Pena, et al, Imaging lipid bodies in cells and tissues using third-harmonic generation microscopy, Nat. Methods, 2006, 3: 47-53.

[16] H J Humecki, Practical Spectroscopy vol 19, ed. E G Brame Jr, New York: Dekker, 1995.

[17] G Turrell and J Corset, Raman Microscopy Development and Applications, Academic, San Diego, Calif., 1996.

[18] G J Puppels, F F M De Mul, C Otto, et al, Studying single living cells and chromosomes by confocal Raman microspectroscopy. Nature, 1990, 347: 301-303.

[19] N M Sijtsema, S D Wouters, C J De Grauw, et al, Confocal direct imaging Raman microscope: design and applications in biology, Appl. Spectrosc., 1998, 52(3): 348-355.

[20] C W Freudiger, W Min, B G Saar, et al, Label-Free Biomedical Imaging with High Sensitivity by Stimulated Raman Scattering Microscopy, Science, 2008, 322(5909): 1857-1861.

[21] M D Duncan, J Reintjes and T J Manuccia, Scanning coherent anti-Stokes Raman microscope, Opt. Lett., 1982, 7(8): 350-352.

[22] A Zumbusch, G R Holtom and X S Xie, Three-Dimensional Vibrational Imaging by Coherent Anti-Stokes Raman Scattering, Phys. Rev. Lett., 1999, 82(20): 4142-4145.

[23] M Hashimoto, T Araki and S Kawata, Molecular vibration imaging in the fingerprint region by use of coherent anti-Stokes Raman scattering microscopy with a collinear configuration, Opt. Lett., 2000, 25(24): 1768-1770.

[24] E O Potma, W P de Boeij and D A Wiersma, Nonlinear coherent four-wave mixing in optical microscopy, J. Opt. Soc. Am. B, 2000, 17(10): 1678-1684.

[25] E O Potma, W P de Boeij, P J M van Haastert, et al, Real-time visualization of intracellular hydrodynamics in single living cells, Proc. Natl. Acad. Sci. U.S.A., 2001 98(4): 1577-1582.

[26] J X Cheng, A Volkmer and X S Xie, Theoretical and experimental characterization of coherent anti-Stokes Raman scattering microscopy, J. Opt. Soc. Am. B, 2002, 19(6): 1363-1375.

[27] P D Maker and R W Terhune, Study of optical effects due to an induced polarization third order in the electric field strength, Phys. Rev., 1965, 137(3A): 801-818.

[28] R F Begley, A B Harvey and R L Byer, Coherent anti-Stokes Raman spectroscopy, Appl. Phys. Lett., 1974, 25(7): 387-390.

[29] Shen Y R, The Principles of Nonlinear Optics, New York: Wiley, 1984.

[30] M D Levenson and S S Kano, Introduction to Nonlinear Laser Spectroscopy, San Diego, CA: Academic, 1988.

[31] S Maeda, T Kamisuki and Y Adachi, Advances in Non-Linear Spectroscopy, ed. R J H Clark and R E Hester, New York: Wiley, 1988.

[32] J Nibler, Advances in Non-Linear Spectroscopy, ed. R J H Clark and R E Hester, New York: Wiley, 1988.

[33] J X Cheng, E O Potma and X S Xie, Coherent anti-Stokes Raman scattering correlation spectroscopy: Probing dynamical processes with chemical selectivity, J. Phys. Chem. A, 2002, 106(37): 8561-8568.

[34] S A Akhmanov, N I Koroteev and A I Kholodnykh, Excitation of the coherent optical
      phonons of E_g-type in calcite by means of the active spectroscopy method, J. Raman
      Spectrosc., 1974, 2(3): 239-248.
[35] C Otto, A Voroshilov, S G Kruglik, et al, Vibrational bands of luminescent zinc(II)-
      octaethylporphyrin using a polarization-sensitive 'microscopic' multiplex CARS
      technique, J. Raman Spectrosc., 2001, 32(6): 495-501.
[36] L Ujj, B L Volodin, A Popp, et al, Picosecond resonance coherent anti-Stokes Raman
      spectroscopy of bacteriorhodopsin: spectra and quantitative third-order
      susceptibility analysis of the light-adapted BR-570, Chem. Phys., 1994, 182(2-3):
      291-311.
[37] B N Toleutaev, T Tahara and H Hamaguchi, Broadband (1000 cm^{-1}) multiplex
      CARS spectroscopy: Application to polarization sensitive and time-resolved
      measurements, Appl. Phys. B, 1994, 59(4): 369-375.
[38] A Voroshilov, C Otto and J Greve, Secondary structure of bovine albumin as studied
      by polarization-sensitive multiplex CARS spectroscopy, Appl. Spectrosc., 1996,
      50(1): 78-85.
[39] P St J Russell, Photonic Crystal Fibers, Science, 2003, 299(5605): 358-362.
[40] H Kano, H Hamaguchi, Near-infrared coherent anti-Stokes Raman scattering
      microscopy using supercontinuum generated from a photonic crystal fiber, Appl.
      Phys. B, 2005, 80(2): 243-246.
[41] T W Kee, H X Zhao and M T Cicerone, One-laser interferometric broadband coherent
      anti-Stokes Raman scattering, Opt. Expr., 2006, 14(8): 3631-3640.
[42] Lee Y J, Liu Y X and M T Cicerone, Characterization of three-color CARS in a two-
      pulse broadband CARS spectrum, Opt. Lett., 2007, 32(22): 3370-3372.
[43] Lee Y J and M T Cicerone, Vibrational dephasing time imaging by time-resolved
      broadband coherent anti-Stokes Raman scattering microscopy, App. Phys. Lett.,
      2008, 92(4): 041108.
[44] Yu L Y, Yin J, Niu H B, et al, Study on the method and experiment of time-resolved
      coherent anti-Stokes Raman scattering using supercontinuum excitation, Acta
      Phys. Sin., 2010, 59(8): 5406-5411 (in Chinese).
[45] D C Harris and M D Bertolucci, Symmetry and Spectroscopy: An Introduction to
      Vibrational and Electronic Spectroscopy, Dover Publications, 1989.
[46] A Smekal, Zur Quantentheorie der Dispersion, Naturwissenschaften, 1923, 11: 873-875.
[47] C V Raman and K S Krishnan, A New Type of Secondary Radiation, Nature, 1928,
      121(3048): 501-502.
[48] C V Raman, A New Radiation, Indian Journal of Physics, 1928, 2: 387-398.
[49] G Placzek, Rayleigh-Streuung und Raman-Effekt. In: Handbuch der Radiologie. ed. E
      Marx, Akademische Verlagsgesellschaft, Leipzig, 1934.
[50] H A Kramers and W Heisenberg, Uber die streuung von strahlen durch atome, Z.
      Phys., 1925, 31: 681.
[51] R Loudon, The Quantum Theory of Light, Oxford University Press, Oxford, 1978.
[52] M O Scully and M S Zubairy, Quantum Optics, Cambridge University Press,
      Cambridge, 1996.
[53] E Garmire, F Pandarese and C T Townes, Coherently Driven Molecular Vibrations and
      Light Modulation, Phys. Rev. Lett., 1963, 11(4): 160-163.

[54] J X Cheng, A Volkmer, L D Book, and X S Xie, An epi-detected coherent anti-Stokes Raman scattering (E-CARS) microscope with high spectral resolution and high sensitivity, J. Phys. Chem. B, 2001, 105(7): 1277-1280.

[55] A Volkmer, J X Cheng and X S Xie. Vibrational imaging with high sensitivity via epi-detected coherent anti-Stokes Raman scattering microscopy, Phys. Rev. Lett., 2001, 87(2): 023901-023904.

[56] S A Ahkmanov, A F Bunkin, S G Ivanov and N I Koroteev. Coherent ellipsometry of Raman scattered light, JETP Lett., 1977, 25(9): 416-420.

[57] J L Oudar, R W Smith and Y R Shen, Polarization-sensitive coherent anti-Stokes Raman spectroscopy, Appl. Phys. Lett., 1979, 34(11): 758-760.

[58] R Brakel and F W Schneider, Polarization CARS spectroscopy In Advances in Nonlinear Spectroscopy, ed Clark RJH and Hester RE, JohnWiley & Sons Ltd., New York, 1988.

[59] D A Kleinman, Nonlinear dielectric polarization in optical media, Phys. Rev., 1962, 126(6): 1977-1979.

[60] J X Cheng, L D Book and X S Xie, Polarization coherent anti-Stokes Raman scattering microscopy, Opt. Lett., 2001, 26(17): 1341-1343.

[61] A Laubereau and W Kaiser, Vibrational dynamics of liquids and solids investigated by picosecond light pulses, Rev. Mod. Phys., 1978, 50(3): 607-665.

[62] F M Kamga and M G Sceats, Pulse-sequenced coherent anti-Stokes Raman scattering spectroscopy: a method for the suppression of the nonresonant background, Opt. Lett., 1980, 5(3): 126-127.

[63] M Fickenscher, M G Purucker and A Laubereau, Resonant vibrational dephasing invetsigated with high-precision femtosecond CARS, Chem. Phys. Lett., 1992, 191(1-2): 182-188.

[64] A Volkmer, L D Book and X S Xie, Time-resolved coherent anti-Stokes Raman scattering microscopy: imaging based on Raman free induction decay, Appl. Phys. Lett., 2002, 80(9): 1505-1507.

[65] D Oron, N Dudovich and Y Silberberg, Single-pulse phase-contrast nonlinear Raman spectroscopy, Phys. Rev. Lett., 2002, 89(27): 273001-273004.

[66] N Dudovich, D Oron and Y Silberberg, Single-pulse coherently controlled nonlinear Raman spectroscopy and microscopy, Nature, 2002, 418(8): 512-514.

[67] D Oron, N Dudovich and Y Silberberg, Femtosecond Phase-and-polarization control for background-free coherent anti-Stokes Raman spectroscopy, Phys. Rev. Lett., 2003, 90(21): 213901-213904.

[68] G Marowsky and G Luepke, CARS-background suppression by phase-controlled nonlinear interferometry, Appl. Phys. B, 1990, 51(1): 49-50.

[69] Y Yacoby and R Fitzgibbon, Coherent cancellation of background in four-wave mixing spectroscopy, J. Appl. Phys., 1980, 51(6): 3072-3077.

[70] C L Evans, E O Potma and X S Xie, Coherent anti-Stokes Raman scattering spectral interferometry: determination of the real and imaginary components of nonlinear suscepstibility for vibrational microscopy, Opt. Lett., 2004, 29(24): 2923-2925.

[71] D L Marks and S A Boppart, Nonlinear interferometric vibrational imaging, Phys. Rev. Lett., 2004, 92(12):123905.

[72] C Vinegoni, J S Bredfeldt, D L Marks and S A Boppart, Nonlinear optical contrast enhancement for optical coherence tomography, Opt. Express, 2004, 12(2):331-341.

[73] E O Potma, W P de Boeij and D A J Wiersma, Nonlinear coherent four-wave mixing in
      optical microscopy, Opt. Soc. Am. B, 2000, 17(10): 1678-1684.
[74] G C Bjorklund, Effects of focusing on third-order nonlinear processes in isotropic
      media, IEEE J. Quantum Electron., 1975, 11(6): 287-296.
[75] E O Potma, X S Xie, L Muntean, et al, Chemical Imaging of Photoresists with Coherent
      Anti-Stokes Raman Scattering (CARS) Microscopy, J. Phys. Chem. B, 2004, 108(4):
      1296-1301.
[76] J X Cheng, Y K Jia, G Zheng, and X S Xie, Laser-scanning coherent anti-Stokes Raman
      scattering microscopy and application to cell biology, Biophys. J., 2002, 83(1): 502-
      509.
[77] J X Cheng, L D Book and X S Xie, Polarization coherent anti-Stokes Raman scattering
      microscopy, Opt. Lett., 2001, 26(17): 1341-1343.
[78] A Zumbusch, G R Holtom and X S Xie, Three-dimensional vibrational imaging by
      coherent anti-Stokes Raman scattering, Phys. Rev. Lett., 1999, 82(20): 4142-4145.
[79] X L Nan, J X Cheng and X S Xie, Vibrational imaging of lipid droplets in live fibroblast
      cells with coherent anti-Stokes Raman scattering microscopy, J. Lipid. Res., 2003,
      44(11): 2202-2208.
[80] X L Nan, E O Potma and X S Xie, Nonperturbative Chemical Imaging of Organelle
      Transport in Living Cells with Coherent Anti-Stokes Raman Scattering Microscopy,
      Biophys. J., 2006, 91(2): 728-735.
[81] E O Potma, W P d Boeij, P J M van Haastert and D A Wiersma, Real-time visualization
      of intracellular hydrodynamics in single living cells, Proc. Natl. Acad. Sci. U.S.A.,
      2001, 98(4): 1577-1582.
[82] H Wang, Y Fu, P Zickmund, R Shi and J X Cheng, Coherent anti-Stokes Raman
      scattering imaging of axonal myelin in live spinal tissues, Biophys. J., 2005, 89(1):
      581-591.
[83] C L Evans, E O Potma, M Puoris'haag, et al, Chemical imaging of tissue in vivo with
      video-rate coherent anti-Stokes Raman scattering (CARS) microscopy, Proc. Natl.
      Acad. Sci. USA. 2005. 102(46): 16807-16812.
[84] Yin Jun, Yu Ling-yao, Niu Han-Ben, et al, Simultaneous measurements of global
      vibrational spectra and dephasing times of molecular vibrational modes by
      broadband time-resolved coherent anti-Stokes Raman scattering spectrography,
      Chin. Phys. B, 2011, 20(1): 014206.
[85] R R Alfano and S L Shapiro, Emission in the region 4000 to 7000 Å via four-photon
      coupling in glass, Phys. Rev. Lett., 1970, 24(11): 584-587.
[86] R R Alfano and S L Shapiro, Observation of self-phase modulation and small-scale
      filaments in crystals and glasses, Phys. Rev. Lett., 1970, 24(11): 592-594.
[87] R R Alfano, The Supercontinuum Laser Source, ed. R. Alfano, Springer, Berlin, 1989.
[88] J K Ranka, R S Windeler and A J Stentz, Visible continuum generation in air-silica
      microstructure optical fibers with anomalous dispersion at 800 nm, Opt. Lett., 2000,
      25(1): 25-27.
[89] J C Knight, Photonic crystal fibres, Nature, 2003, 424: 847-851.
[90] J C Knight and P St J Russell, New ways to guide light, Science, 2002, 296(5566): 276-
      276.
[91] T A Birks, J C Knight and P St J Russell, Endlessly single-mode photonic crystal fiber,
      Opt. Lett., 1997, 22(13): 961-963.

[92]   J C Knight, J Broeng, T A Birks, et al. Photonic band gap guidance in optical fibers, Science, 1998, 282(5393): 1476-1478.

[93]   R F Cregan, B J Mangan, J C Knight, et al. Single-mode photonic band gap guidance of light in air, Science, 1999, 285(5433): 1537-1539.

[94]   J M Dudley, G Genty and S Coen, Supercontinuum generation in photonic crystal fiber, Rev. Mod. Phys., 2006, 78 (4): 1135-1184.

[95]   G P Agrawal, Nonlinear Fiber Optics (4th ed.), Elsevier, 2007.

[96]   Yin Jun, Ph D. Thesis, 2010.

[97]   Q Cao, X Gu, E Zeek, et al, Measurement of the intensity and phase of supercontinuum from an 8-mm-long microstructure fiber, Appl. Phys. B, 2003, 77(2-3):239-244.

[98]   Liu Xing, Yin Jun, Niu Han-Ben, et al, Optimization of Supercontinuum Sources for Ultra-Broadband T-CARS Spectroscopy, Chin. Phys. Lett., 2011, 28(3): 034202.

[99]   M Muller and J M Schins, Imaging the Thermodynamic State of Lipid Membranes with Multiplex CARS Microscopy, J. Phys. Chem. B, 2002, 106(14): 3715-3723.

[100] J X Cheng, A Volkmer, L D Book and X S Xie, Multiplex Coherent Anti-Stokes Raman Scattering Microspectroscopy and Study of Lipid Vesicles, J. Phys. Chem. B, 2002, 106(34): 8493-8498.

[101] T W Kee and M T Cicerone, Simple approach to one-laser, broadband coherent anti-Stokes Raman scattering microscopy, Opt. Lett., 2004, 29(23): 2701-2703.

[102] H N Paulsen, K M Hilligsoe, J Thogersen, et al, Coherent anti-Stokes Raman scattering microscopy with a photonic crystal fiber based light source, Opt. Lett., 2003, 28(13): 1123-1125.

[103] H Kano and H Hamaguchi, Femtosecond coherent anti-Stokes Raman scattering spectroscopy using supercontinuum generated from a photonic crystal fiber, Appl. Phys. Lett., 2004, 85(19): 4298-4300.

[104] A F Pegoraro, A Ridsdale, D J Moffatt, et al, Optimally chirped multimodal CARS microscopy based on a single Ti:sapphire oscillator, Opt. Expr., 2009, 17(4): 2984-2996.

[105] S O Konorov, D A Akimov, A A Ivanov, et al, Microstructure fibers as frequency-tunable sources of ultrashort chirped pulses for coherent nonlinear spectroscopy, Appl. Phys. B., 2004, 78(5): 565-567.

[106] H A Rinia, M Bonn and M Muller, Quantitative multiplex CARS spectroscopy in congested regions, J. Phys. Chem. B., 2006, 110(9): 4427-4479.

[107] Yin Jun, Yu Lingyao, Niu Hanben, et al, Theoretical Analysis of Time-resolved Coherent Anti-Stokes Raman Scattering Method for Obtaining the Whole Raman Spectrum of Biomolecules, ACTA OPTICA SINICA, 2010, 30 (7): 2136-2141 (in Chinese).

[108] M Fickenscher, H G Purucker and A Laubereau, Resonant vibrational dephasing investigated by high-precision femtosecond CARS, Chem. Phys. Lett., 1992, 191(1-2): 182-188.

[109] H Okamoto, R Inaba, M Tasumi and K Yoshihara, Femtosecond vibrational dephasing of the CN stretching in alkanenitriles with long alkyl chains. Dependence on the chain length and hydrogen bonding, Chem. Phys. Lett., 1993, 206(1-4): 388-392.

[110] W Kiefer, A Materny and M Schmitt, Femtosecond time-resolved spectroscopy of elementary molecular dynamics, Naturwissenschaften, 2002, 89(6): 250-258.

[111] M Schmitt, G Knopp, A Materny and W Kiefer, Femtosecond time-resolved coherent
       anti-Stokes Raman scattering for the simultaneous study of ultrafast ground and
       excited state dynamics: iodine vapour, Chem. Phys. Lett., 1997, 270(1-2): 9-15.
[112] T Joo and A C Albrecht, Femtosecond time-resolved coherent anti-Stokes Raman
       spectroscopy of liquid benzene: A Kubo relaxation function analysis, J. Chem.
       Phys., 1993, 99(5): 3244-3251.
[113] J C Kirkwood, D J Ulness and A C Albrecht, On the mechanism of vibrational
       dephasing in liquid benzene by coherent anti-Stokes Raman scattering using
       incoherent light, Chem. Phys. Lett., 1998, 293(3-4): 167-172.
[114] A Fendt, S F Fisher and W Kaiser, Vibrational lifetime and Fermi resonance in
       polyatomic molecules, Chem. Phys., 1981, 57(1-2): 55-64.
[115] D Pestov, M Zhi, Z E Sariyanni, et al, Femtosecond CARS of methanol-water
       mixtures, J. Raman Spectrosc., 2006, 37(1-3): 392-396.
[116] Y Huang, A Dogariu, Y Avitzour, et al, Discrimination of dipicolinic acid and its
       interferents by femtosecond coherent Raman spectroscopy, J. Appl. Phys., 2006,
       100(12): 124912.
[117] M Fickenscher and A Laubereau. High-precision femtosecond CARS of simple
       liquids. J. Raman Spectrosc., 1990, 21(12): 857-861.
[118] G Beadie, M Bashkansky, J Reintjes and M O Scully, Towards a FAST-CARS anthrax
       detector: Analysis of cars generation from DPA, J. Mod. Opt., 2004, 51: 2627-2635.
[119] S Mishra, R K Singh and A K Ojha, Investigation on bonding interaction of
       benzonitrile with silver nano particles probed by surface enhanced Raman
       scattering and quantum chemical calculations, Chem. Phys., 2009, 355(1): 14-20.
[120] J X Cheng and X S Xie, Coherent Anti-Stokes Raman Scattering Microscopy:
       Instrumentation, Theory, and Applications, J. Phys. Chem. B, 2004, 108(3): 827-840.
[121] R Inaba, K Tominaga, M Tasumi, et al, Observation of homogeneous vibrational
       dephasing in benzonitrile by ultrafast Raman echoes, Chem. Phys. Lett., 1993,
       211(2-3): 183-188.
[122] H Okamoto, R Inaba, K Yoshihara and M Tasumi, Femtosecond time-resolved
       polarized coherent anti-Stokes Raman studies on reorientational relaxation in
       benzonitrile, Chem. Phys. Lett., 1993, 202(1-2): 161-166.
[123] Y J Lee, S H Parekh, Y H Kim and M T Cicerone, Optimized continuum from a
       photonic crystal fiber for broadband time-resolved coherent anti-Stokes Raman
       scattering, Opt. Express, 2010, 18(5): 4371-4379.
[124] E Abbe, Beitraezur Theorie des Mikroskops und der mikroskopischen
       Wahrnehmung, Arch. f. Mikr. Anat., 1873, 9(1): 413-418.
[125] E Betzig, G H Patterson, R. Sougrat, et al., Imaging Intracellular Fluorescent Proteins
       at Nanometer Resolution, Science, 2006, 313(5793): 1642-1645.
[126] M J Rust, M Bates and X Zhuang, Sub-diffraction-limit imaging by stochastic optical
       reconstruction microscopy (STORM), Nat. Methods, 2006, 3: 793-796.
[127] S W Hell and J Wichmann, Breaking the diffraction resolution limit by stimulated
       emission: stimulated-emission-depletion fluorescence microscopy, Opt. Lett., 1994,
       19(11): 780-782.
[128] S W Hell, Far-Field Optical Nanoscopy, Science, 2007, 316(5828): 1153-1158.
[129] W P Beeker, P Groß, C J Lee, et al., A route to sub-diffraction-limited CARS
       Microscopy, Opt. Express, 2009, 17(25): 22632-22638.

[130] V V Krishnamachari and E O Potma, Interferometric switching of coherent anti-Stokes Raman scattering signals in microscopy, Phys. Rev. A, 2009, 79(1): 013823.

[131] K M Hajek, B Littleton, D Turk, et al., A method for achieving super-resolved widefield CARS microscopy, Opt. Express, 2010, 18(18): 19263-19272.

[132] W Liu and H B Niu, Diffraction barrier breakthrough in coherent anti-Stokes Raman scattering microscopy by additional probe-beam induced phonon depletion, Phys. Rev. A, 2011, 83(2): 023830.

Part 3

Photonic Crystal Waveguides and Plasmonics

11

Label-Free Biosensing Using Photonic Crystal Waveguides

Jaime García-Rupérez, Veronica Toccafondo and Javier García Castelló
Nanophotonics Technology Center, Universidad Politécnica de Valencia
Spain

1. Introduction

The development of fast, efficient, and reliable sensing devices to be used for the detection, identification, and quantification of substances and biological material is one of the main current investigation fields. Among the substances and analytes to be detected we can find gases, liquids, proteins, hormones, bacteria, or DNA. These sensing devices can find applications in many fields, such as medical diagnostics, food safety control, environmental control, or drug detection.

Most of the currently available sensing devices base their detection on labelling the target analytes because of the difficulty to directly detect very small size analytes in a low concentration. By performing a proper treatment of the sample in which the target analyte is contained, it is possible to attach a label to it. The label consists of a material with certain physical properties such as fluorescence, radioactivity, metallic material, etc. so that the analyte detection is indirectly carried out by detecting the physical properties of the label that has been attached to it. However, although analyte labelling is an effective way to detect small size analytes in a small concentration, it requires previous sample preparation which needs to be performed by skilled personnel and is time consuming. Moreover labelling presents the difficulty of finding a proper method to specifically attach the labels only to the target analyte to be detected. It is therefore important to have more sensitive sensing devices or mechanisms at our disposal, which allow performing the detection without the need of labels.

One of the main candidates for the development of these highly sensitive sensing devices able to perform a label-free detection is photonic technology (Fan et al., 2008), and more specifically, integrated planar photonic devices, which have been attracting an increasing interest in the last few years. Although more mature photonic sensing technologies, such as those based on surface plasmon resonances (Homola et al., 1999) or fiber Bragg gratings (Kersey et al., 1997), have even become commercially available some years ago, devices based on integrated planar photonic structures are envisaged as a highly promising alternative for future lab-on-a-chip (LoC) devices.

First demonstrations of the possibility of using these devices for sensing applications were reported by the end of the past century, as for example in (Luff et al., 1996) and (Schubert et al., 1997), but it has not been until 5-6 years ago that many researchers have focused their investigations on the application of already developed integrated planar photonic devices for biosensing purposes. The transduction principle of these devices is based on the high dependence of their response to changes in the refractive index (RI) of the surrounding medium. Therefore, when the target analyte is deposited on the top (or generally, the surroundings) of the photonic structure, it induces a RI variation which can be directly detected by means of the change in the device's response.

The two main advantages of integrated planar photonic devices for sensing applications are their high sensitivity due to the high confinement of the electromagnetic field in the photonic structure which enhances the interaction with the target analyte, and their reduced size, which makes it possible both to detect very small analytes and to integrate many of these devices on a single chip to perform a multi-analyte detection. Other advantages of integrated photonic sensing devices are derived from these two, such as a short analysis time, the possibility to perform a label-free detection avoiding a costly previous sample preparation procedure (in terms of time and money), or the requirement of very low volumes of analyte and reagents to perform the analysis. Moreover, if these integrated planar photonic devices are based on CMOS-compatible materials such as Silicon-On-Insulator (SOI), the possibility of using mass manufacturing techniques for the fabrication opens the door to a low cost and high volume production of these devices.

Many different integrated planar photonic structures are widely used for sensing applications, among which resonant structures, Mach-Zehnder Interferometers (MZIs), and photonic crystal structures are the most common. Results reported using these structures range from basic characterization of the sensing device to refractive index variations to more complex bio-sensing experiments where the presence of a specific analyte such as proteins, bacteria or even DNA strands is detected and quantified without the need of any label.

The most popular integrated planar photonic sensing devices are probably those based on resonant structures such as rings or disks, which are coupled to an access waveguide (De Vos et al., 2007, 2009; Barrios et al., 2008; Ramachandran et al., 2008; Carlborg et al., 2010; Iqbal et al., 2010; Xu et al., 2010a). In these structures, variations on the surrounding media provoke a spectral shift of the transmission resonances, which is used to perform the sensing. These kind of sensing structures exhibit high sensitivity and low detection limit values, with a small footprint on the order of a few hundreds of μm^2.

Other groups work on the development of photonic biosensors using Mach-Zehnder Interferometers (MZIs) (Schubert et al., 1997; Sepulveda et al., 2006; Densmore et al., 2009), where changes in the refractive index in the proximity of one of the arms provoke an additional phase shift over the propagated wave which is translated into an amplitude variation at the output. Although these sensing structures exhibit performances similar to or even better than resonant structures, their main problem is that very long arm lengths (of some millimetres) are required in order to provide enough interaction to detect the presence of the analyte, thus limiting their integration level. Nevertheless, this issue has been greatly overcome by folding the arms of the MZI, thus significantly reducing the final footprint of the device (Densmore et al., 2009).

The third main alternative for the development of integrated planar photonic biosensors are photonic crystal structures, on which this chapter will focus. Several configurations are typically used when developing photonic crystal based sensing structures: bulk photonic crystals (Huang et al., 2009), photonic crystal waveguides (Skivesen et al., 2007; Buswell et al., 2008; Di Falco et al., 2009; García-Rupérez et al., 2010; Toccafondo et al., 2010; Scullion et al., 2011), and photonic crystal cavities (Loncar et al., 2003; Chow et al., 2004; Lee & Fauchet, 2007; Sünner et al., 2008; Dorfner et al., 2009; Zlatanovic et al., 2009; Xu et al., 2010b). For bulk photonic crystal sensors, the periodic dielectric structure does not present any defects and it is directly used to perform the sensing. For these structures, a variation of the refractive index of the surrounding medium provokes a shift of the position of the photonic bandgap (PBG) edge, which is used to perform the sensing. However, when using bulk photonic crystals, light is not well localized on the in-plane direction of the structure, so that configurations where linear or punctual defects are introduced on the bulk structure are usually preferred in order to enhance the localization of the optical field in an active region well below 100 μm^2 and thus increase the interaction with the target analyte.

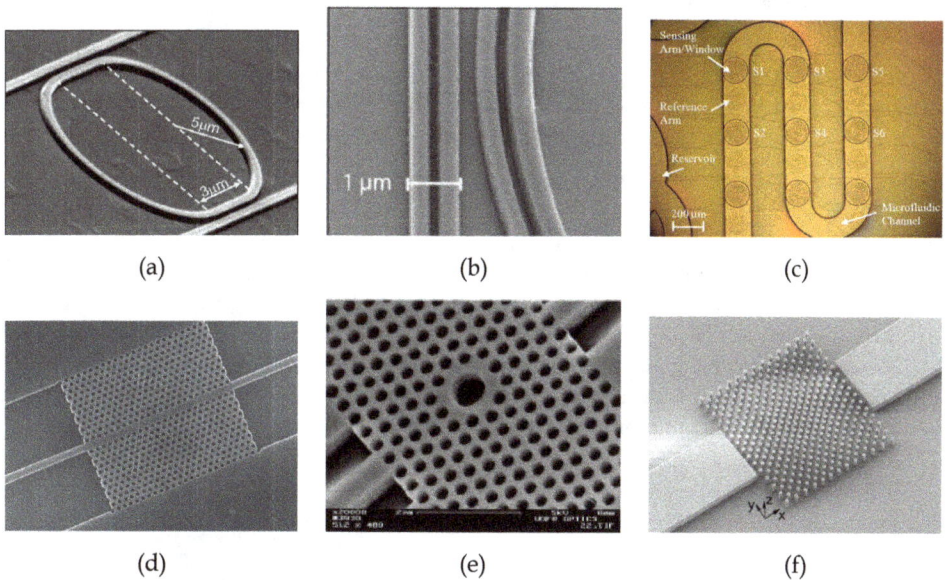

Fig. 1. SEM (Scanning Electron Microscope) pictures of several integrated planar photonic structures for sensing purposes. (a) SOI ring resonator (De Vos et al., 2007); (b) coupling region of a silicon nitride slot ring resonator (Carlborg et al., 2010); (c) array of SOI MZIs with folded arms (Densmore et al., 2009); (d) SOI photonic crystal waveguide (Buswell et al., 2008); (e) holes-based SOI photonic crystal cavity (Lee & Fauchet, 2007); (f) pillar based silicon photonic crystal cavity (Xu et al., 2010b).

For photonic crystal waveguide (PCW) based sensors, a linear defect is introduced in the propagation direction of the optical wave. In this way, light is localized in the region of the linear defect (and its surroundings), leading to an increased interaction with the target analyte. When the linear defect is created within the bulk photonic crystal, a guided mode appears inside the PBG of the photonic crystal structure. The position of the guided band edge (either lower band edge or upper band edge) is used for sensing, as it is dependent on the refractive index variations of the surrounding media.

For photonic crystal cavity based sensors, a punctual or area defect is created in the bulk photonic crystal, leading to the formation of a resonator with a strong localization of the optical field within this region. This can be achieved for instance by changing the dimensions of one or several elements of the periodic structure (see Figs. 1.(e) and 1.(f)), or by eliminating then. The coupling of light to this defect is usually achieved in two different manners:

- The defect is created in the bulk photonic crystal, and light is directly coupled to/from it through the access dielectric waveguides. In this configuration, a resonant mode appears inside the PBG of the photonic crystal structure, with a central wavelength determined by the defect profile.
- The defect is created next to the linear defect of a photonic crystal waveguide. In this configuration, light matching the resonance wavelength of the cavity is extracted from the PCW, leading to the appearance of a notch-like response in the transmission spectrum.

For both cases, the resonance position depends on the refractive index of the medium surrounding the cavity region, leading to a very high sensitivity to variations produced in a very small active region, which can even be smaller than 1 μm^2.

Finally, it is also worth noting that planar 1D periodic waveguides or cavities, such as the ones shown in Fig. 2, which are not usually referred to as photonic crystal structures although they are based in the same principles as them, are also a very promising alternative for the development of integrated planar photonic biosensors (García et al., 2008; Kauppinen et al., 2009; Goddard et al., 2010; Castelló et al., 2011). When using a planar 1D periodic structure, the position of its guided bands or resonant modes is also highly dependant on the refractive index of the surrounding media, so shifts in their position can be used to perform the sensing. Moreover, these structures can be simpler to fabricate and have a smaller footprint than 2D photonic crystal based structures.

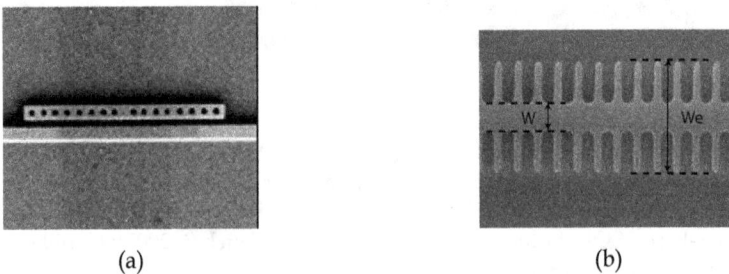

(a)   (b)

Fig. 2. SEM pictures of two planar 1D periodic structures used for sensing purposes. (a) 1D SOI holes cavity (Goddard et al., 2010); (b) 1D SOI corrugated waveguide (Castelló et al., 2011).

The next sections of this chapter will focus on the use of photonic crystal waveguides for the development of label-free photonic biosensors, as they are very promising thanks to their high sensitivity and small footprint. However, the performance of these sensors may be limited by the difficulty to accurately determine the position of the edge of the guided band of the PCW as this feature is not sharp enough to provide very low uncertainty. This problem can be overcome by using the sharp Fabry-Perot fringes appearing in the slow-light regime near the edge of the guided band, whose position can be determined very accurately, as will be explained in section 2. Moreover, their use ensures that we are working in the slow-light regime of the PCW, which allows a stronger interaction of the optical field with the target substance. In sections 3, 4 and 5, the application of PCW-based biosensors for the detection of refractive index variations, label-free antibody detection, and label-free DNA detection will be shown. Finally, section 6 will describe an alternative method for the indirect measurement of the shift of the guided band's edge, which will allow having a low-cost and real-time sensing system.

2. Photonic crystal waveguide sensor design and sensing principle

Planar PCWs are one of the preferred photonic structures for the development of biosensing devices because of their reduced size and the high confinement of the optical field in the linear defect region. In this text, we will consider planar SOI PCWs, due to the compatibility of this technological platform with CMOS fabrication techniques.

When designing a PCW for biosensing there are several factors to be taken into account. We will focus on five of them:

- First, we have to select between a pillars-on-air or a holes-on-dielectric photonic crystal structure. For pillars-on-air photonic crystals (Xu et al., 2010b), we can have a higher interaction with the target analyte flowing through the pillars; however, light coupling to these structures is less efficient, and their fabrication process can be more complex. On the other hand, for holes-on-dielectric photonic crystals (Buswell et al., 2008), the fabrication process is easier and the light coupling is more efficient, but a weaker interaction with the target analyte is produced. The photonic crystal configuration will also determine the working polarization, as pillars-on-air photonic crystals present PBGs for TM-polarization and holes-on-dielectric photonic crystals present PBGs for TE-polarization.
- We have to select the working wavelength, which is usually determined by the available readout system and light sources, so that typically used wavelengths are those from telecom applications: around 1310 nm and around 1550 nm. A working wavelength around 1310 nm allows having lower water absorption losses, however this is not really a critical point because power is not a major issue for integrated photonic biosensors. On the other hand, working at higher wavelengths, such as 1550 nm, makes the evanescent field wider and thus provides a higher sensitivity.
- The thickness of the patterned silicon layer will also affect the sensitivity of the device; the thinner the silicon layer, the higher the sensitivity because of the larger amount of evanescent field traveling through the upper cladding and interacting with the target analytes. This parameter is usually set by the SOI substrates used to fabricate the devices.
- The PCW configuration will also affect the sensitivity of the sensing device. Again, the configuration with a higher amount of evanescent field traveling through the cladding will provide a higher sensitivity due to an enhanced interaction between the optical field and the target analyte.

- Finally, the interface between the PCW and the access waveguide is also something to be taken into account. As previously commented, the shift of the guided band's edge is used in PCW to perform the sensing. Therefore, it is important to have a PCW with a sharp band edge in order to be able to accurately determine its position.

PCWs are designed by means of simulations to determine their optimal parameters and match the previously commented criteria. The most widely used simulation techniques for the design of photonic crystals are FDTD (Finite-Differences Time-Domain), which is used to calculate the propagation of the electromagnetic wave through the structure, and the plane-wave expansion (PWE) method, which is used to determine the band diagram of the perfect infinite photonic crystal structure.

Now the design and experimental characterization of a PCW sample for biosensing will be described. We will consider a PCW in a holes-on-dielectric photonic crystal fabricated on a SOI wafer with a 250nm-thick silicon layer on a 3μm-thick buried oxide layer. The PCW will be a W1-type, where one row of holes is removed in the Γ-K direction to create the waveguide. The PCW was designed to have a working wavelength located around $\lambda = 1550$ nm.

By PWE simulations, the lattice constant and the hole radius which will yield a guided band's edge located around 1550 nm can be determined. In case we decide to have the upper edge (in terms of wavelength) of the guided band around this wavelength, these parameters must be 390 nm and 110 nm, respectively. Fig. 3 shows the band diagram for TE-modes calculated for the holes-on-silicon W1-PCW using the PWE method, where the fundamental mode of the PCW with its upper edge located around 1550 nm can be seen.

The next step is the fabrication of the PCW on a SOI substrate. For the fabrication of holes-on-dielectric photonic crystal structures, the lithographic process is usually carried out using a positive resist such as PMMA, where the holes area and the trenches defining the structure need to be exposed by e-beam or DUV lithography. After developing the resist, the pattern is transferred to the silicon layer by an etching process such as inductively coupled plasma (ICP) etching.

Fig. 4 shows a scanning electron microscopy (SEM) picture of a W1-PCW fabricated to be used for biosensing. The structural parameters of this PCW were determined by the theoretical simulations, and its length is 20 μm (≈ 52 periods), which provide enough periods for achieving a strong photonic bandgap effect while keeping the structure size as compact as possible. For coupling, 450nm-wide single-mode access waveguides are used to couple/collect light to/from the PCW. The interface between the PCW and the single-mode access waveguides is the one shown in Fig. 4, where the structure ends next to the inner row of holes of the PCW. This interface presents a sharper band edge than other interfaces.

For the optical characterization of the PCW, which is essentially determined by its transmission spectrum, light from an external source must be coupled to the chip. The two main coupling mechanisms used for chip's characterization are butt coupling, where light from a lensed fiber is laterally coupled to the chip, and grating couplers, where light is vertically coupled to the chip using grated structures. For the characterization of these PCW sensors, light from a tunable laser is TE-polarized using a polarization controller and butt-coupled into the PCW using a lensed fiber. Output light is then collected using an objective and detected using an optical powermeter after passing through a free-space polarizer configured for TE polarization. Since the fabricated PCWs are designed for biosensing applications, and these require the use of the sensors in a wet environment, the design was optimized for these working conditions. Fig. 5 thus shows the transmission spectrum of one of the fabricated PCWs when deionized water (DIW, $n = 1.3173$) is used as upper cladding. Sharp peaks can be observed at the edge of the guided band (located around 1563 nm).

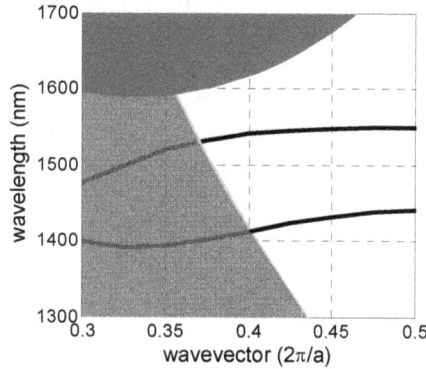

Fig. 3. Band diagram for TE modes of the SOI W1-PCW with a silicon thickness of 250 nm, a hole radius of 110 nm and a lattice constant of 390 nm. Dark and light shaded areas depict modes going into the bulk photonic crystal and the silicon oxide lower cladding, respectively.

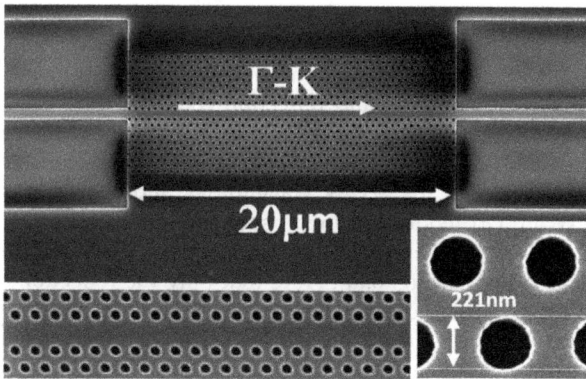

Fig. 4. SEM image of the SOI photonic crystal waveguide used for the sensing, with close-up view of the sensor area and photonic crystal holes (insets)

Fringes appearing near the band edge of the transmission spectrum of the PCW are Fabry-Perot (FP) fringes of the cavity defined by the interface between the PCW and the access waveguides (García et al., 2006), as schematically depicted in Fig. 6. A good power coupling between modes in the access waveguides and the PCW is achieved for wavelengths in the transmission band, making the FP cavity effect almost negligible. However, mode mismatching increases between the two waveguides as we get closer to the edge of the guided band, thus reducing the coupling efficiency and increasing the reflected power, so higher amplitude FP fringes begin to appear. Not only is there an increase of the reflection coefficient at the interfaces (and thus a reduction of the transmission coefficient), but also a reduction of the group velocity of the guided mode of the PCW as we get closer to the edge of the Brillouin zone. The reduction of the group velocity makes the optical length of the FP cavity longer, thus increasing the frequency of the FP fringes of the transmission spectrum in the region of the band edge. The main point of using these fringes to perform the biosensing is that we are working in the slow-light regime of the PCW, so we will have a higher interaction of the electromagnetic field with the target analyte.

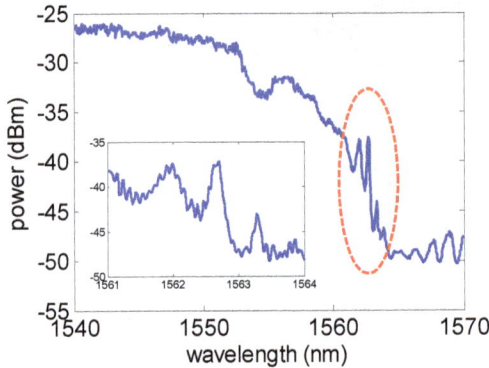

Fig. 5. Spectrum of the PCW in the region of the band edge when having DIW as upper cladding. Transmission fringes at the band edge are marked with dashed red line and enlarged in the inset.

From expressions shown in Fig. 6.(b), it can be seen that the transmission spectrum is dependent on both the transmission and reflection coefficients at the interfaces and the propagation constant of the guided mode in the PCW, which determines the optical length of the structure. Fig. 7.(a) shows these three parameters (transmission and reflection coefficients between the access waveguides and the PCW, and propagation constant of the PCW guided mode) obtained using the numerical tool CAMFR, which is based in the eigenmode expansion (EME) method, for the PCW with interface with the access waveguides as shown in Fig. 6.(a). Note that only the reflection coefficient for the incidence from the PCW to the access waveguide (r_{21}, r_{23}) is needed for the calculations and that the transmission coefficient is the same for the incidence from the PCW to the access waveguide and vice versa due to reciprocity properties ($t_{ij} = t_{ji}$). Although this software does not allow 3D calculations, a generic 2D modelling (i.e., infinite height of the structure) using an effective refractive index of 2.8 for the silicon slab and 1.33 for the cladding surrounding the structure is useful to realistically predict the appearance of FP fringes in the transmission spectrum. It can be seen that the transmission coefficient decreases to zero as we get closer to the band edge (the opposite is true for the reflection coefficient, which tends to one), thus the amount of power coupled into the PCW decreases and a stronger cavity effect is created. Concerning the propagation constant, it can be seen that it gets flatter as the edge of the guided band is approached, which means a reduction on the group velocity (which tends to zero) and makes the fringes narrower and closer between them. This reduction in the group velocity will also provoke a higher interaction of the optical field with the target analytes.

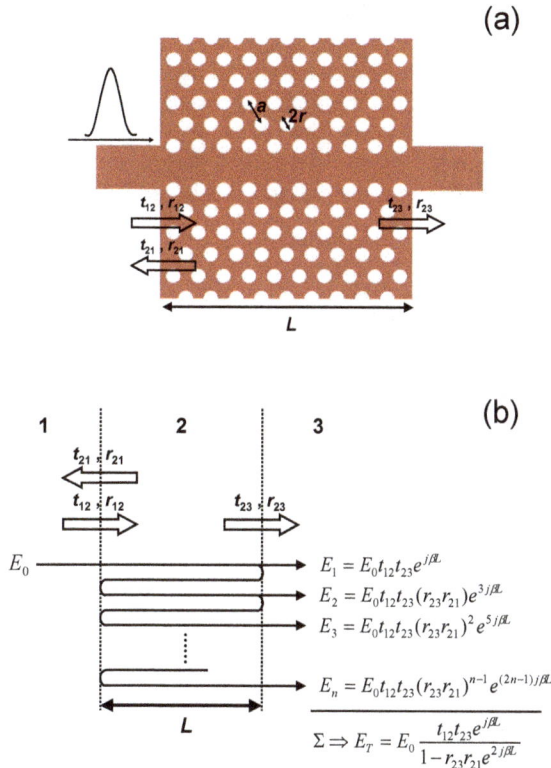

Fig. 6. (a) Schematic picture of the PCW with its access waveguides. The interface with the access waveguides is that of the fabricated PCW. The structural parameters of the PCW (i.e., lattice constant, hole radius and PCW length), and the transmission and reflection coefficients at each interface (t_{ij}, r_{ij}) are depicted. (b) All signal contributions generated in the PCW because of the reflections at the interfaces with the access waveguides are combined at the output, being responsible of the appearance of FP oscillations in the transmission spectrum. Media 1 and 3 represent the access waveguides and medium 2 represents the PCW.

When these three parameters are combined using the expression shown in Fig. 6.(b), the transmission response shown in Fig. 7.(b) for a 20µm-long 2D-PCW is obtained. One can see how FP resonances appear because of the cavity created inside the PCW, and how these FP fringes get stronger and narrower as we approach the edge of the guided band because of the increase of the reflection coefficient and the decrease of the group velocity.

To corroborate the appearance of these FP fringes in the edge of the guided mode of the PCW and analyze the potential for using them for biosensing purposes, 3D-FDTD simulations of the PCW with the parameters of the fabricated structure and two different upper claddings (n_1 = 1.3173 and n_2 = 1.3200; corresponding to the refractive indices of DIW and a ethanol-DIW 4% dilution) can be performed (in this case, RSoft's FullWAVE software was used). Simulation results for each upper cladding are shown in Fig. 8, where the FP fringes at the band edge are observed again and where the position of the guided band is dependent of the refractive index of the upper cladding. Table 1 shows the wavelength shift and the calculated sensitivity in terms of refractive index units (RIU) for each fringe near the band edge. As predicted from the previous theoretical modeling, an increase in the sensitivity is obtained as we get closer to the guided band edge because of the reduction in the group velocity of the guided mode, obtaining a sensitivity ~40% higher for the fringe closest to the band edge compared to the first of the selected fringes (78.70 nm/RIU vs. 57.15 nm/RIU).

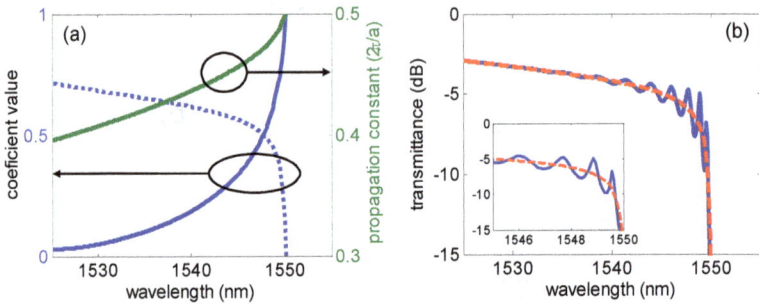

Fig. 7. (a) Theoretical transmission (blue dotted line) and reflection coefficients (blue solid line), and propagation constant of the PCW (green solid line) calculated with CAMFR. (b) Transmission response of the 20µm-long PCW when the created FP cavity is considered (blue solid line). The transmission response when no cavity is considered (only the direct contribution propagating into the PCW) is also depicted (red dashed line). The inset shows a detail of the transmission band edge.

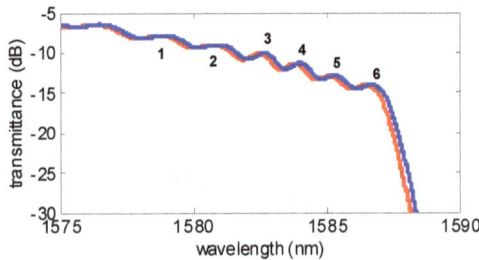

Fig. 8. 3D-FDTD simulations of the transmission spectrum of the 20µm-long PCW for two different upper claddings: n_1 = 1.3173 (red line) and n_2 = 1.3200 (blue line). FP fringes near the band edge have been labeled from 1 to 6. The band edge fringes are slightly smoothed because of the stop time used to make the duration of the simulation reasonable.

Fringe #	1	2	3	4	5	6
Wavelength shift (nm)	0.1543	0.1865	0.192	0.2002	0.2039	0.2125
Sensitivity (nm/RIU)	57.15	69.07	71.11	74.15	75.52	78.70

Table 1. Wavelength shift and sensitivity for each FP fringe near the guided band edge for the 3D-FDTD simulations.

3. Refractive index sensing using photonic crystal waveguides

A straight-forward way to to test the possibility of using these band edge fringes to perform biosensing, is to carry out a simple refractive index sensing experiment, for instance using several dilutions of ethanol in DIW. Dilution concentrations are (in mass %): pure DIW, ethanol 2% in DIW, and ethanol 4% in DIW, whose refractive indices at $\lambda \approx 1550$ nm and T = 25°C are 1.3173, 1.3186, and 1.3200, respectively (García-Rupérez et al., 2010).

For carrying out the RI sensing experiments, a flow cell is required in order to flow the target substances over the chip. In this case, a 2-port flow cell with a fluidic cavity of size 5.5mm x 2mm x 0.5mm (length x width x depth) is placed on the top of the chip. For pumping the liquid passing through the flow cell, an automatic syringe pump in withdrawal mode connected to one of the ports of the flow cell using silicone tubing is used. The liquid is flowed at a rate of 15 µl/min. Tubing from the second port of the flow cell is placed into a vial with the liquid to be flowed over the chip. This configuration is used in order to avoid having to replace the syringe to change the liquid to be flowed: with this configuration, the liquid is drawn from vials, enabling an easier handling of the tubes when manual changing between them is performed. The TE transmission spectrum of the PCW in the vicinity of the guided band edge (shown in Fig. 5) is continuously acquired using a tunable laser with a sweep resolution of 10 pm, and a cubic interpolation is used to increase the wavelength accuracy on the determination of the position of the peak's maximum. Fig. 9 shows the temporal evolution of the position of the maximum of the FP peak located around 1563.3 nm for the different ethanol-DIW dilutions flowed.

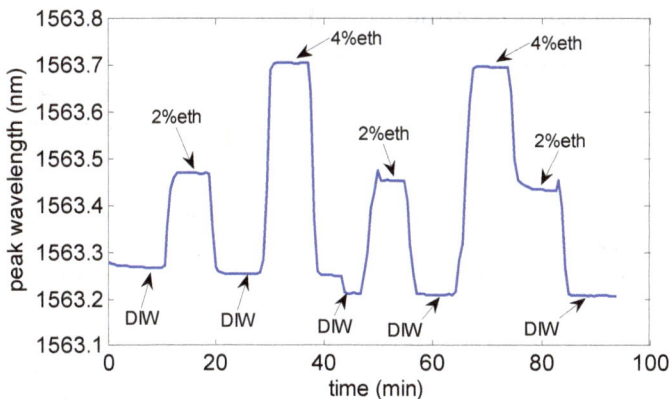

Fig. 9. Temporal evolution of the FP peak for the different ethanol-DIW dilutions flowed.

Fig. 10 shows the relative peak shift for the three ethanol-DIW concentrations used in the experiments, where a linear behavior is observed, with a sensitivity of 118 pm/% calculated with respect to the ethanol concentration. Considering the almost linear RI variation for this ethanol concentration range ($\Delta n \approx 6.75 \times 10^{-4}$ /%), a sensitivity to RI variations of $S = \Delta\lambda/\Delta n = 174.8$ nm/RIU is obtained. This value is 2.2x the value obtained in 3D-FDTD simulations, and this difference can be attributed to the discretization step of the PCW used for the FDTD simulations, which may have lead to a non accurate modelling of the optical field which senses the variation of the refractive index of the cladding. The reason for this high sensitivity value is also the higher interaction with the target analyte due to the reduction of the group velocity in the band edge, where the FP fringes used to sense are located.

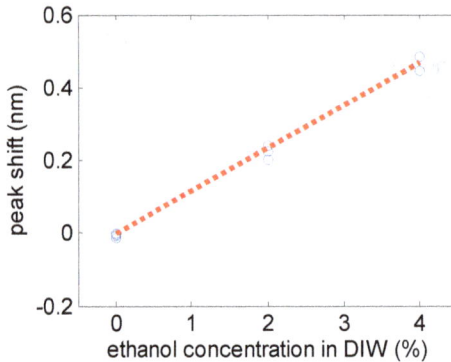

Fig. 10. Blue circles depict the wavelength shift of the position of the maximum of the peak for the different ethanol-DIW dilutions. The linear fit of the data is depicted with dashed red line.

We can then use the standard deviation σ of the peak position for a continuous flow of an ethanol-DIW dilution, which is 0.6 pm, as the noise level, and use it to calculate the theoretical detection limit DL, which is defined as $DL = \sigma/S$. A value as low as $DL = 3.5 \times 10^{-6}$ RIU is obtained.

4. Antibody sensing using photonic crystal waveguides

Once the origin of band edges fringes is modeled and the possibility of using them for sensing purposes is checked by performing RI sensing experiments, they are used for label-free antibody sensing. A different PCW sample (with the same nominal parameters) from another fabrication run is used for this. In order to give specificity to the antibody sensing, the PCW sensing device needs to be bio-functionalized with proper antigen probes which will act as receptors for the target analyte. In this case, anti-BSA (bovine serum albumin) antibody is used for the sensing experiments, so BSA antigen probes need to be immobilized on the surface of the photonic sensor. This bio-functionalization process consists in the activation of the surface of the chip with pure ICPTS (3-isocyanatepropyl triethoxysilane) vapour, the deposition and incubation of BSA antigen 10 µg/ml in PBS 0.1x, and a final blocking step with ovoalbumin protein (OVA) 1% in PBS 0.1x (García-Rupérez et al., 2010). The flow cell and the tubing used for the experiments are also blocked with OVA to avoid the absorption of the flowed molecules.

The transmission spectrum for the TE polarization is obtained for the bio-functionalized PCW when having PBS 0.1x as upper cladding, and is shown in Fig. 11, where the band edge is now located around 1542 nm. A closer look at the spectral area close to the band edge is given in Fig. 11.(b), where sharp peaks appearing in this region are once again observed.

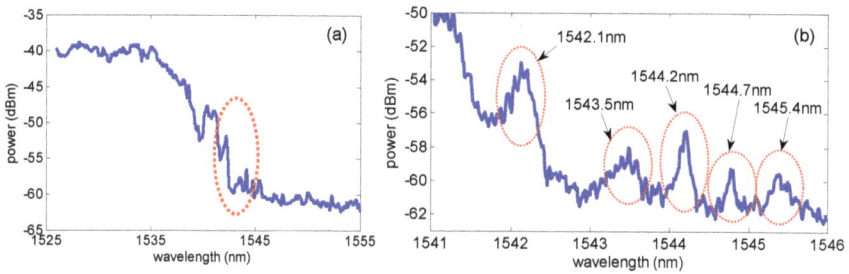

Fig. 11. (a) Spectrum of the PCW in the region of the band edge when PBS 0.1x is flowed. (b) More detailed view of the transmission spectrum close to the band edge. Transmission peaks in this region are marked with dashed red circles and their approximated wavelength positions are depicted.

Fig. 12 shows the temporal evolution of the peak shift for the first four peaks depicted in Fig. 11.(b) (peaks at 1542.1nm, 1543.5nm, 1544.2nm and 1544.7nm) during the whole experiment. The initial baseline is obtained by continuously flowing the PBS 0.1x buffer. Once a stable baseline is achieved, anti-BSA antibody at a concentration of 10 µg/ml in PBS 0.1x solution is flowed over the chip, which will bind to the BSA probes attached to the PCW surface thus inducing a shift in the peak position. The anti-BSA is flowed long enough to achieve a monolayer on the top of the BSA-functionalized chip, as indicated by the saturation in the shift of the PCW response. Then, the flow is switched back to the PBS 0.1x buffer to remove any anti-BSA which has not specifically bound to probes on the surface of the PCW, thus only the net shift due to the binding of the anti-BSA to the BSA probes is obtained. Finally, a control step is done by flowing an anti-digoxigenin (anti-DIG) antibody 15 µg/ml in PBS 0.1x dilution to check that the shift previously obtained for the anti-BSA antibody flow is only due to a specific binding to the BSA probes and not due to absorption or any other mechanism. Because of the low affinity between the anti-DIG and the BSA probes, a very slight peak shift is observed during this flow. Later, PBS 0.1x is flowed again to finish the experiment.

One can see from Fig. 12 that the temporal evolution is almost the same for all the tracked peaks, reaching a plateau for the anti-BSA flow, indicating the formation of a monoloayer. Concerning the total shift when the anti-BSA is flowed, it is slightly different for each peak, as shown in Table 2. The shift is higher as we move to peaks closer to the band edge (i.e., from peak #1 to peak #4). This is due to the reduction of the group velocity of the guided mode and was already observed in the 3D-FDTD simulations for RI sensing. However, peak #3 does not follow this trend and shows a wavelength shift smaller than peak #2 (this fact will be commented later).

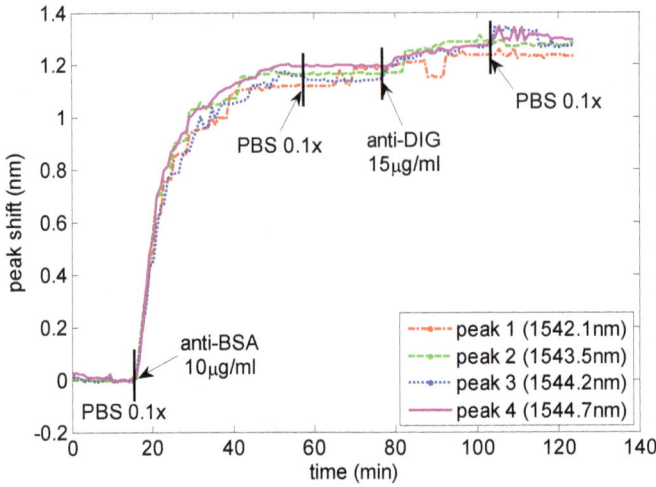

Fig. 12. Wavelength shift vs time for the different solutions flowed in the experiment. Each line (color and style identified in the legend) correspond to the relative shift of each tracked peak respect its initial wavelength position. The time instants when the flowed solution is switched are depicted in the figure.

The surface density for a close-packed anti-BSA monolayer when considering a 100% binding efficiency is $\rho_{anti\text{-}BSA}$ = 1.7 ng/mm^2, which is calculated from the molecular mass and the size of the anti-BSA molecule (as described in (Barrios et al., 2008)), and will give us an upper limit for the detection limit of the device (in the real situation there is no close-pack monolayer as binding efficiency is below 100%). The sensitivity for the anti-BSA detection is given by $S_{anti\text{-}BSA} = \Delta\lambda_{anti\text{-}BSA}/\rho_{anti\text{-}BSA}$; calculated values for each peak are shown in Table 2.

Peak	$\lambda_{initial}$ (nm)	$\Delta\lambda_{anti\text{-}BSA}$ (nm)	$S_{anti\text{-}BSA}$ (nm/ng/mm^2)	σ_{PBS} (pm)	$DL_{anti\text{-}BSA}$ (pg/mm^2)	$DL_{anti\text{-}BSA}$ (fg)	$\Delta\lambda_{anti\text{-}DIG}$ (nm)
#1	1542.128	1.123	0.661	1.5	2.3	0.23	0.114
#2	1543.471	1.17	0.688	1.5	2.2	0.22	0.108
#3	1544.235	1.142	0.672	3.1	4.6	0.46	0.128
#4	1544.75	1.2	0.706	1.5	2.1	0.21	0.1

Table 2. Parameters characterizing the performance of the FP resonances used for the sensing experiments.

The noise level values obtained in the tracking of each peak, which correspond to the standard deviation of the peak position (σ) for the stable PBS 0.1x cycle flowed after the anti-BSA, are shown in Table 2. It can be seen that the noise level is very low (σ = 1.5pm), except for peak #3, for which the noise level is twice this value (σ = 3.1pm). With these noise values and the sensitivities previously calculated for each peak, the surface mass density detection limits (given by $DL_{anti\text{-}BSA} = \sigma_{PBS}/S_{anti\text{-}BSA}$) are calculated, which are shown in Table 2 for each peak. A surface mass density detection limit of <2.1 pg/mm^2 is obtained from the calculations.

The total mass detection limit can also be obtained from the surface mass density detection limit. If the active region of the PCW is considered for the calculations, since it is where the optical field is confined and where the interaction with the target anti-BSA is actually taking place (around two-three rows of holes at each side of the linear defect, which is around 100 μm^2 considering the internal surface of the holes too), a total mass detection limit of ~0.2 fg is obtained.

Concerning the improvement because of the reduction of the group velocity, Table 2 shows how calculated values for the sensitivity and the detection limit slightly improve as we move closer to the band edge (peak #3 is the only which does not follow this trend and also has a higher noise level, suggesting that it has a poorer quality than the other peaks used in the experiments), although only a 10% improvement is obtained when moving from peak #1 to #4. This is the same increase that was predicted by the 3D-FDTD simulations when taking into account only the four peaks closest to the band edge.

The high sensitivity values which are achieved when working in the wavelength region close to the band edge of the PCW, together with the device's small footprint, make it suitable for the detection of very small amounts of analyte.

5. DNA sensing using photonic crystal waveguides

For label-free DNA sensing, the experimental protocol is similar to the one previously described for antibody sensing, but now the goal of the experiment is to detect DNA hybridization events occurring on the sensor surface, so single DNA strands (ssDNA) need to be immobilized on the PCW. To do so, the chip's surface is still activated with pure ICPTS as described for the antiBSA sensing in the previous section, then an intermediate layer of streptavidin is deposited on the chip (a concentration of 0.1 mg/ml in 0.1x PBS is used), and finally biotinylated ssDNA probes 10 μM in PBS 0.1x are incubated on the sample, which will bind to the streptavidin layer thanks to the high affinity between the streptavidin and the biotin molecules (Toccafondo et al., 2010). Fig. 13 shows the TE transmission spectrum of the PCW with a PBS 0.1x upper cladding, where FP fringes are shown. In this case, the peak marked with dashed red line has been used for sensing. Fig. 14 shows the steps to be carried out in the experiment and Fig. 15 shows the temporal evolution of the peak position for the different solutions flowed. First of all PBS 0.1x is flowed to obtain the baseline of the measurement. Then a solution containing the complementary ssDNA to that immobilized on the chip, with a concentration of 0.5 μM, is flowed. Binding of the complementary DNA is effectively shown, which induces a shift in the peak position of $\Delta\lambda_{DNA}$ = 47.1 pm. The noise level in this experiment is estimated to be σ = 1.865 pm, thus giving an estimated detection limit of 19.8nM for ssDNA hybridization detection.

The strand-end of the complementary ssDNA chosen for this experiment is marked with digoxigenin, in order to allow to perform a control step and confirm the hybridization events. Therefore, after the complementary ssDNA is detected, anti-DIG 10ppm, which has a high affinity with DIG, is flowed. Its binding to the DIG marker of the target ssDNA causes a permanent shift in the peak position of 0.246 nm, which confirms the specific binding of the target DIG-marked ssDNA on the chip.

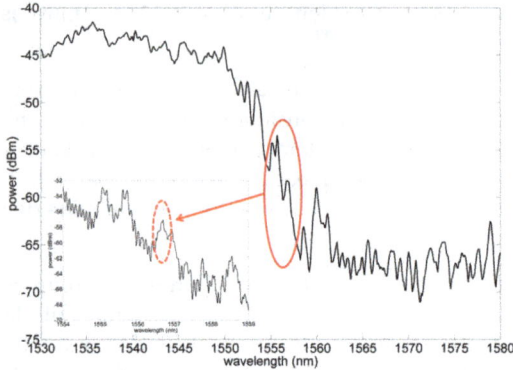

Fig. 13. Spectrum of the photonic crystal waveguide in the region of the band edge with a PBS 0.1x upper cladding. The transmission peak used for sensing is marked with dashed red line in the inset, where a higher resolution sweep has been made.

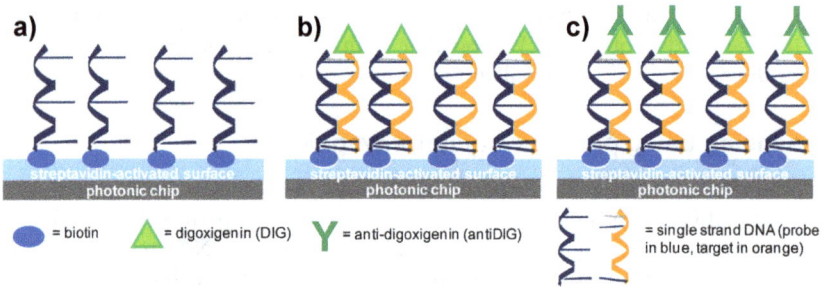

Fig. 14. Scheme of the experiment steps for the ssDNA detection and anti-DIG control.

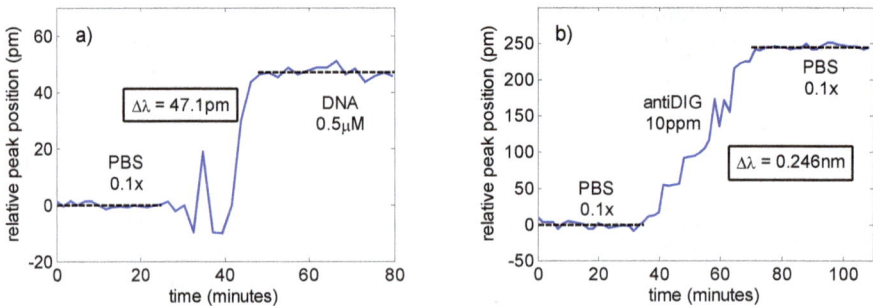

Fig. 15. (a) Wavelength shift vs time for single strand DNA 0.5 μM sensing, and (b) for anti-DIG 10 ppm sensing. Relative wavelength shift from the initial baseline is represented.

6. Low-cost sensing technique using photonic crystal waveguides

One of the problems of using PCWs for biosensing, as well as of other photonic structures like ring resonators, is that the detection is based on the measurement of the shift of the structure's spectral response, as previously discussed in this chapter. Therefore, these systems require the use of either a tunable laser source or an optical spectrum analyzer (OSA) to perform the readout of the device, making the total cost of the system significant (above 20.000-30.000 Euros or even higher). Moreover, sweeping times of the order of several seconds up to minutes are needed to acquire each spectrum, preventing an instantaneous observation of the interactions of the target analyte with the sensor.

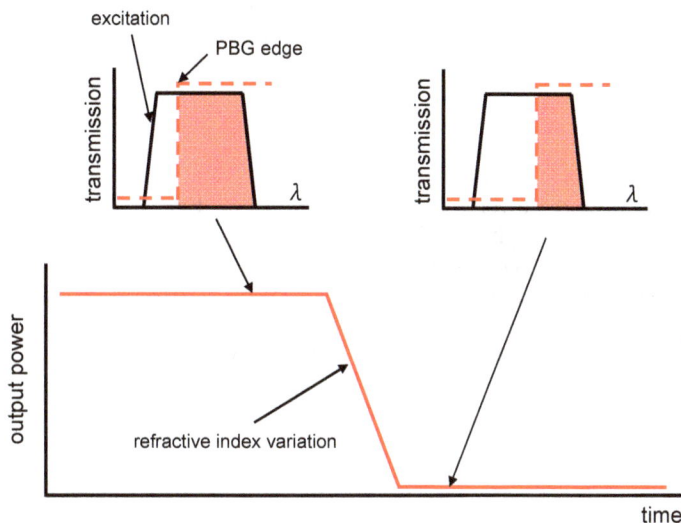

Fig. 16. Output power evolution using the proposed sensing technique. In the initial state, the source spectrum (black solid line) is filtered by the PBG of the sensing structure (red dashed line) and only a certain amount of power is transmitted (shaded area). When a refractive index variation occurs, the PBG shifts, and the amount of input power filtered changes (decreases in this case).

An alternative technique for the development of real-time and low-cost integrated photonic sensing devices using photonic bandgap (PBG) structures is schematically described in Fig. 16. Using this technique, the PBG shift is indirectly determined by using a filtered broadband optical source as excitation instead of a tunable laser, and a power meter at the output instead of an eventual OSA, thus significantly simplifying the system and reducing its cost to values around 2.000-3.000 Euros (or even lower), making it competitive with other sensing systems based on other transduction mechanisms (e.g., electro-chemical based sensors). If the PBG edge is located within the source wavelength range (for the case shown in Fig. 16, the upper edge of the PBG is used for the sensing), the PBG filters the optical source and the overlap of both spectra can be directly measured at the output using a simple power meter. When a variation of the refractive index of the surrounding medium occurs, it induces a shift in the spectral response of the photonic sensing structure. This is translated into a shift in the position of the PBG edge, and thus into a change in the optical power measured at the output, as is illustrated in Fig. 16 where an increase in the RI of the surrounding medium occurs. This power variation is directly used to perform the sensing, without the need to obtain the transmission spectrum of the structure using expensive tunable elements. Moreover, since the output power can be continuously monitored (several power values per second can be measured), a real-time sensing is performed, which allows an instantaneous observation of the interactions taking place in the sensing structure.

The initial spectral alignment between the source and the sensor will determine the sensitivity and the linearity of the device. Eq. 1 describes the relative power variation at the output (in dB) as a function of the initial spectral overlap between source and sensor (BW) and the shift of the guided band's edge due to a change in the refractive index ($\Delta\lambda$).

$$10 \times \log_{10}\left(1 - \frac{\Delta\lambda}{BW}\right) \qquad (1)$$

Fig. 17 shows the output power variation depending on the initial overlap between the source and the sensor. A high initial overlap leads to a linear response of the sensor, although a lower sensitivity is obtained. On the other hand, as the initial overlap is reduced, the sensitivity increases but a more non-linear behaviour is observed. However, a proper modeling and calibration of the sensor response will allow working in the non-linear regime, with a significant increase in the sensor sensitivity.

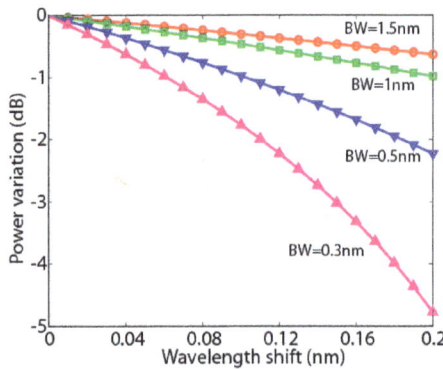

Fig. 17. Power variation versus wavelength shift for different initial alignments between the source and the sensor. BW indicates the initial overlapping bandwidth.

Another advantage of this readout technique is the possibility of continuously acquire the output power of the device (rigorously talking, many output power samples are taken each second). This will not only allow us to instantaneously observe any interaction taking place within the sensing device, but it will also allow to perform a temporal averaging of the power values and reduce the noise, thus leading to a significant reduction of the detection limit of the device.

7. Conclusion

Integrated planar photonic structures are one of the main candidates for the development of label-free biosensing devices, and among them, photonic crystal based structures. In this chapter, we have shown how PCWs can be used for biosensing purposes and how the use of Fabry-Perot fringes appearing at the guided band's edge instead of the edge itself allows performing the sensing with a higher accuracy and leading to a reduced detection limit. Experimental results have been shown for the detection of refractive index variations (with a detection limit of 3.5×10^{-6} RIU), label-free antibody sensing (with a surface mass density detection limit below 2.1 pg/mm^2 and total mass detection limit below 0.2 fg), and label-free ssDNA sensing (with a detection limit of 19.8 nM).

A technique for the indirect tracking of the guided band's edge shift has also been presented. This technique avoids the use of expensive tunable sources or detectors, which are required when carrying out a direct tracking of the spectral response shift, as usually done when using photonic sensing devices such as ring resonators or photonic crystal based structures.

8. Acknowledgment

Financial support from the Spanish MICINN under contract TEC2008-06333, the Universidad Politécnica de Valencia through program PAID-06-09, and the Conselleria d'Educació through program GV-2010-031 is acknowledged.

9. References

Barrios, C. A., Bañuls, M. J., González-Pedro, V., Gylfason, K. B., Sánchez, B., Griol, A., Maquieira, A., Sohlström, H., Holgado, M., & Casquel, R. (2008). Label-free optical biosensing with slot-waveguides. *Optics Letters*, Vol. 33, No. 7, (April 2008), pp. 708-710, ISSN 0146-9592

Buswell, S. C., Wright, V. A., Buriak, J. M., Van, V., & Evoy, S. (2008). Specific detection of proteins using photonic crystal waveguides. *Optics Express*, Vol. 16, No. 20, (September 2008), pp. 15949-15957, ISSN 1094-4087

Carlborg, C. F., Gylfason, K. B., Kaźmierczak, A., Dortu, F., Bañuls Polo, M. J., Maquieira Catala, A., Kresbach, G. M., Sohlström, H., Moh, T., Vivien, L., Popplewell, J., Ronan, G., Barrios, C. A., Stemme, G., & van der Wijngaart, W. (2010). A packaged optical slot-waveguide ring resonator sensor array for multiplex label-free assays in labs-on-chips. *Lab on a Chip*, Vol. 10, No. 3, (February 2010), pp. 281-290, ISSN 1473-0197

Castelló, J.G., Toccafondo, V., Pérez-Millán, P., Losilla, N.S., Cruz, J.L., Andrés, M.V., & García-Rupérez, J. (2011). Real-time and low-cost sensing technique based on photonic bandgap structures *Optics Letters*, Vol. 36, No. 14, (July 2011), pp. 2707-2709, ISSN 0146-9592

Chow, E., Grot, A., Mirkarimi, L. W., Sigalas, M., & Girolami, G. (2004). Ultracompact biochemical sensor built with two-dimensional photonic crystal microcavity. *Optics Letters*, Vol. 29, No. 10, (May 2004), pp. 1093-1095, ISSN 0146-9592

De Vos, K., Bartolozzi, I., Schacht, E., Bienstman, P., & Baets, R. (2007). Silicon-on-Insulator microring resonator for sensitive and label-free biosensing. *Optics Express*, Vol. 15, No. 12, (June 2007), pp. 7610-7615, ISSN 1094-4087

De Vos, K., Girones, J., Popelka, S., Schacht, E., Baets, R., & Bienstman, P. (2009). SOI optical microring resonator with poly(ethylene glycol) polymer brush for label-free biosensor applications. *Biosensors and Bioelectronics*, Vol. 24, No. 8, (April 2009), pp. 2528-2533, ISSN 0956-5663

Di Falco, A., O'Faolain, L., & Krauss, T. F. (2009). Chemical sensing in slotted photonic crystal heterostructure cavities. *Applied Physics Letters*, Vol. 94, No. 6, (February 2009), pp. 063503, ISSN 0003-6951

Dorfner, D., Zabel, T., Hürlimann, T., Hauke, N., Frandsen, L., Rant, U., Abstreiter, G., & Finley, J. (2009). Photonic crystal nanostructures for optical biosensing applications. *Biosensors and Bioelectronics*, Vol. 24, No. 12, (August 2009), pp. 3688-3692, ISSN 0956-5663

Fan, X. D., White, I. M., Shopova, S. I., Zhu, H., Suter, J. D., & Sun, Y. (2008). Sensitive optical biosensors for unlabeled targets: A review. *Analytica Chimica Acta*, Vol. 620, No. 1-2, (July 2008), pp. 8-26, ISSN 0003-2670

García, J., Sanchis, P., & Martí, J. (2006). Detailed analysis of the influence of structure length on pulse propagation through finite-size photonic crystal waveguides. *Optics Express*, Vol. 14, No. 15, (July 2006), pp. 6879-6893, ISSN 1094-4087

García, J., Sanchis, P., Martínez, A., & Martí, J. (2008). 1D periodic structures for slow-wave induced non-linearity enhancement. *Optics Express*, Vol. 16, No. 5, (March 2008), pp. 3146-3160, ISSN 1094-4087

García-Rupérez, J., Toccafondo, V., Bañuls, M. J., Castelló, J. G., Griol, A., Peransi-Llopis, S., & Maquieira, Á. (2010). Label-free antibody detection using band edge fringes in SOI planar photonic crystal waveguides in the slow-light regime. *Optics Express*, Vol. 18, No. 23, (November 2010), pp. 24276-24286, ISSN 1094-4087

Goddard, J.M., Mandal, S., Nugen, S.R., Baeumner, A.J., & Erickson, D. (2010). Biopatterning for label-free detection. *Colloids and Surfaces B: Biointerfaces*, Vol. 76, No. 1, (March 2010), pp. 375-380, ISSN 0927-7765

Homola, J., Yee, S. S., & Gauglitz, G. (1999). Surface plasmon resonance sensors: review. *Sensors and Actuators, B: Chemical*, Vol. 54, No. 1, (January 1999), pp. 3-15, ISSN 0925-4005

Huang, M., Yanik, A. A., Chang, T.-Y., & Altug, H. (2009). Sub-wavelength nanofluidics in photonic crystal sensors. *Optics Express*, Vol. 17, No. 26, (December 2009), pp. 24224-24233, ISSN 1094-4087

Iqbal, M., Gleeson, M. A., Spaugh, B., Tybor, F., Gunn, W. G., Hochberg, M., Baehr-Jones, T., Bailey, R. C., & Gunn, L. C. (2010). Label-Free Biosensor Arrays Based on Silicon Ring Resonators and High-Speed Optical Scanning Instrumentation. *IEEE Journal on Selected Topics in Quantum Electronics*, Vol. 16, No. 3, (May 2010), pp. 654-661, ISSN 1077-260X

Kauppinen, L. J., Hoekstra, H. J. W. M., & de Ridder, R. M. (2009). A compact refractometric sensor based on grated silicon photonic wires. *Sensors and Actuators, B: Chemical*, Vol. 139, No. 1, (May 2009), pp. 194-198, ISSN 0925-4005

Kersey, A. D., Davis, M. A., Patrick, H. J., LeBlanc, M., Koo, K. P., Askins, C. G., Putnam, M. A., & Friebele E. J. (1997). Fiber Grating Sensors. *Journal of Lightwave Technology*, Vol. 15, No. 8, (August 1997), pp. 1442-1462, ISSN 0733-8724

Lee, M. R., & Fauchet, P. M. (2007). Nanoscale microcavity sensor for single particle detection. *Optics Letters*, Vol. 32, No. 22, (November 2007), pp. 3284-3286, ISSN 0146-9592

Loncar, M., Scherer, A., & Qiu, Y. (2003). Photonic crystal laser sources for chemical detection. *Applied Physics Letters*, Vol. 82, No. 26, (June 2003), pp. 4648-4650, ISSN 0003-6951

Luff, B. J., Harris, R. D., Wilkinson, J. S., Wilson, R., & Schiffrin, D. J. (1996). Integrated-optical directional coupler biosensor. *Optics Letters*, Vol. 21, No. 8, (April 1996), pp. 618-620, ISSN 0146-9592

Ramachandran, A., Wang, S., Clarke, J., Ja, S. J., Goad, D., Wald, L., Flood, E. M., Knobbe, E., Hryniewicz, J. V., Chu, S. T., Gill, D., Chen, W., King, O., & Little, B. E. (2008). A universal biosensing platform based on optical micro-ring resonators. *Biosensors and Bioelectronics*, Vol. 23, No. 7, (February 2008), pp. 939-944, ISSN 0956-5663

Schubert, Th., Haase, N., Kück, H., & Gottfried-Gottfried, R. (1997). Refractive-index measurements using an integrated Mach-Zehnder interferometer. *Sensors and Actuators, A: Physical*, Vol. 60, No. 1-3, (May 1997), pp. 108-112, ISSN 0924-4247

Scullion, M.G., Di Falco, A., & Krauss, T.F. (2011). Slotted Photonic Crystal Cavities with Integrated Microfluidics for Biosensing Applications. *Biosensors and Bioelectronics*, Vol. 27, No. 1, (September 2011), pp. 101-105, ISSN 0956-5663

Sepúlveda, B., Río, J. S., Moreno, M., Blanco, F. J., Mayora, K., Domínguez, C., & Lechuga, L. M. (2006). Optical biosensor microsystems based on the integration of highly sensitive Mach–Zehnder interferometer devices. *Journal of Optics A: Pure and Applied Optics*, Vol. 8, No. 7, (July 2006), pp. S561-S566, ISSN 1464-4258

Skivesen, N., Têtu, A., Kristensen, M., Kjems, J., Frandsen, L. H., & Borel, P. I. (2007). Photonic-crystal waveguide biosensor. *Optics Express*, Vol. 15, No. 6, (March 2007), pp. 3169-3176, ISSN 1094-4087

Sünner, T., Stichel, T., Kwon, S.-H., Schlereth, T. W., Höfling, S., Kamp, M., & Forchel, A. (2008). Photonic crystal cavity based gas sensor. *Applied Physics Letters*, Vol. 92, No. 26, (July 2008), pp. 261112, ISSN 0003-6951

Toccafondo, V., García-Rupérez, J., Bañuls, M. J., Griol, A., Castelló, J. G., Peransi-Llopis, S., & Maquieira, A. (2010). Single-strand DNA detection using a planar photonic-crystal-waveguide-based sensor. *Optics Letters*, Vol. 35, No. 21, (November 2010), pp. 3673-3675, ISSN 0146-9592

Xu, D.-X., Vachon, M., Densmore, A., Ma, R., Delâge, A., Janz, S., Lapointe, J., Li, Y., Lopinski, G., Zhang, D., Liu, Q. Y., Cheben, P., & Schmid, J. H. (2010). Label-free biosensor array based on silicon-on-insulator ring resonators addressed using a WDM approach. *Optics Letters*, Vol. 35, No. 16, (August 2010), pp. 2771-2773, ISSN 0146-9592

Xu, T., Zhu, N., Xu, M. Y.-C., Wosinski, L., Aitchison, J. S., & Ruda, H. E. (2010). Pillar-array based optical sensor. *Optics Express*, Vol. 18, No. 6, (March 2010), pp. 5420-5425, ISSN 1094-4087

Zlatanovic, S., Mirkarimi, L. W., Sigalas, M. M., Bynum, M. A., Chow, E., Robotti, K. M., Burr, G. W., Esener, S., & Grot, A. (2009). Photonic crystal microcavity sensor for ultracompact monitoring of reaction kinetics and protein concentration. *Sensors and Actuators, B: Chemical*, Vol. 141, No. 1, (August 2009), pp. 13-19, ISSN 0925-4005

12

Plasma Photonic Crystal

Rajneesh Kumar

Plasmonics and Metamaterials Lab, Department of Physics
Indian Institute of Technology Kanpur
India

1. Introduction

Recently, there has been a rapid growth in the use of plasma for industrial applications where the use of plasma-based technologies offer distinct advantage over the conventional technologies (Kumar, 2011c). A number of spin-off plasma based technologies have spawned in the area of plasma-microwave interactions and plasma stealth technology. Plasma is the fourth state of matter and its material properties (electric permittivity and magnetic permeability) can be tuned by changing plasma parameters for electromagnetic radiations. As it is well known that electric permittivity (ε) and a magnetic permeability (μ) are the fundamental characteristics which determine the propagation of electromagnetic waves in matter. Here, it may be quite interesting to study the electromagnetic wave propagation in the plasma. Plasma can be used as metamaterials for negative refrction of electromagnetic waves (Kumar, 2010a; Kumar 2011d).

2. Photonic crystals (PCs) for negative refraction

Photonic crystals (PCs) are structures with periodic arrangement of dielectrics or metals, which provide the ability to manipulate the propagation of electromagnetic waves. In fact the lattice constant of common materials is 0.2-1 nm, i.e., much shorter than the wavelength of visible light (a few 100 nm). This is the reason why the response of such materials on the electrical and magnetic fields of light wave can be described by macroscopic parameters ε and μ. In 1987, anomalous refraction properties of PCs were reported based on a numerical analysis with transfer matrix method (Yablonovich, 1987). The light propagation in the PCs can not be considered as an average effect of atoms as in common crystals. In contrary, light propagation in PCs is the result of Bragg diffraction for each atom. Hence the periodic structure of the PCs is very important. The macroscopic constant ε and μ can not describe the light propagation in PC and the light refraction at the PC boundary. More precisely, light waves in PCs should be considered as the Bloch waves but in the so-called envelope function approximation they may be considered as plane waves.

An effective index of refraction for the crystal is used to describe the overall reflectivity form the photonic crystal:

$$\eta = c\frac{d\omega}{dk} \tag{1}$$

Thus, calculating band structure of a PC numerically leads to calculation of η. From the experimental point of view η can be calculated by Snell's law. Hence, the negative refraction can be realized also with PCs that is in contrast to the composite metamaterials pave inhomogeneous media with a lattice constant comparable to the wavelength. Although both ε and μ are positive in dielectric PCs and metallic photonic crystals (MPCs), phenomenon of negative refraction and super resolution can be expected from peculiarities of the dispersion characteristics of certain PCs. The main advantage of PCs over composite metamaterials (CMMs) currently is that they can be more easily scaled to 3D and adapted to visible frequencies (Parimi et al, 2004). Negative refraction at microwave frequencies was observed in both dielectric and metallic PCs, for example, using a square array of alumina rods in air (Cubukcu et al, 2003). 2D and 3D PCs consisting of alumina rods were used for the demonstration of negative refraction in the microwave and millimeter wave range. Two techniques namely, manual assembly of alumina rods and rapid phototyping were used in this study for fabricating low-loss PCs (investigated in the wave range form 26 GHz to 60 GHz). The negative refraction in a metallic PC with hexagonal lattice acting as a flat lens with out optical axis at microwave frequencies was reported at 10.4 GHz for TM mode (Parimi et. al, 2004). Such PC contains cylindrical copper rods, are in triangular lattice, in which negative refraction was found for both TM and TE mode propagation between 8.6 and 11 GHz (TM mode) and between 6.4 and 9.8 GHz (TE mode). Hence, extensive experimental and simulation results were achieved, which pave the way to a variety of well tailored PCs structures. However, the advantages of metallic PC were reported to be highest dielectric constant, low attenuation, and the possibility of focusing. Most of efforts have been dedicated to the engineering and extension of the functionalities of metamaterials or PCs at terahertz (Yen et al, 2004 ; Padilla et al, 2006, Chen et al, 2006) and optical frequencies (Linden et al, 2004; Soukoulis et al, 2007). Negative refraction of surface plasmons was also demonstrated but was confined to a two-dimensional waveguide (Lezec et al, 2007). Three dimensional optical metamaterials have come into focus recently, including the realization of negative refraction in semiconductor metamatetrials and a 3D magnetic metamaterial in the infra red frequencies. However neither of these had a negative index of refraction (Liu et al, 2008 ; Hoffman et al, 2007). Three dimensional optical metamaterial with a negative refractive index has been also demonstrated recently (Valentine et al, 2008). Negative and positive refaction tubability of x-band microwve in MPC have been achieved recently by making defults or holes (Kumar, 2011a).

3. Plasma photonic crystal (PPC)

In previous section, we have seen that negative refraction of electromagentic wave is possible by photonic crystal. We now interested to study the plasma photonic crystal due to its applications over the convetional PCs. The plasma photonic crystals (PPCs) are artificially periodic array composed of alternating thin unmagnetised or magnetized plasmas and dielectric materials or vacuum (Hojo and Mase, 2004). It is well known that nonmagnetised plasma can be characterized by a complex frequency-dependent permittivity medium. On the other hand, the unmagnetised plasma is frequency dispersive medium. The refractive index of collisionless unmagnetised plasma that is determined by electromagnetic wave frequency and plasma frequency is less than one. Dispersion relation of propagating electromagnetic waves in nonmagnetised plasma can be modified if bulk plasma is replaced by a microplasma array (Park et al, 2002), which is analogically understood from the extensive studies of photonic crystal. Hence in plasma photonic crystal, array of periodic micro plasmas are used at the place of array of dielectrics or metals in the conventional photonic crystals. One or two dimensional layers of array of micro plasmas make forbidden bands for wave propagation are formed beyond the bulk cut of frequency (electron plasma frequency) due to periodicity, where one can refer to such a functional structures as plasma photonic crystal. A photo of plasma photonic crystal is given below in Fig.1.

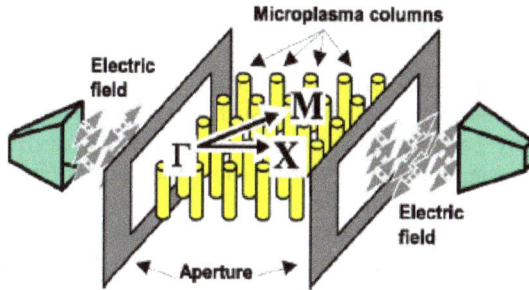

Fig. 1. Plasma photonic crystal

We know that plane-wave-expansion method has been widely used to analytically derive photonic band diagram of two-and three-dimensional dielectric periodic structures (Ho et al, 1990; Phihal et al, 1991). Dielectric constant of plasma can be obtained considering the field components in electromagnetic waves proportional to $\exp.[j(\omega t - k.x)]$, where k and x are the complex wave number and spatial position vector, respectively. The dielectric constant as a function of frequency $\varepsilon_p(\omega)$ inside a cold plasma column with electron plasma frequency $\left(\omega_{pe}\right)$ is written as

$$\varepsilon_p = 1 - \left(\frac{\omega_{pe}}{\omega}\right)^2 \frac{1}{1 - j\left(\upsilon_m / \omega\right)} \tag{2}$$

where υ_m is the electron elastic collision frequency determined by neutral gas pressure and elastic collision cross section. In metal cases, a similar value (γ) to υ_m was used as an inverse of electron relaxation time, and γ was much smaller than ω and ω_{pe} (Kuzmiak and 1997) it is also possible that υ_m is comparable to ω and ω_{pe} where electron density is around 10^{13} cm^{-3} at a gas pressure around atmospheric pressure. Therefore plane-wave expansion method with Drude model in collision plasma has been studied (Sakai et al, 2007) for plasma photonic crystals. Experimental demonstrations have also been performed (Sakai et al, 2005 ; Sakaguchi et al, 2007) those typical parameters are summarized below,

- One and two dimensional periodic structures have been studied.
- In 2 D structures, a mesh type DBD (Dielectric Barrier Discharge) electrode assembly, mounted at 6 mm separation from third electrode (Micro-hollow-cathode-discharge MHCD like configuration).
- Array forms a 4.4 x 4.4 cm^2 squire lattice of plasma columns.
- Array size 33 x 33 lattices, where 33 rows of a micro plasma column with diameter of 0.6 mm.

- Lattice constant = 2.1 mm to 2.5 mm.
- Squire hole with opening 1.4 mm x 1.4 mm.
- He, N_2, and Ne gas is used.
- For generation of microwaves, signal generator of 33-50 GHz and 50 – 75 GHz are used.
- Pyramidal horn antennas are used for transmitting and receiving microwaves.

Several experimental studied have been conducted with the help of given experimental set-up and parameters. Important results which emerged from the studied are listed below,

- Lattice structure of micro-plasma arrays behaves as a photonic crystal similar to solid dielectrics.
- A millimeter wave at 33-110 GHz was injected into two-dimensional plasma column array, and the transmitting signal through such array attenuated less than 20%.
- Band gap forms by periodic dielectric constant above the electron plasma frequency $\left(\omega > \omega_{pe}\right)$ and propagation of flat bands below the plasma frequency $\omega < \omega_{pe}$. Hence structure of photonic crystal play role rather then cut-off conditions.
- Band gap frequency could be varied by changing the lattice constant, leading to a function of dynamic and time-controllable band-stop filter in millimeter and sub terahertz regions.
- 30 rows of plasma columns are similar in the case of 17 rows of metal due to a lower ratio of dielectric constant between plasma and vacuum region, plasma photonic crystals require more rows than case of an ordinary photonic crystal.

4. Plasma crystal as photonic crystal

So far we have studied about PCs and plasma photonic crysals. It is quite interesting to describe the plasma crystal as photonic crystal due to its applications over the PCs and PPCs. Infact coulomb lattices of charged dust grains are called plasma crystals, which can be generated in laboratory in dusty plasma experiments (Morfill et al, 1997). Plasma crystals are composed dust grains that become electrically charged to very high charge states while thermal kinetic energy of the grains remains low. In plasma crystal composed of negatively charged dust grains, the dust grain radius is typically in the order of a few micrometers, while the average intergrain distance is on the order of a few hundreds micrometers, with variations in different experiments (Morfill et al, 2002). Therefore plasma crystals generated in the laboratory generally have only a few layers in the vertical dimension owing to gravitational compression, although recently a three-dimensional structure has been demonstrated (Zuzic et al, 2000). However, plasma crystal has some similarities to colloidal crystal, which are colloidal suspensions of ordered charged particles in solvents. Due to Bragg scattering properties, colloidal crystals have applications as narrow band rejection filters in optics. Tunable colloidal crystal in the optical, ultraviolet and infrared have also been demonstrated in which the particle size or spacing charge with temperature to tune the diffraction (Weissman et al, 1996). Recently, a magnetically tunable optical filter comprising a ferro-fluid based emulsion cell has been discussed (Philips et al, 2003). Viewing above it can be noticed that if dust plasma crystal can be generated in sufficient large multilayer closer-packed configuration, they may have similar use as filters in the longer wavelength terahertz (THz) regime. A possible experimental prototype of magnetically controlled and tuned plasma crystal in dusty plasma is shown in Fig. 2. With the help of above study, THz refraction or scattering can be studied.

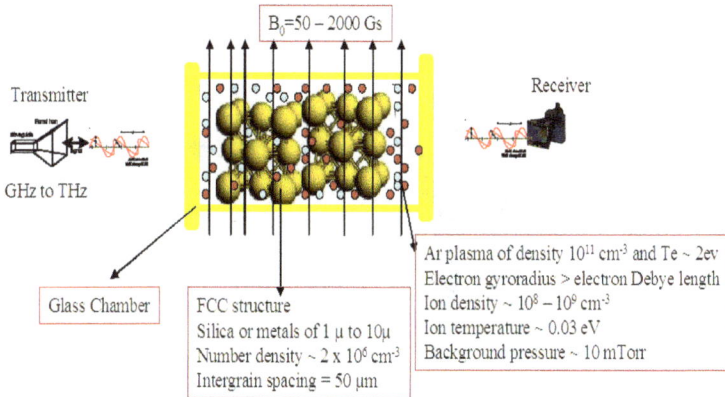

Fig. 2. A prototype to present EM wave transition from plasma crystal

5. Plasma in photonic crystal

In this section experimental study on plasma added photonic crystal (hybrid plasm photonic crystal) is presented. We have seen that negative refraction at microwave frequencies   was also observed in both dielectric and metallic PCs. Although, research indicates that the MPC is suitable for negative refraction (Kumar 2011a), however, to date tunability in the fabricated MPC to control the wave propagation has not been achieved. From this aspect, plasma can be a good candidate to replace metal or dielectrics from PCs because plasma is a frequency dependent dispersive medium and its refractive index can be determined by electromagnetic wave frequency and plasma frequency (Ginzberg, 1970; Hojo et.al 2003; Hojo and Mase, 2004) has proposed that plasma photonic crystals (PPCs) are artificially made periodic arrays composed of alternating discharge plasma and other dielectric materials (including vacuum). On the bases of different approaches, two types of PPC are being studied. In the first type of PPC, cylindrical glass rods or dielectrics forming a crystal lattice are immersed in discharge background plasma (Laxmi and Parmanand, 2005; Liu, et al, 2006; Hojo et al, 2006; LIU et al, 2009) while the second type consists of cylindrical rods of discharge plasma that constitutes a crystal lattice in vacuum or air (Sakai, et al, 2005; Sakai et al, 2005; Sakai and Tachibana, 2007). It can also be composed of plasma with spatially periodic density variation, which can be induced naturally in plasmas i.e in the presence of laser pulses in underdense plasmas (Botton and Ron, 1991 ; Zhang et al, 2003; Wu et al, 2005 ; Yin et al, 2009), dust plasma crystals (Rosenberg et al, 2006), self-organised small plasma blobs or patterns (Fan et al, 2009 ; Kumar and Bora, 2010a ; Kumar, 2011c), etc. However theoretical and experimental studies have been going on since last few years to find out the possible applications of PPC over the conventional PCs, although there is no strong evidence of negative refractive index or metamaterial properties of PPC. Meanwhile, difficulties in the construction of PPC have been experienced during experimental realizations. Even after a long research history of MPC and PCC, a number of problems related to controllability and fabrications in both PCs are still unresolved. There is, however, plenty of scope to work on a hybrid PCs (Kumar , 2009a; 2009b, 2011b, 2011d) of MPC and PPC in such a manner so that properties of both PCs can be utilized. Hence, the motivation of this study is to investigate the effect of a plasma column to control the microwave propagation through MPC as presented in Fig 3.

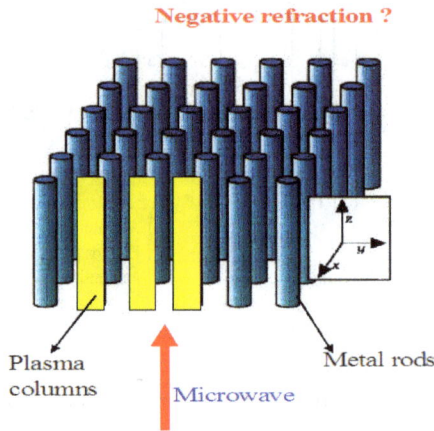

Fig. 3. Hybrid plasma photonic crystal

5.1 Selection of plasma parameters

Our motivation for the study to use plasma column in place of copper rods lies in the fact that by switching OFF and ON plasma, which can be formed and destroyed rapidly, and microwave refraction can be controlled. Hence, to accomplish such purpose it is required to study the characteristics of plasma at X-band frequency microwave (18 GHz) to obtain its behavior close to the metallic copper rods. Hence phase difference in 18 GHz microwave is calculated using reflection coefficient of copper and plasma with the help of relative reflective index ($\varepsilon_r = -0.40 - 1.26i$) and conductivity ($\sigma = 1.24 - 1.4i$). Reflection behavior of microwave from metal and plasma is shown in Fig. 4. The phase difference and path difference are achieved as 2.5 radian and 6 mm respectively.

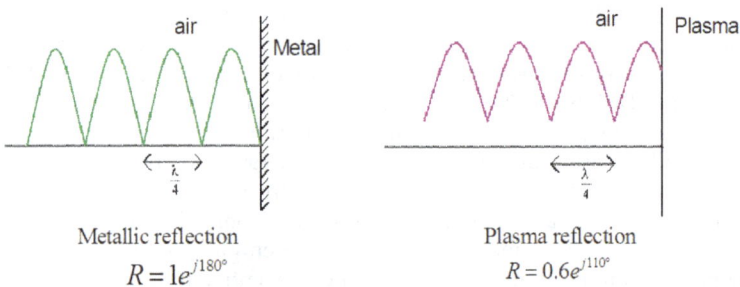

Fig. 4. Reflection of 18 GHz microwave from metal and plasma medium

The plasma is considered as a homogeneous cylindrical channel with complex relative electric permittivity given as

$$\varepsilon(r) = 1 - \left(\frac{\omega_{pe}}{\omega}\right)^2 \left(1 - i\frac{\upsilon_m}{\omega}\right)^{-1} \tag{3}$$

where ω_{pe} is the electron plasma angular frequency, ω is the angular frequency of the incident wave and υ_m is the electron-neutral collision frequency. We see that in order to obtain a significant effect of the plasma for reasonable values of the plasma channel diameter, the plasma density should be in the range of 5 x 10^{12} cm^{-3}. Plasma parameters are precisely chosen to full fill the conditions. Possible precautions for cut-off criteria, collisionality and skin depth effect have been considered. The calculated values of these are given below in Fig. 5.

5.2 Experimental set-up

To accomplish the motivation of this study an experimental set-up is made (This experiemnt was performed in GREPHE, LAPALCE, Toulouse, France). A schematic of experimental set-up for measuring the microwave transmission through metallic photonic crystal (MPCs) with and without plasma column is shown in Fig.6, where a MPC, glass chamber, horn antennas, etc. are presented. A triangular MPC is made by 33 copper rods of diameter 2 mm and length of 100 mm, which are fixed with lattice constant of 10mm in two flanges or discs made by dielectrics in such a way so that different structures of MPC can be configured. Metallic photonic crystal is placed in a glass chamber to carry out experiments with and without plasma column. MPC is kept at the centre of the glass chamber in such a manner that transmitter and receivers can be aliened with the MPC and plasma columns. Two suitable canonical horn antennas are used as transmitter and receiver for X-band microwave. Transmitting horn antenna is fixed and receiving horn antenna can be moved around the MPC form 0^0 to 360^0. Both antennas are used in far-field region form the MPC. The heights of both the antennas are same from the ground. A microwave generator is used to generate X-band microwave (2 GHz-28GHz). A vector network analyzer (VNA) is attached with both the antennas to measure the transmission coefficient. An angle chart is made on the ground level to measure the angle of the position of the receiver from the origin of the MPC. A microwave absorber is used to absorb the microwave so that reflected microwave cannot affect the measurements. In order to investigate the application of plasmas in MPC experiments are conducted with and without plasma column at different places; thus, experimental set-up is modified as follows. In the first set of experiments in which plasma column is formed at the place of central rod of the front row in MPC and in the second set of experiments plasma column is formed between MPC and position of transmitter.

Same discharge mechanism can be used to form the plasma column in all the experiments, so the details of discharge mechanism and formation of plasma columns are common. It is well known that micro-discharge can be used to produce large volume plasma columns up to atmospheric pressure (Kunhard, 2000 ; Park et al, 2003; JING and WANG, 2006). Therefore, micro-discharge is also used to form plasma column in and out of MPC (Kumar, 2009b). For this purpose, three electrodes are made of molybdenum foil and alumina is used as a dielectric to make sandwich of electrodes with the hole diameter of 0.5 mm to 1 mm. High temperature glue is used to pack the electrodes and alumina. Two DC power supplies are used to produce voltage differences between electrodes. Length of the plasma column is equal to the separation between electrodes and, of course, length of plasma column can be varied by changing the separation of electrodes. Typical cathode voltage is 800 V and anode voltage varies form 1 KV to 2 KV maintaining current up to 15 mA. Argon and helium gases are used as background gases in the glass chamber. Experiments are carried out with different cathode-anode configurations at different background pressures. Turbo pumps and needle valves are used to control the gas pressure inside the glass chamber respectively. For transmitting microwave a horn antenna is fixed at the flange of one of the ports of glass chamber and properly aligned according to the position of plasma column and MPC. Flange of the second port of glass chamber is used to take the electrical connections between power supplies and electrodes. For receiving the transmitted microwave power, another horn antenna is fixed on a stand outside the glass chamber. Such horn antenna can be moved on the angle chart form $+90^0$ to -90^0. Microwave is fed to the transmitting antenna using microwave generator and receiving antenna is fitted to a spectrum analyzer. Transmitter, MPC and plasma column are arranged in a glass chamber in such a way that the experiments can be carried out for different positions of plasma column.

Cuts-off density for 18 GHz frequency

$$n_c = 1.24 \times 10^4 \times \{f(MHz)\}^{\frac{1}{2}} cm^{-3} \quad n_c = 4 \times 10^{12} cm^{-3} \text{ for } f = 18 GHz$$

At low working pressure (up to 100 Torr)

$$\omega(1.13 \times 10^{11}) \gg v_m (3 \times 10^{10}) \quad \text{Collisionless plasma for 18 GHz}$$

At high working pressure (up to atmospheric pressure 760 Torr)

$$v_m (5 \times 10^{11}) \geq \omega(1 \times 10^{11}) \quad \text{Collisional plasma for 18 GHz}$$

For collisionless plasma of density 5 X 10^{12} cm^{-3} at working pressure 10 Torr

Electric conductivity of plasma,
$$\sigma = 2.82 \times 10^{-4} \times \frac{n_e}{\omega^2} v_m$$
$$\sigma = 1.1 \times 10^{-3} Ohm^{-1}.Cm^{-1}$$

Skin depth in given plasma for 18 GHz

$$\delta = \frac{5.03}{\{\sigma(Ohm^{-1}.cm^{-1}).f(MHz)\}^{\frac{1}{2}}} cm$$

$$\delta = 11 \text{ mm for } f = 18 \text{ GHz}$$

Fig. 5. Calculations for plasma parameters

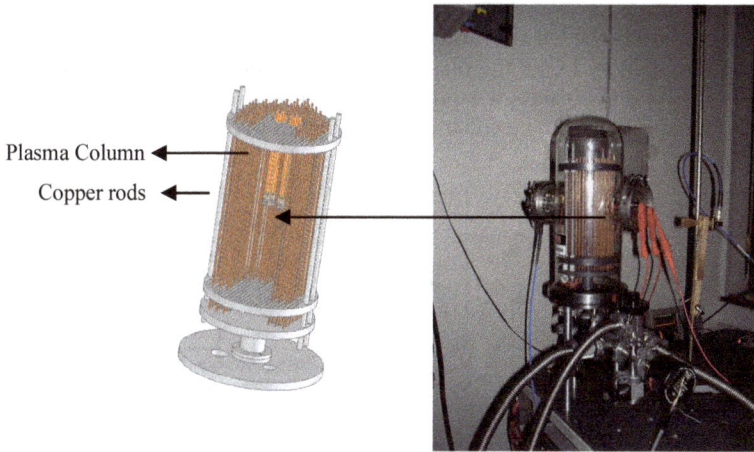

Fig. 6. A photo of experimental set up for plasma added MPC

5.3 Measurements and results

Experiments are carried out to study the electromagnetic bands gaps (EBGs) of X-band microwave through different configurations of triangle structure of MPC with and without plasma columns. Measurement method is shown in Fig.7.

5.3.1 Without and with plasma in central default

It has been studied that flat and forbidden bands at 18 GHz can be formed by MPC and defaulted MPC (Kumar 2011a, Kumar 2011b). Although this research work has potential to improve the tunability of PCs, it seems that enhancement in the tunability and controllability in this MPC is required because for tuning the MPC one needs to physically remove the metallic rods by mechanical effort (Kumar 2011a). For this concern, attention is paid to use a plasma column in the hole or default of MPC because plasma can be created and destroyed by switching ON and OFF. With the help of this approach, tunability of MPC can be increased as rapidly as the plasma can be formed and destroyed.

Hence experiments are carried out to measure the transmitted power at 18 GHz through MPC with and without plasma column. For this purpose electrodes of separation 20 mm are kept at the centre hole and well connected with the power-supplies. Finally, a plasma column of density of 5×10^{12} cm^{-3} and electron temperature of 2eV is formed around atmospheric pressure. Transmitted power of microwave is measured at different angles. Plasma is characterized [Kumar 2009b] as a collisional medium, which shows cut-off for 18 GHz microwave. A schematic of measurement method with electrode and with plasma in electrodes is shown in Fig. 8 (a) and (b) respectively.

Measurements of transmitted power of 18 GHz with electrodes and with plasma column are presented in Fig.9. Results of this figure show that transmitted power –38dBm at + 45^0 for electrodes at the separation of 20 mm are fixed and transmitted power becomes –48 dBm when plasma is formed between electrodes. Negative and positive refraction is also studied by forming the plasma in the left side and the right side default from the centre in front row. Hence by switching ON and OFF the plasma, flat and forbidden bands can be achieved at 45^0. Due to these strong evidences, it can be concluded that tunability and controllability of MPC over the PCs and PPCs can be enhanced by using a plasma column in a MPC.

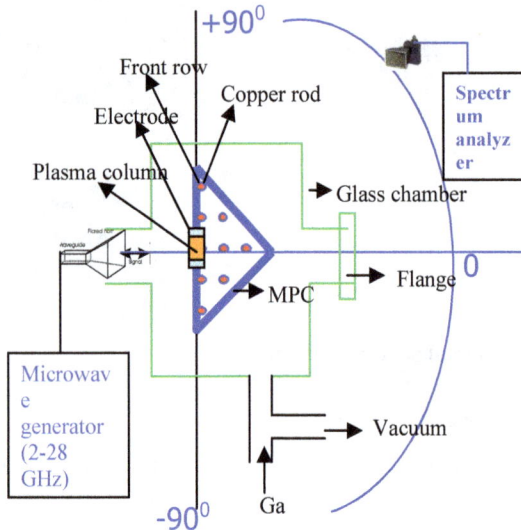

Fig. 7. A schematic diagram of measurement method showing MPC in triangular shape inside a glass chamber for measuring microwave transmission through MPC with and without plasma columns.

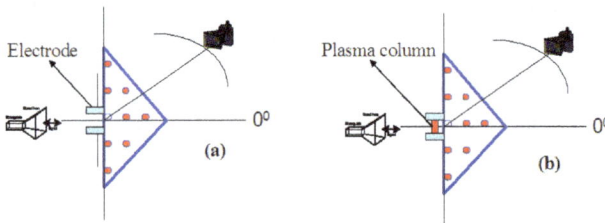

Fig. 8. (a) Measurement method of transmission of 18 GHz with electrode only (b) with plasma column at the central default of MPC.

So far, it has been successfully demonstrated how forbidden bands and flat bands can be formed by hybrid plasma photonic crystal. Although this study reveals importance and applications of plasma to control the microwave propagation in PCs, some minor problems are realized during the experiment e.g. adjustment of the electrodes, initiation of breakdown for discharge and sustaining plasma for long time because plasma column is formed inside the MPC where it is surrounded by metallic rods, which creates capacitive effects. Here it may be quite interesting to use plasma column between transmitter and MPC. Hence in the next section, attention is paid to the study of microwave propagation when plasma column is formed between transmitter and front row of MPC.

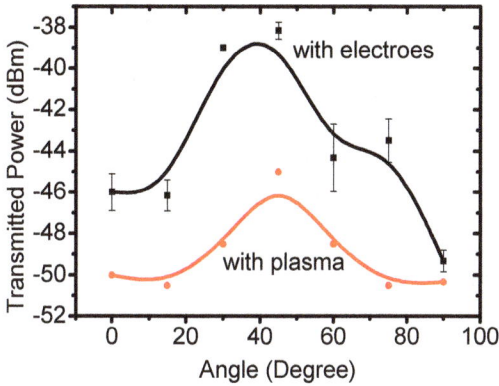

Fig. 9. Variation in the microwave transmitted power at different angles with electrodes only and plasma column within electrodes at the centre hole or default.

5.3.2 Plasma column between transmitter and metallic photonic crystal

In this experiment, a plasma column of density around 10^{12} cm^{-3} is formed at 40 mm to 100 mm away from the MPC and transmitted power of microwave is measured at different angles. Measurements with and without plasma column are presented in Fig.10. Findings of this study suggest that transmitted power is –50dBm (forbidden band) at every angle around MPC when there is no plasma column within electrodes of separation 15 mm. When plasma is formed between electrodes, transmitted power –35 dBm (flat band) is received at +45^0. Thus, negative refraction can also be controlled using plasma column. During this experiment it is also noticed that when the plasma column is situated between 70 mm to 100 mm away from the front row of MPC towards the transmitter, flats bands are measured and if plasma column is situated at a distance of 10 mm to 40 mm from front row of MPC, forbidden bands are measured at same angle. With this experiment, it can also be pointed out that position of plasma column can also control the propagation of microwave.

6. Conclusion

Negative refraction by the photonic crystal has been achieved for microwave to optical range. Plasma photonic crystals are used to enhance the controllability and tunability of microwave propagation. Plasma crystals can be also used for microwave to terahertz frequency filter. However plasma in metallic photonic crystal is a suitable technique to control the microwave in such a way that negative and positive refraction can be achieved. For this purpose micro-discharged mechanism is used to form plasma column at atmospheric pressure in a metallic photonic crystal. Argon, helium, xenon and their mixtures are used as a background gas. Transmitter, MPC and plasma column are arranged in a glass chamber in such way that the experiments can be carried out for different positions of plasma columns. A 20 mm long plasma column of electron density around 10^{13} cm^{-3} is formed. Experiments are conducted to study the electromagnetic band gaps of X-band microwave through different configurations of triangular MPC with and without plasma column. Transmitted power of 18 GHz with electrodes (without plasma between electrodes) and with plasma in electrodes are measured. Results reveal that transmitted power of – 38 dBm is received at + 45⁰ for electrodes at a separation of 20 mm which becomes -48 dBm when plasma is formed between electrodes. In another experiment, a plasma column is formed between MPC and transmitter and transmitted power is measured for every angle. Findings of the study suggest that when plasma column of length 15 mm is formed in electrodes, which are fixed 70 mm to 100 mm away from the MPC, flat band of power level -35 dBm is received at + 45⁰ while forbidden band of power level – 50 dBm is noticed when plasma column is formed at a diastase of 10 mm to 40 mm from the MPC.

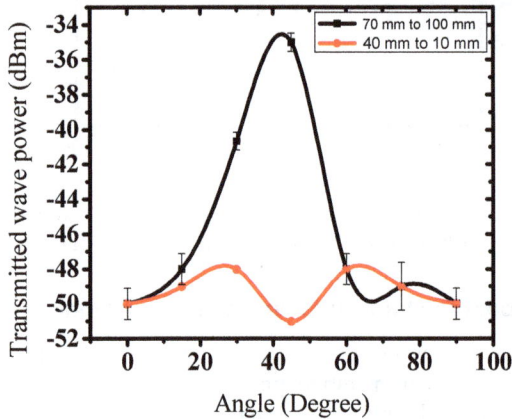

Fig. 10. Variations in the transmitted microwave power at different angles in the presence of plasma column at different places in between transmitter and metallic photonic crystal.

Therefore, by switching ON and OFF the plasma column, propagation of microwave in metallic photonic crystal can be controlled in such a way that positive and negative refraction can be achieved. This chapter can be concluded from the fact that plasma can be used to form tunable / controllable photonic crystals.

7. Acknowledgment

Author acknowledges the RTRA STAE, Toulouse, France for financial support to this work under the PLASMAX project.

8. References

Botton M. and Ron A. (1991), Efficiency enhancement of a plasma-filled backward-wave oscillator by self-induced distributed feedback. *Phys. Rev. Lett,* vol. 66, no.19, pp. 2468-2472.

Chen H. T. et al (2006) Active terahertz metamaterial devices. *Nature,* vol. 444, pp.597-600.

Cubukcu E. (2003), Negative refraction by photonic crystals. *Nature (Lomdon),* vol.423, pp 604-605.

Fan Weili et al (2009), A potential tunable plasma photonic crystal : Applications of Atmospheric patterned gas discharge. *IEEE Trans. On Plas. Sci.,* vol. 37, no.6, pp. 1016-1020.

Ginzberg V. L., *The propagation of electromagnetic waves in plasmas,* Pergammon press, New York (1970).

Ho K. M. et al (1990), Existence of a photonic gap in periodic dielectric structures. *Phys. Rev. Lett.*, vol. 65, no. 25, pp. 3152-3155.

Hoffman A. et.al (2007), Negative refraction in semiconductor metamaterials.*Nature Mater*, vol.6,pp. 946 – 950.

Hojo H., Akimoto K., Mase A. (2003), Enhanced wave transmission in one-dimensional plasma photonic crystals. presented at Conference Digest on 28th Int. Conf. On infrared and millimeter waves, (Otsu, Japan),pp. 347.

Hojo H. and Mase A. (2004), Dispersion relation of electromagnetic waves in one-dimensional plasma photonic crystals. *J. Plasma Fusion Res*, vol. 80, pp. 89- 95.

Hojo,H. et al (2006), Beaming of millimeter waves form plasma photonic crystal wave guides. *Plasma and Fusion Research*, vol. 1 , pp. 021-1021-2.

Jing Chao and Wang Youqing (2006), Formation of large-volume high pressure plasma in triode-configuration discharge device. *Plas. Sci. and Tech.*, vol. 8, no.2, pp.185-189.

Kumar Rajneesh (2009a), Hybrid plasma photonic crystal. Proc. 12th International Symposium on Microwave and Optical Technology, New Delhi, India, 16-19 December 2009. p 890.

Kumar Rajneesh (2009b), Microdischarges and applications in photonic crystals. 24th National symposium on plasma science and technology, Hamirpur, India 8-11 December 2009 (unpublished).

Kumar Rajneesh and Bora Dhiraj (2010a), A reconfigurable plasma antenna. *J. Appl. Phys.*, vol. 107, pp. 053303-1-9.

Kumar Rajneesh (2010b), Possible application of ultra cold plasmas as tunable matamaterial for microwave, 22-25 Feb 2010, 2nd International Conference on Metamaterials, Photonic Crystals and Plasmonics (META'10) Cairo, Egypt.

Kumar Rajneesh (2011a), Study of negative and positive refractions for microwave using metallic photonic crystals, *International Journal for Microwave and Optical Technology*, vol. 6, no. 2, pp.80-84.

Kumar Rajneesh (2011b), Experimental study of electromagnetic band gaps using plasmas or defaults in metallic photonic crystals, *Microwave and Optical Technology Letter*, vol. 53, no. 5, pp1109-1113.

Kumar Rajneesh (2011c), Plasma Antenna, LAP LAMBERT Academic Publishing GmbH & Co. KG, Dudweiler Land., Germany.ISBN: 978-3-8443-0530-2.

Kumar Rajneesh (2011d), Plasma as Metamaterials, LAP LAMBERT Academic Publishing GmbH & Co. KG, Dudweiler Landstraße, Germany, ISBN: 978-3-8443-1811-1.

Kunhard E. E. (2000), Generation of large-volume, atmospheric pressure nonequilibirum plasmas. *IEEE Tras. On Plas. Scie.*, vol. 28, no. 1, pp. 189-200.

Kuzmiak V. and Maradudin A. A. (1997), Photonic band structures of one- and two-dimensional periodic systems with metallic components in the presence of dissipation. *Phys. Rev. B*, vol. 55, no.12, pp 7427–7444.

Laxmi S. and Parmanand M. (2005), Photonic band gap in one-dimensional plasma dielectric photonic crystals. *Solid state communication*, vol. 138, pp.160-164.

Lezec H. J. et al (2007), Negative refraction at visible frequencies. *Science*, vol. 316, pp. 430-432.

Linden S. et al (2004)., Magnetic Response of Metamaterials at 100 Terahertz. *Science*, Vol. 306 no. 5700, pp. 1351-1353.

Liu N. et al (2008), Three-dimensional photonic metamaterials at optical frequencies. *Nature Mater*, vol. 7, pp. 31-

Liu Shaobin, Hong Wei and Naichang (2006), Finite-Difference Time-domain analysis of unmagnetised plasma photonic crystals. *Int. Jr. of Infra. And Millimeter wave*, vol. 27, no.3, pp. 403-423.

Liu Song, Shuangying and LIU Sanqiu (2009), A study of properties of the photonic band gap of unmagnetised plasma photonic crystals. *Plasma Sci. Tech.*, vol.11, no.1, pp. 14-17.

Morfill G. E. et al (1997), Plasma Crystal-A review, in Advances in Dusty Plasmas, P. K. Shukla, Ed, Singapur: Word Scientific, 99.

Morfill G. E. et al (2002), A review of liquid and crystalline plasmas — new physical states of matter?. *Plasma Phys. Control. Fusion*, vol. 44, no. 12 B, pp. B 263 .

Padilla W.J. et.al (2006), Dynamical Electric and Magnetic Metamaterial Response at Terahertz Frequencies. *Phys. Rev. Lett.*, vol, 96, no.10, pp 107401-107405.

Parimi P. V. et al (2004), Negative refraction and left-handed electromagnetism in microwave-photonic crystal, *Phys. Rev. Lett.*, vol. 92, no. 12, pp. 127401-127404.

Park Hae II et al (2003), Formation of large-volume, high pressure plasmas in microhollow cathode discharge. *Appli. Phys. Lett.*, vol. 82, no. 19, pp. 3191-3193.

Philips J. et al (2003), A tunable optical filter, Means. Sci. Technol.vol. 14, pp 1289–1294.

Rosenberg M. et al (2006), A note on the use of dust plasma crystal as tunable THz filters. *IEEE Trans. On Plas. Scie.* pp.1-4.

Sakaguchi T. et al (2007), Photonic bands in two-dimensional microplasma arrays. II. Band gaps observed in millimeter and subterahertz ranges. *J. Appli. Phys.*, vol. 101, no.7, pp. 073305 – 073312.

Sakai O, Sakaguchi T, Ito Y and Tachibana K. (2005), Interaction and controll of millimeter-wave with microplasm array. *Plasma Phy. Control. Fusion*, vol 47, pp. B617-B627.

Sakai O. et al (2005), Verification of a plasma photonic crystal for microwaves of millimeter wavelength range using two-dimensional array of columnar microplasmas. *Appli.. Lett.*, vol. 87, pp. 241505-241505.

Sakai O. and Tachibana K. (2007), Properties of electromagnetic wave propagation emerging in 2-D periodic plasma structures. *IEEE Trans. On Plasma Scie.*, vol. 35, no. 5, pp.1267-1273.

Sakai O. et al (2007), Photonic bands in two-dimensional microplasma arrays. I. Theoretical derivation of band structures of electromagnetic waves. *J. Appl. Phys.*,vol. 101, no.7,pp 073304-073313.

Soukoulis C.M. et al (2007), Negative refraction index at optical frequencies. *Science*, vol. 315, pp47-49.

Valentine J. et al (2008), Three-dimensional optical metamaterial with a negative refractive index. *Nature*, vol.455, pp. 376-379.

Weissman J. M. et al (1996), Thermally Switchable Periodicities and Diffraction from Mesoscopically Ordered Materials. *Science*, vol. 274, no.5289,pp. 959-963.

Wu H. C., Sheng Z.-M. and Zhang Q.-J. (2005), Manipulating ultrashort intense laser pulses by plasma bragg gratings. *Phys. Plasmas*, vol.12, pp. 113103-1-113103-5.

Yablonovitch E. (1987), Inhibited spontaneous emission in solid-state physics and electronics. *Phys. Rev. Lett.*, vol. 58, no. 20, pp 2059-2062.

Yen T. J. et.al (2004), Terahertz Magnetic Response from Artificial Materials. *Science*, Vol. 303 no. 5663 pp. 1494-1496.

Yin Yan et al (2009), Bandgap characteristics of one-dimensional plasma photonic crystals. *Phys. Plasmas*, vol. 16, pp. 102103-1-102103-5.

Zhang P. et al (2003), An optical trap for relativistic plasma. *Phys. Plasmas*, vol.10, no.5, pp. 2093-2099.

Zuzic M. et al (2000), Three-Dimensional Strongly Coupled Plasma Crystal under Gravity Conditions. *Phys. Rev. Lett.*, vol.85, no.19, pp.4064–4067.

13

Photonic Crystal for Polarization Rotation

Bayat and Baroughi
South Dakota State University
USA

1. Introduction

Due to the unique guiding properties of photonic crystal (PC) structure, such as sharp low loss bends, it is considered one of the main contenders of a compact optical integrated circuit (OIC). PC is foreseen as the next generation of hybrid photonic-electronic integrated circuit. However, one of the main issues in implementation of a PC based OIC is its strong polarization dependence guiding behavior.

The components of optical integrated circuit exhibit strong polarization dependence behavior which translates into their random response to random polarizations. One approach to render the polarization sensitivity of an optical integrated circuit is to eliminate the randomness of the input polarization by splitting it into two orthogonal polarizations (TE, TM) and rotating one of the polarizations; thus, single polarization is realized on the chip (Barwicz et al., 2007). The focus of this chapter is to implement PC based polarization rotator which is capable of rotating the polarization of light to an arbitrary angle. The large birefringence in PC structure leads to a small optical path difference between the two polarizations which can result in realization of ultra-compact polarization rotators. Ease of fabrication and its compatibility with integrated PC technology is considered another main advantage of the PC based polarization rotator. This chapter is organized as follows. In Sec.2, an overview on the passive polarization rotators is given. In Sec. 3, a novel polarization rotator structure is introduced and designed. Fabrication and characterization of the PC based polarization rotator are discussed in Sec. 4 and Sec. 5, respectively.

2. An overview on passive polarization rotators

Passive polarization rotator structures are mostly composed of geometrically asymmetric structures where the symmetry of the structure is disturbed so that two orthogonal polarizations could be coupled to each other. Imposing asymmetry into a symmetric waveguide structure leads to a perturbation in the primary waveguide axes, depicted in Fig.1, where E_f and E_s are projected fields on fast and slow axes called fast and slow normal modes, respectively. They travel with different speeds resulting in phase delay between the two components. For the phase delay of 180°, the power conversion between the two components has reached to its maximum and the propagating distance is called the half-beat length L_π, defined as (Mrozowski, 1997):

$$L_\pi = \frac{\pi}{\beta_s - \beta_f} = \frac{\pi}{(n_s - n_f)k_0} = \frac{\lambda}{2(n_s - n_f)} \qquad (1)$$

Where β_s and β_f are the propagation constants of slow and fast modes, and n_s and n_f are corresponding effective refractive indices, respectively. The process of polarization rotator in geometrically asymmetric structure can be explained with more details as follows. According to Fig. 1(b), the transverse component of the normal modes of for example the asymmetric loaded rib waveguide (Shani et al., 1991) can be expressed as following, where φ is the optical perturbation angle and E_s and E_f are slow and fast modes, respectively:

$$E_s = \cos\varphi\hat{x} - \sin\varphi\hat{y} \tag{2.a}$$

$$E_f = \sin\varphi\hat{x} + \cos\varphi\hat{y} \tag{2.b}$$

Assuming that the input wave is x-polarized, it can be expressed as the combination of E_s and E_f as following:

$$E_i = \cos\varphi\, e^{-j\beta_s z}E_s + \sin\varphi e^{-j\beta_f z}E_f \tag{3}$$

At half-beat length long, the slow and fast modes become out of phase resulting in destructive interference between the two modes. Thus, at $z=L_\pi$, the total field become:

$$E_{o1} = \cos\varphi E_s - \sin\varphi E_f \tag{4}$$

Substituting equ. (2) into equ. (4) results in:

$$E_{o1} = \cos 2\varphi\hat{x} - \sin 2\varphi\hat{y} \tag{5}$$

Thus, at $z=L_\pi$ the input wave has been rotated by 2φ with respect to x-axis; depicted in Fig. 1 (c). To avoid the reversal of power conversion and synchronize the power conversion, where $\varphi<45°$, the top loaded layer of the asymmetric loaded layer rib waveguide must be inverted w.r.t the center of rib waveguide at $z=L_\pi$ where z is the propagation direction, Fig. 1(a). At the end of the next brick, the polarization of the input signal has been rotated by 4φ; depicted in Fig 1(d). The top loaded layers will be arranged periodically and repeated until the total phase shift becomes 90° (Snyder & Love, 1983).

In single section polarization rotator structures, φ is adjusted to 45°; so that, 90° polarization rotation could be achieved by only one section.

Several structures of longitudinally variable passive polarization rotators have been reported in literature including: periodic asymmetric loaded rib waveguide (Shani et al., 1991), periodic tilted waveguiding section (Heidrich et al., 1992), periodically loaded strip waveguide (Mertens et al., 1998), cascaded bend waveguides (Van Dam et al., 1996; Liu et al., 1997, 1998a, 1998b) and a mode-evolution-based polarization rotator structure (Watts et al., 2005).

Periodic asymmetric loaded rib waveguide was experimentally demonstrated by Shani. The asymmetric loading of the waveguide would perturb the axes of the primary waveguide. By periodically alternating the loaded layer in longitudinal direction the polarization conversion or rotation will be accumulated coherently. The total length of the device was more than 3 mm. Haung and Mao employed coupled mode theory based on scalar modes to analyze the structure theoretically (Haung & Mao, 1992). Later on, Obayya and at el.

Employed full vectorial analysis based on versatile finite element beam propagation method (VFEBPM) to improve the design and reduce the polarization conversion length to 400 μm at operating wavelength of 1.55 μm (Obayya et al., 2000). Due to the huge size of the device, the design could not be verified with the aid of rigorous numerical methods, three dimensional finite difference time domain method (3D-FDTD).

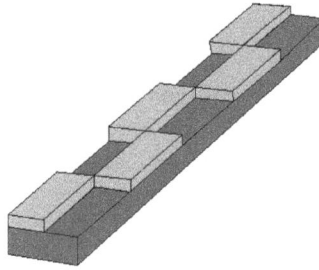

(a)

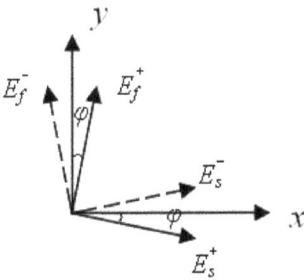

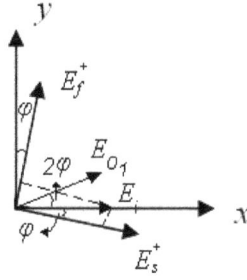

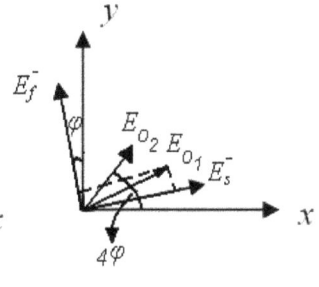

(b)   (c)   (d)

Fig. 1. The sketch of asymmetric loaded rib waveguide and its normal modes (a) asymmetric loaded rib waveguide (b) normal modes of perturbed optical axis (E_s^+, E_f^+), (E_s^-, E_f^-) (c) the input electrical field, E_i, and the rotated output field, E_{o1}, after propagation distance of L_π (d) the output field after another L_π, E_{o2} (E_s and E_f are the slow and fast modes, respectively).

Based on the same idea, tilted waveguiding polarization converter was introduced first by Heidrich (Heidrich et al., 1992). The first device was implemented by laterally tilted InP/GaInAsP rib waveguide on stepped substrate. The total length of the device was more than 7 mm . Later on by improving the design using coupled mode theory and BPM, the more compact device with the total length of 0.9 mm was implemented by Van der Tol. Besides being bulky, the device undertakes huge coupling loss at the junctions between the adjacent periodic sections. To eliminate the loss, single section devices were proposed (Tzolov & Fontaine, 1996; Huang et al., 2000; Rahman et al., 2001). By using angled single section waveguides in InP/InGaAs material system, another type of short polarization rotator was designed and fabricated (El-Refaei & Yevick, 2003).

Silicon based polarization rotators are more attractive in the sense that fabrication process is more compatible with the complementary metal-oxide semiconductor (CMOS) technology. Chen and et al. introduced silicon slanted rib waveguide for polarization rotation (Chen et al., 2003). Deng and et al. implemented slanted wall in Si by wet etching of Silicon <100>; thus, the side wall angle (52°) was not a flexible parameter. The total length of the fabricated device was more than 3 mm which was considered bulky (Deng, 2005). Moreover, the fabrication process of slanted-wall ridge waveguide is not compatible with planar optics circuit.

Recently, Wang and Dai proposed Si nanowire based polarization rotator with asymmetrical cross section, depicted in Fig. 2. The side wall is vertical; thus, it could be realized utilizing dry etching, reactive ion etching (RIE) (Wang & Dai, 2008). They were able to design asymmetric si nanowire device as small as 10 µm. Single mode guiding is required to avoid multimode interface that leads to lower polarization conversion efficiency. However, single mode silicon nanowires are so small that makes the fabrication very difficult and challenging. The fabrication tolerance is very small; thus, the proposed structure is not a robust device in the sense that small fabrication error could diminish the performance of the device. Moreover, to achieve a compact polarization rotator, the height of the loading (h) is 240 nm that is almost half of the thickness of the nanowire (H=500 nm) leading to a huge coupling loss.

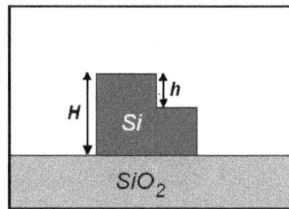

Fig. 2. The sketch of the cross section of asymmetric Si nanowire for polarization rotation application (Wang & Dai, 2008).

Having studied the existing polarization rotators, the major issues are either size or complexity of the structure. To tackle these issues, a PC based polarization rotator is introduced in the following section.

3. PC based polarization rotator

3.1 Introduction

The two main criteria are compactness and compatibility with planar PC circuit. PC structure was emerged as the best candidate to meet both requirements. PC slab waveguide is highly birefringent and compact; moreover, the fabrication process of the asymmetric loaded PC slab waveguide and integrated planar optical circuits are compatible.

Fig. 3(a) shows the schematic of the proposed polarization rotator. It consists of a single defect line PC slab waveguide. The geometrical asymmetry that is required to couple two orthogonal polarizations to each other was introduced to the upper layer of the defect line. The upper layer is made of the same material as the slab layer etched asymmetrically with respect to z-axis (propagation direction). Power conversion reversal happens at half beat lengths along the line. In order to avoid power conversion reversal and synchronize the coupling, the upper layer that is half beat length long is alternated on either side of the z-axis with the given period. The proposed structure is described as periodic asymmetric loaded PC slab waveguide. Because of the large birefringence of PC structures, the PC based polarization rotator is expected to be very compact as opposed to periodic asymmetric loaded rib waveguide. Compact structure requires smaller number of loading layers; hence the radiation loss at the junctions between different sections will be reduced.

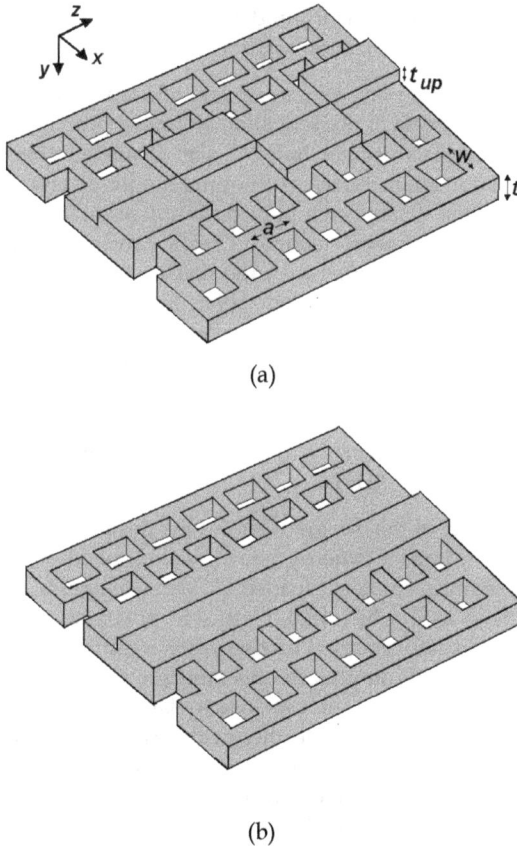

(a)

(b)

Fig. 3. The sketch of (a) periodic asymmetric loaded triangular PC slab waveguide (b) asymmetric loaded PC slab waveguide.

Due to the compactness of the structure, a rigorous numerical method, 3D-FDTD can be employed for analysis and simulation. However, for preliminary and quick design an analytical method that provides the approximate values of the structural parameters is preferred. Coupled-mode theory is a robust and well-known method for the analysis of perturbed waveguide structures. Thus, the coupled-mode theory based on semi-vectorial modes was developed for PC structures (Bayat et al., 2009). However, the frequency band of the modes of the asymmetric loaded PC slab waveguide must be determined prior to the coupled mode analysis. Plane wave expansion method (PWEM) was employed for modal analysis of the asymmetric loaded PC slab waveguide. For coupled mode analysis, the semi-vectorial modes of the asymmetric loaded PC slab waveguide, Fig. 3 (b), were calculated using semi-vectorial beam propagation method (BPM) of RSOFT, version 8.1. Coupled mode theory was employed to calculate the cross coupling between x-polarized and y-polarized waves. To simplify the problem for analytical calculations, instead of circular-hole PC pattern, square-hole PC pattern was employed. The coupled mode theory is an approximate method that provides an estimation of the structural parameters. The combination of the coupled mode theory and PWEM provides the frequency band over which low loss high efficiency polarization rotation is expected to be achieved.

Although, coupled mode theory presents a quick and efficient design methodology; it poses a tedious and error-prone process of discretization along the propagation direction for more complicated geometries such as circular hole PC slab. Thus, another design methodology based on vector-propagation characteristics of normal modes of asymmetric loaded PC slab waveguide was implemented. It can be employed for any arbitrary shape PC slab structure. Normal mode analysis of asymmetric loaded PC slab waveguide provides with fast and slow normal modes of the structure; so that, the half-beat length can be easily calculated (Mrozowki, 1997). The vector-propagation characteristics of normal modes of the asymmetric loaded PC slab waveguide were calculated using 3D-FDTD analysis combined with spatial fourier transform (SFT) of the electric field along the propagation direction. Both coupled-mode analysis and normal mode analysis led to the same results for square shape PC slab waveguide based polarization rotator. To verify the design parameters obtained using coupled mode theory and normal mode analysis, polarization rotator structure was simulated using 3D-FDTD.

Fig. 4 presents the flow chart of the design. It shows the design consists of three main steps. In the first step, the operational frequency band is determined using PWEM analysis. In this step, the thickness of the asymmetric loaded PC slab waveguide is optimized to provide maximum frequency band over which highly efficient polarization conversion is expected to take place. In the second step, coupled mode theory is employed for preliminary design of the polarization rotator. The outputs of this step are the length of top loaded layers (half-beat length) and total number of top loaded layers. Finally, to verify the design parameters obtained using coupled mode theory the 3D-FDTD simulation is performed. Coupled mode theory was developed for square hole geometries. For circular hole PC structure, another design methodology based on 3D-FDTD was developed, shown by 2' in the figure. As it was explained before, 3D-FDTD analysis of asymmetric loaded PC slab waveguide combined with SFT was employed to obtain the vector-propagation characteristics of the normal modes (slow and fast modes) of the structure. The accuracy of this method can be examined by 3D-FDTD simulation of the polarization rotator structure. In Sec. 3.2, the design methodology has been elaborated. The design and simulation results are presented in Sec. 3.3.

3.2 Theory

The schematic of the asymmetric square-hole PC slab polarization rotator is shown in Fig. 3(a). In this structure, the unit cell, the width of the square holes, the thicknesses of silicon

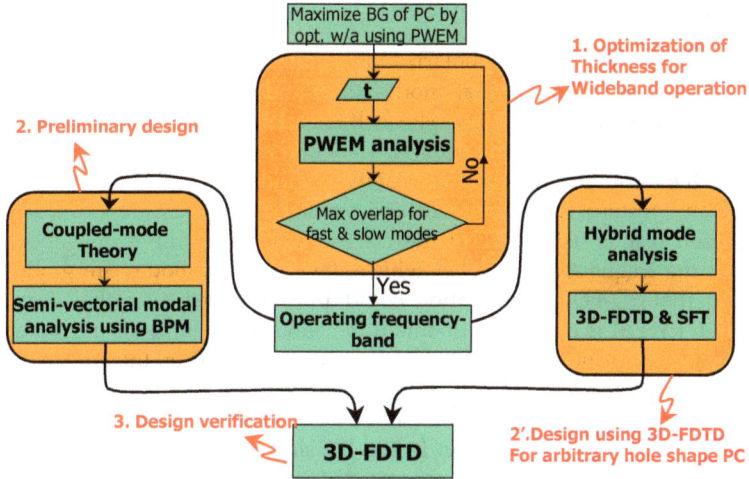

Fig. 4. The flow chart of the design methodology of PC slab waveguide based polarization rotator.

PC slab waveguide and top loaded layer are represented by a, w, t and t_{up}, respectively. The top cladding layer is asymmetric with respect to the z-axis (propagation direction) and alternates periodically throughout the propagation direction to synchronize the coupling between the two polarizations. The vector wave equation for the transverse electric field (x-y and z are the transverse and propagation directions, respectively) is given by (Haung et al., 1992):

$$\frac{\partial^2 E_x}{\partial z^2} + \nabla_t^2 E_x + n^2 k^2 E_x = -\frac{\partial}{\partial x}\left(E_x \frac{1}{n^2}\frac{\partial n^2}{\partial x}\right) - \frac{\partial}{\partial x}\left(E_y \frac{1}{n^2}\frac{\partial n^2}{\partial y}\right)$$ (6.a)

$$\frac{\partial^2 E_y}{\partial z^2} + \nabla_t^2 E_y + n^2 k^2 E_y = -\frac{\partial}{\partial y}\left(E_x \frac{1}{n^2}\frac{\partial n^2}{\partial x}\right) - \frac{\partial}{\partial y}\left(E_y \frac{1}{n^2}\frac{\partial n^2}{\partial y}\right)$$ (6.b)

where, n is the refractive index distribution of the waveguide and ∇_t^2 is the transverse differential operator defined as:

$$\nabla_t^2 = \frac{\partial^2}{\partial x^2} + \frac{\partial^2}{\partial y^2}$$

The vector properties are manifested on the right hand side of equ. (6.a), equ. (6.b); which indicates that the two orthogonal polarizations may be coupled to each other as a result of geometrical asymmetry.

Two approaches including normal mode analysis of the asymmetric loaded PC slab waveguide and coupled mode theory based on semi-vectorial modes were employed to design the polarization rotator structure. Both methods are explained in the following subsections.

3.2.1 Normal mode analysis using 3D-FDTD

The first approach to design the polarization rotator structure is to calculate vector-propagation characteristics of normal modes (fast and slow modes) of the asymmetric loaded PC slab waveguide (Fig. 3(b)). The half-beat length and the total number of the loaded layers can be calculated using equ. (1). To obtain modal characteristics of the fast and slow modes of the asymmetric loaded PC slab waveguide, 3D-FDTD method is employed. The propagation constants of x-polarized and y-polarized waves are extracted from 3D-FDTD simulation results using SFT of the electric field along the propagation direction. However, first the frequency band over which slow and fast modes are guided must be determined so that 3D-FDTD simulation could be performed over the aforementioned frequency band. PWEM is employed to obtain the band diagram of the asymmetric loaded PC slab waveguide. To calculate the birefringence of the structure, the effective frequency-dependent index of refraction of the normal modes is to be calculated. To obtain the aforementioned data, the accurate dispersion analysis is carried out, which is based on SFT of the electromagnetic field distribution in the PC slab waveguide along the propagation direction at any point on defect line cross section, the plane normal to the propagation direction (y, z). To employ SFT, it is assumed that the electromagnetic field in the PC slab waveguide can be expressed as a modal expansion at the normal plane, as following:

$$E_\omega(x,y,z) = \sum_{n,m} E_{n,m,\omega} e^{j\beta_{n,m,\omega} z} \qquad (7)$$

where $E_{n,m,\omega}$ and $\beta_{n,m,\omega}$ represent the electric field component and the propagation constant of the $(n,m)^{th}$ mode at frequency ω. The peaks of the SFT spectrum describe the propagating modes of the structure. These peaks are independent of the location, (x_0, y_0) and the electromagnetic field components. The effective refractive indices of the modes can be determined by locating these peaks.

Having determined the effective refractive indices of the fast and slow modes, L_π can be calculated as well as the total number of top loaded layers. The design methodology presented in this subsection requires finding the vector-propagation characteristics of normal modes of asymmetric loaded PC slab waveguide. It is a very general methodology and can be extended to any air hole geometry of PC slab structure as opposed to coupled-mode theory that is more efficient for simple air hole geometry PC structures such as square hole PC slab based polarization rotator. In the following subsection, coupled-mode theory based on semi-vectorial modes of the structure is presented.

3.2.2 Coupled-mode theory

Huang and Mao employed similar coupled mode theory based on the scalar modes to analyze polarization conversion in a periodic loaded rib waveguide. In a PC slab waveguide, the propagation characteristics strongly depend on the polarization of the propagating wave leading to a large birefringence (Genereux et al., 2001). However, scalar modal analysis completely ignores the polarization dependence of the wave propagation; thus, it is too simplified to represent the wave propagation inside a PC slab waveguide. Here, coupled mode theory based on semi-vectorial modes of a PC structure was developed to analyze the asymmetric loaded PC slab waveguide. Using semi-vectorial modal analysis, the polarization dependence of wave propagation has been partially taken into account; thus, the coupling between the two x-polarized and y-polarized waves can be modeled more accurately using coupled mode analysis.

In a triangular lattice PC structure, cross-section varies along the propagation direction within one unit cell. Employing square holes instead of circular holes simplifies the problem of modeling of such structures. According to Fig. 5, the unit cell can be divided into two regions with designated coupling coefficients. Thus, the problem boils down to calculating the coupling coefficients for regions 1 and 2. Semi-vectorial BPM (BPM package of RSOFT) was employed to calculate the semi-vectorial modes of the asymmetric PC slab waveguide shown in Fig. 3(b). The output of BPM analysis were the profile and the propagation constants of the x-polarized and y-polarized modes of the asymmetric loaded PC slab waveguide that were used to calculate the coupling coefficients of the x-polarized and y-polarized waves. Assuming that the profile of the total transverse field in the asymmetric loaded PC slab waveguide is represented as following:

$$E = E_x \hat{x} + E_y \hat{y} = a_x(z) e_x(x,y) e^{-j\beta_x z} + a_y(z) e_y(x,y) e^{-j\beta_y z}, \quad (8)$$

Where $e_x(x,y) e^{-j\beta_x z}$ and $e_y(x,y) e^{-j\beta_y z}$ are x- and y-components of electric field of the semi-vectorial solution of wave equation for x-polarized and y-polarized waves, respectively. β_x and β_y are propagation constants along x and y directions, respectively. Substituting equ. (8) into equ. (6) and multiplying both side of equ. (6.a), and equ. (6.b) by $e^{j\beta_x z}$ and $e^{j\beta_y z}$, respectively, and assuming that the amplitude of the field are slowly varying along z-direction (propagation direction); the following equation is obtained:

$$-j2\beta_x e_x \frac{da_x(z)}{dz} + a_x(z)\nabla_t^2 e_x + n^2 k^2 a_x(z) e_x - \beta_x^2 a_x(z) e_x =$$
$$-a_x(z)\frac{\partial}{\partial x}(e_x \frac{1}{n^2}\frac{\partial n^2}{\partial x}) - a_y(z) e^{-j\Delta z}\frac{\partial}{\partial x}(e_y \frac{1}{n^2}\frac{\partial n^2}{\partial y}) \qquad (9.a)$$

$$-j2\beta_y e_y \frac{da_y(z)}{dz} + a_y(z)\nabla_t^2 e_y + n^2 k^2 a_y(z) e_y - \beta_y^2 a_y(z) e_y =$$
$$-a_y(z)\frac{\partial}{\partial y}(e_y \frac{1}{n^2}\frac{\partial n^2}{\partial y}) - a_x(z) e^{j\Delta z}\frac{\partial}{\partial y}(e_x \frac{1}{n^2}\frac{\partial n^2}{\partial x}) \qquad (9.b)$$

Where: $\Delta = \beta_y - \beta_x$,

By invoking the following assumption:

$$\nabla_t^2 e_x + (n^2 k^2 - \beta_{ave}^2) e_x = 0$$
$$\nabla_t^2 e_y + (n^2 k^2 - \beta_{ave}^2) e_y = 0 \qquad (10)$$

Where, $\beta_{ave} = \dfrac{\beta_x + \beta_y}{2}$,

A simplified form of equ. (9) is obtained:

$$-j2\beta_x e_x \frac{da_x(z)}{dz} + (\beta_{ave}^2 - \beta_x^2)a_x(z)e_x =$$
$$-a_x(z)\frac{\partial}{\partial x}(e_x \frac{1}{n^2}\frac{\partial n^2}{\partial x}) - a_y(z)e^{-j\Delta z}\frac{\partial}{\partial x}(e_y \frac{1}{n^2}\frac{\partial n^2}{\partial y})$$

(11.a)

$$-j2\beta_y e_y \frac{da_y(z)}{dz} + (\beta_{ave}^2 - \beta_y^2)a_y(z)e_y =$$
$$-a_y(z)\frac{\partial}{\partial y}(e_y \frac{1}{n^2}\frac{\partial n^2}{\partial y}) - a_x(z)e^{j\Delta z}\frac{\partial}{\partial y}(e_x \frac{1}{n^2}\frac{\partial n^2}{\partial x})$$

(11.b)

Multiplying both sides of equ. (11.a), and equ. (11.b) by e_x^* and e_y^* (*- conjugate), respectively and integrating over the cross-section, the following coupled mode equations are obtained:

$$\frac{da_x(z)}{dz} = -j\kappa_{xx}a_x(z) - j\kappa_{xy}a_y(z)$$
$$\frac{da_y(z)}{dz} = -j\kappa_{yy}a_y(z) - j\kappa_{yx}a_x(z)$$

(12)

Where:

$$\kappa_{xx} = \frac{(\beta_{ave}^2 - \beta_x^2)\iint e_x^*.e_x dxdy + \iint e_x^*.\frac{\partial}{\partial x}(e_x \frac{1}{n^2}\frac{\partial n^2}{\partial x})dxdy}{2\beta_x \iint e_x^*.e_x dxdy}$$

(13.a)

$$\kappa_{xy} = \frac{e^{-j\Delta z}\iint e_x^*.\frac{\partial}{\partial x}(e_y \frac{1}{n^2}\frac{\partial n^2}{\partial y})dxdy}{2\beta_x \iint e_x^*.e_x dxdy}$$

(13.b)

$$\kappa_{yy} = \frac{(\beta_{ave}^2 - \beta_y^2)\iint e_y^*.e_y dxdy + \iint e_y^*.\frac{\partial}{\partial y}(e_y \frac{1}{n^2}\frac{\partial n^2}{\partial y})dxdy}{2\beta_y \iint e_y^*.e_y dxdy}$$

(13.c)

$$\kappa_{yx} = \frac{e^{j\Delta z}\iint e_y^*.\frac{\partial}{\partial y}(e_x \frac{1}{n^2}\frac{\partial n^2}{\partial x})dxdy}{2\beta_y \iint e_y^*.e_y dxdy}$$

(13.d)

κ_{xx} and κ_{yy} are the self-coupling coefficients; whereas, κ_{xy} and κ_{yx} refer to cross-coupling coefficients. In equ. (13.a), and equ. (13.c), the second terms were negligible in comparison with the first terms. The coupling coefficients must be solved for both regions 1 and 2 (see Fig. 5), using equ. (13). The distribution of the electric fields in both regions are the same;

where as, the refractive index profile is different as depicted in Fig. 5 leading to different values of coupling coefficients for regions 1 and 2. If the cross-coupling coefficients in both regions 1 and 2 were assumed to be equal ($\kappa_{xy}= \kappa_{yx}= \kappa$), the coupled-mode equations could be solved analytically as presented in equ. (14) below. Nonetheless, numerical methods could be easily implemented for general cases where the cross-coupling coefficients were not equal. Given the exact analytical solution as $A(z)=MA(0)$; where A is a column vector for coefficients a_x and a_y; the transfer matrix (M) is expressed as following:

$$M_{i\pm} = \begin{pmatrix} \cos(\Omega_i z_i) - j\cos(\varphi_i / 2)\sin(\Omega_i z_i) & \mp j\sin(\varphi_i / 2)\sin(\Omega_i z_i) \\ \mp j\sin(\varphi_i / 2)\sin(\Omega_i z_i) & \cos(\Omega_i z_i) + j\cos(\varphi_i / 2)\sin(\Omega_i z_i) \end{pmatrix} \quad i=1,2 \qquad (14)$$

$$\Omega_i = \sqrt{\delta_i^2 + \kappa_i^2}$$

$$\delta_i = \frac{\kappa_{xxi} - \kappa_{yyi}}{2} \qquad (15)$$

$$\tan(\varphi_i / 2) = \frac{\kappa_i}{\delta_i}$$

The $\pm$ signs correspond to the alternative sections of the periodic loading. z_1 and z_2 are the length of regions 1 and 2 shown in Fig. 5. Assuming that w is the width of a square hole, z_1 and z_2 are determined as following:

$$z_1=a-w \qquad (16.a)$$

$$z_2=w \qquad (16.b)$$

Having set M_1 and M_2 as the transfer matrix of regions 1 and 2, the transfer matrix for one unit cell is obtained:

$$M_\pm=M_{1\pm}.M_{2\pm}, \qquad (17)$$

The loading period or half-beat length can be approximated as follows:

$$L_\pi \approx \frac{\pi}{\Omega_1 + \Omega_2}, \qquad (18)$$

Thus, the length of one top silicon brick is L_π and the top cladding layer alternates periodically throughout the propagation length. The simulation results revealed that for our structure, $|k_{xy}| \approx |k_{yx}|$ and $k_{xy} \approx k_{yx}^*$; where the imaginary parts were very small. Numerical, and analytical solutions of the coupled mode theory, equ. (12.a) and (8.b), give us almost the same results. From equ. (14) the preliminary value of the loading period before employing the numerical method to solve the coupled mode equation, equ. (12), was calculated.

In next section, first the band structure of the asymmetric loaded PC slab waveguide is calculated using PWEM. Therefore, frequency band over which the polarization rotator can be operated is determined.

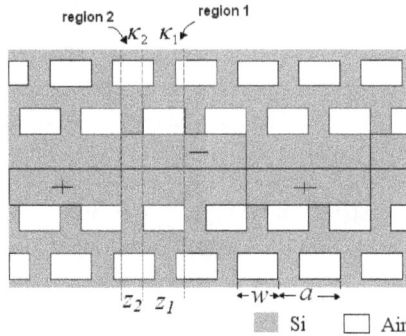

Fig. 5. Top view of the asymmetrically loaded PC based polarization rotator. The top cover layer is marked by the dark solid line in the figure. κ_1 and κ_2 represent the cross-coupling coefficient for regions 1 and 2 inside a unit cell.

3.3 Design of the polarization rotator

In this section the results of the design using both coupled mode theory and normal mode analysis are presented.

3.3.1 Design of the polarization rotator using 3D-FDTD modal analysis

3.3.1.1 PWEM analysis

The first step of the design of the polarization rotator structure is to calculate the frequency band over which lossless propagation takes place for both x-polarized and y-polarized waves and then proceed with the design using both normal mode analysis and coupled mode theory within the aforementioned frequency band. The asymmetric loaded PC slab waveguide shown in Fig. 3(b) was first simulated using PWEM to obtain the band diagram and the frequency band of the modes. The thickness of the PC slab structure plays an important role on the polarization dependent guiding (Bayat et al., 2007). It is possible to obtain maximum overlap between x-polarized and y-polarized waves by optimizing the thickness of the PC slab waveguide.The same design methodology was employed to design the thickness of the asymmetric loaded PC slab waveguide.

Fig. 6(a) shows the super-cell for the asymmetric loaded triangular PC slab waveguide. By including several unit cells in horizontal plane, the defect lines in the super-lattice structure are isolated. In PWEM analysis, the definition of the TE-like and TM-like waves is based on the symmetry planes of the modes. The dominant components of TM-like mode (H_y, E_z, E_x) and the non-dominant components (E_y, H_z, H_x) have even and odd symmetry w.r.t. y=0 plane, respectively. Similarly, the dominant components of TE-like mode (E_y, H_z, H_x) and the non-dominant components (H_y, E_z, E_x) have even and odd symmetry to y=0 plane, respectively. The thickness of the top loaded layer, t_{up}, is an important design parameter. Heuristically speaking, the larger the thickness of the upper layer is, the stronger the geometrical asymmetry is leading to a more compact device. However, to comply with the fabrication constrains the upper limit of the thickness of the loaded layer is restricted to t_{up}=0.2a.

The band diagrams for two different slab thicknesses t=0.6a and t=0.8a are obtained by PWEM and plotted in Fig. 6(b) and (c), respectively. The thickness of the top loaded layer, width of the PC squares and refractive index of silicon are t_{up}=0.2a, w=0.6 and n_{si}=3.48, respectively. There are two modes depicted by dotted and solid lines. The mode graphed by dotted line resembles an index-guided mode except the mini-stop band observed at the zone boundary. We will call it index-guided mode. The other mode depicted by solid line will be called Bloch mode. The index-guided mode is considered to be y-polarized wave for which the dominant electric field component is in y-direction. On the other hand, the Bloch mode is considered x-polarized wave. For t=0.6a, the index-guided mode crosses the Bloch mode and is folded back at the zone boundary at a/λ=0.28. The Bloch mode touches the zone boundary (K) at a/λ=0.259 and crosses TE-like PC slab modes at a/λ=0.274. Since the index-guided mode and Bloch mode cross each other, the difference between the effective refractive indices of the two modes, which is proportional to 1/L_π, varies with frequency significantly. Thus, the polarization converter made of a PC slab with t=0.6a is expected to be narrow band.

On the other hand, by increasing the thickness of the slab to t=0.8a, the index-guided mode has been pushed down to the lower frequencies. The index-guided mode depicted by dotted line in Fig. 6(c) has folded back at the zone boundary at a/λ=0.243. Within the frequency band of the Bloch mode, 0.257-0.267, the two bands are parallel; thus, the variation of the difference between the effective indices of the two modes and L_π with frequency are negligible. In this diagram, the index-guided and Bloch modes also correspond to fast and slow modes, respectively. At normalized frequency of a/λ=0.267, Bloch mode crosses TE-like PC slab mode. In 3D-FDTD simulations, the central normalized frequency of a/λ=0.265 is assigned to f=600 GHz resulting in the unit cell size of a=132.5 μm. The cross-section of E_x, E_y, H_x and H_y components for TE-like input at a/λ=0.267 (f=604.5 GHz) are plotted in Fig. 7. Graphs verify the presence of PC slab modes. E_y and H_x components of PC slab modes own even symmetry w.r.t. y=0 plane as opposed to E_x and H_y components where have odd symmetry. Therefore, PC slab modes are TE-like verifying the PWEM analysis. Moreover, the field distribution inside the defect line indicates that E_y and H_x are the dominant components of electric and magnetic field, respectively indicating that TE-like wave at the input excites y-polarized wave. All four components of y-polarized mode have even symmetry with respect to y=0 and x=0 planes; although, the minor components of input wave, TE-like, (E_x and H_y) have odd symmetry w.r.t y=0 plane. Thus, TE-like input wave has been evolved into the mode of the asymmetric loaded PC slab waveguide. By further increasing the thickness of PC slab waveguide, higher order modes will be pushed down inside the bandgap; which is not suitable for our application. Having compared the band diagram for t=0.6a and t=0.8a; it is seen that t=0.8a suits better for the polarization conversion application.

Now that the overlap between x-polarized and y-polarized guiding are defined, 3D-FDTD simulation for extracting the modal characteristics of asymmetric loaded PC slab waveguide will be limited to the aforementioned frequency band. In following, the modal analysis using 3D-FDTD simulation is presented.

(a)

(b)

(c)

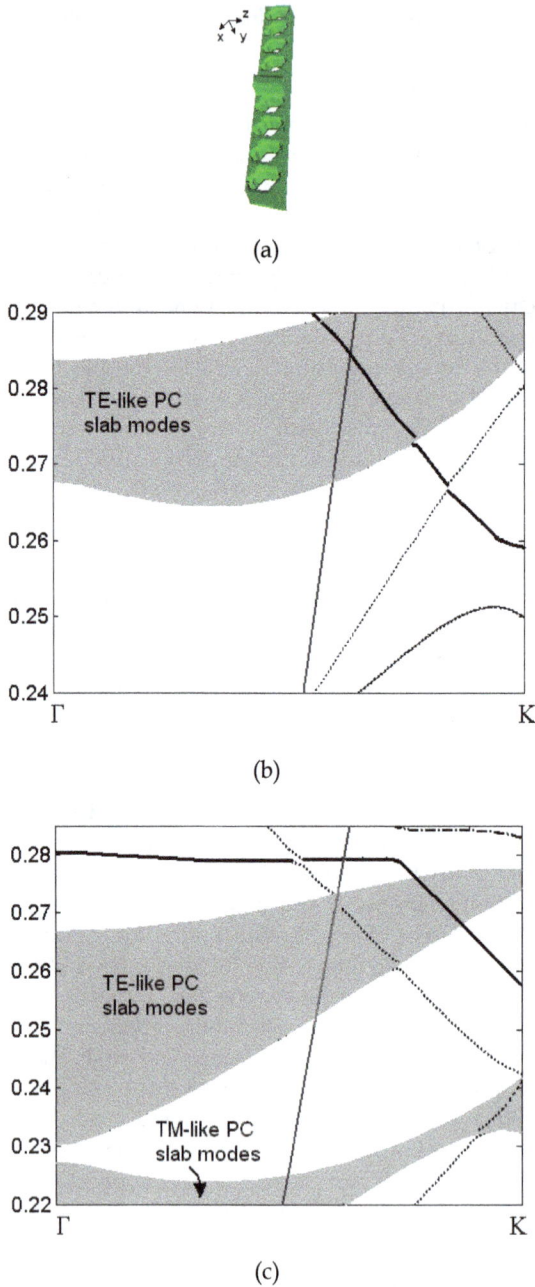

Fig. 6. (a) The supercell of the asymmetric loaded PC slab waveguide for PWEM analysis. The band diagram for the asymmetric loaded PC slab waveguide obtained by PWEM for (b) t=0.6a, t_{up}=0.2a and w=0.6a (c) t=0.8a, t_{up}=0.2a and w=0.6a.

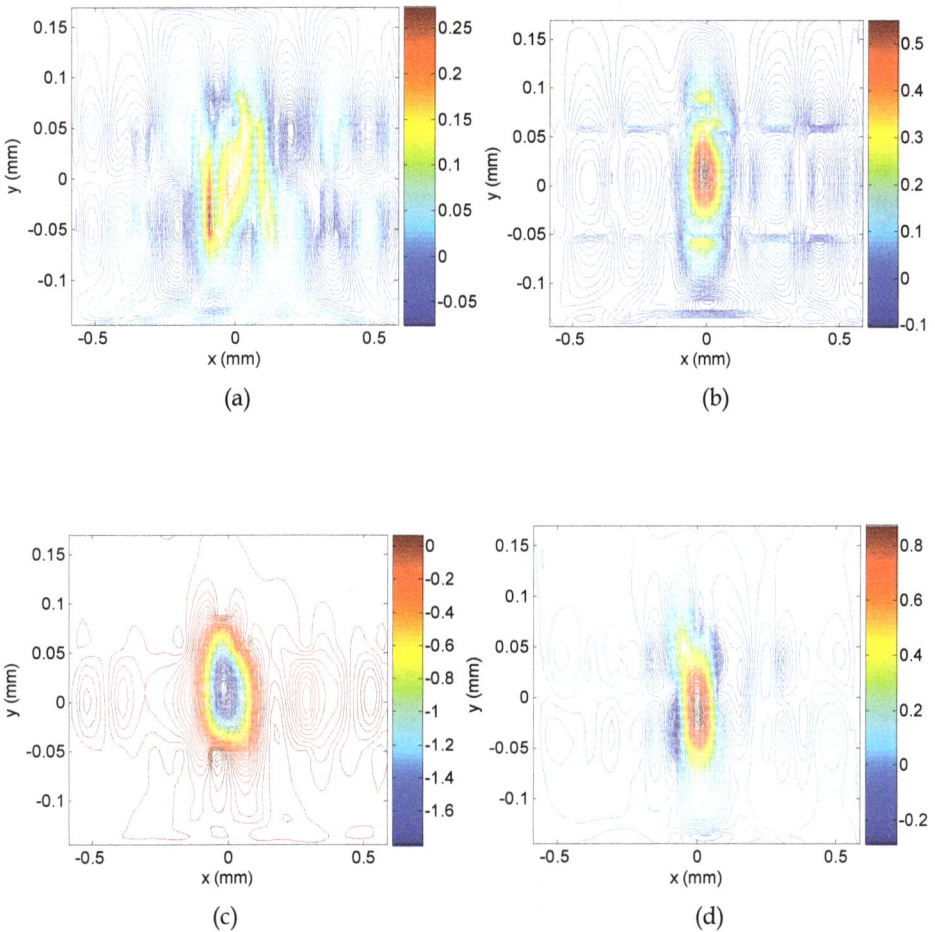

Fig. 7. Cross-section of electromagnetic field components obtained by 3D-FDTD analysis of asymmetric loaded PC slab waveguide at f=604.5 GHz (a/λ=0.267) with t=0.8a, w=0.6a, t_{up}=0.2a, n_{si}=3.48 (a) E_x (b) E_y (c) H_x (d) H_y.

3.3.1.2 3D-FDTD modal analysis

To calculate birefringence and L_{π}, it is required to obtain the vector-propagation characteristics of normal modes of the asymmetric loaded PC slab waveguide, Fig. 3(b), using 3D-FDTD simulation. To calculate the birefringence of the structure, the effective frequency-dependent indices of refraction of the structure for both E_x and E_y components are calculated using SFT analysis of transverse electric field components along the propagation direction at the center point of the defect line (x=0, y=0). For 3D-FDTD simulation, the input is a single frequency sinusoidal with guassian distribution in space. As time proceeds and wave propagates along z-direction inside the defect line, the wave evolves into the modes of the structure. Thus, in the steady state case by applying SFT to the field distribution along the propagation direction at any point inside the defect line, the propagation characteristics of the modes can be obtained.

In our design example, the structural parameters of the PC polarization rotator are determined by assigning the normalized central frequency of the fundamental mode, 0.265, to the operating frequency. For example for f=600 GHz corresponding to the normalized frequency of 0.265, the unit cell size would be 132.5 μm (a= 0.265λ). Fig. 6 (c) shows that the overlap between x-polarized and y-polarized guiding lay within the frequency band of 0.258-0.267. The normalized SFT diagram for the input normalized frequency of a/λ=0.265 is calculated and plotted in Fig. 8. Fig. 8(a) and 8(b) correspond to TE-like and TM-like excitations, respectively. For TE-like excitation, at the input E_y is the dominant component; however, as the wave proceeds, the input evolves into the normal modes of structure that are depicted by the two dominant peaks in SFT spectrum. In SFT spectrum, the horizontal axis is the $1/\lambda_g$ where $\lambda_g = \dfrac{\lambda}{n_{eff}}$. λ and n_{eff} are free space wavelength and effective refractive index of the propagating mode, respectively. For example, in Fig. 8(a) the SFT spectrum of E_y has a peak at $1/\lambda_g$=5.16 where coincides with one the peaks of SFT spectrum of E_x component. The corresponding effective refractive index n_{eff} or n_f (refractive index of fast normal mode) would be 2.58. The other peak of SFT spectrum of E_x component correspond to x-polarized or slow mode. It has been located at $1/\lambda_g$=5.52 resulting in n_{eff} or n_s (refractive index of slow normal mode) of 2.76. It is seen that the asymmetric loaded layer has induced a large birefringence. Having determined the effective refractive indices of the two normal modes, the half-beat length or L_π can be calculated using equ. (1). For the two refractive indices calculated above at λ=500 μm, half-beat length (L_π) would be 1.39 mm which is equivalent to 10.5a, a is the unit cell of PC slab. The same values are obtained for n_f and n_s from graph 7(b) which is plotted for TM-like input wave.

Another important parameter that can be extracted from modal analysis is polarization rotation angel, φ; which is tilted angel of optical axes with respect to Cartesian coordinate. The following expression is used to calculate the polarization rotation angel:

$$\varphi = \tan^{-1}\left(\frac{abs(SFT(E_x))}{abs(SFT(E_y))}\right)\Bigg|_{@peak,TE-like} = \tan^{-1}\left(\frac{abs(SFT(E_y))}{abs(SFT(E_x))}\right)\Bigg|_{@peak,TM-like} \qquad (19)$$

Value of polarization rotation angel (φ) for above example is 6.5º. Polarization rotation angel is important in determining the total number of loaded layers required to achieve 90º polarization rotation. For this example, at the end of fourth loaded layer, the polarization of input wave should be rotated by 104º that exceeds 90º. To compensate the extra rotation angel, the length of the last top loaded layer can be increased. Thus, the normal mode analysis provides with the length and total numbers of top loaded layers.

Below normalized frequency of a/λ=0.257, only y-polarized wave is guided; thus, no polarization rotation is expected as x-polarized wave is not guided. To verify it, normalized SFT diagram of asymmetric loaded PC slab waveguide for the input normalized frequency of a/λ=0.252 is calculated and plotted in Fig. 9. The input wave is y-polarized or TE-like wave. Thus, the dominant component of electric field is E_y. The normalized SFT spectrum of both E_y and E_x components of electric field are plotted in Fig. 9. The SFT diagram of E_x has only one peak verifying PWEM diagram that only y-polarized wave is guided at this frequency.

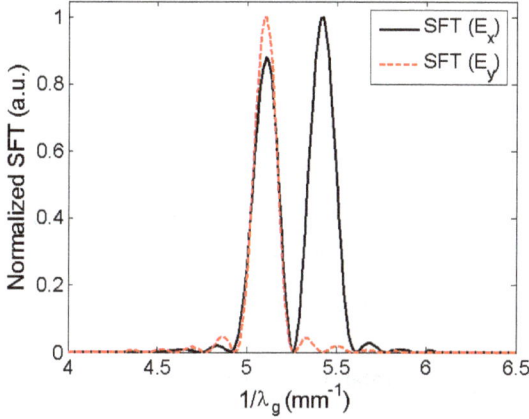

(a)

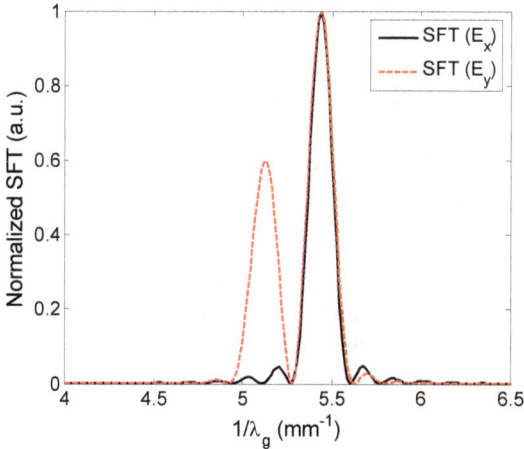

(b)

Fig. 8. The normalized SFT spectrum of transverse electric field components of asymmetric loaded PC slab waveguide for (a) TE-like wave input and (b) TM-like wave input (w=0.6a, t=0.8a and t_{up}=0.2a, λ=500 μm).

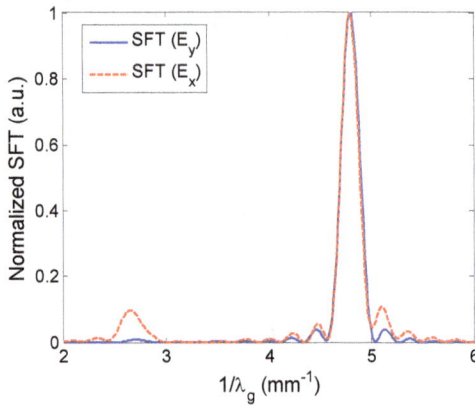

Fig. 9. The normalized SFT spectrum of transverse electric field components of asymmetric loaded PC slab waveguide for TE-like wave input (w=0.6a, t=0.8a and t_{up}=0.2a, a/λ_0=0.252).

In next section, the design of the polarization rotator using coupled-mode theory based on semi-vectorial modes is presented.

3.3.1.3 Design of the polarization rotator using coupled-mode theory

In this section, coupled mode theory discussed earlier was employed to design the asymmetrically loaded PC polarization rotator. In order to employ the coupled mode theory, first the semi-vectorial modes of the asymmetrically loaded PC slab waveguide, Fig. 3(b), must be calculated. Semi-vectorial BPM was employed for semi-vectorial modal analysis of the structure. The normalized electric field for x-polarized (TM-like) and y-polarized (TE-like) waves for the normalized frequency of a/λ=0.265 are shown in Fig. 10(a) and 10(b), respectively. It shows that the electric field distribution is asymmetric in both vertical and lateral directions as a result of the geometrical asymmetry. The propagation constants of the corresponding modes were calculated using semi-vectorial BPM simulation, as well. The effective refractive indices of x-polarized and y-polarized waves were 2.6567 and 2.5007, respectively. A big birefringence was observed as expected in PC slab waveguide structure. For aforementioned parameters, the coupling coefficients of the periodic asymmetric loaded PC polarization rotator (shown in Fig. 3(b)) were calculated using equ. (13) for both regions of 1 and 2, depicted in Fig. 5. Using equ. (15), the loading period was calculated, 10.8a. The value of half-beat length, L_π, computed using coupled-mode theory and normal mode analysis were 10.8a and 10.5a, respectively. Thus, both methods deliver the same results that is a proof of the effectiveness of them. Fig. 11 shows the power exchange between the two polarizations along the propagation distance for a/λ=0.275, 0.265 and 0.255, λ_0=500 μm (600GHz). The length of each top loaded layer is 10a.

Defining the power conversion efficiency (P.C.E.) as following:

$$P.C.E. = \frac{P_{TM}}{P_{TM} + P_{TE}} \times 100 = \frac{a_x^2}{a_x^2 + a_y^2} \times 100 \tag{20}$$

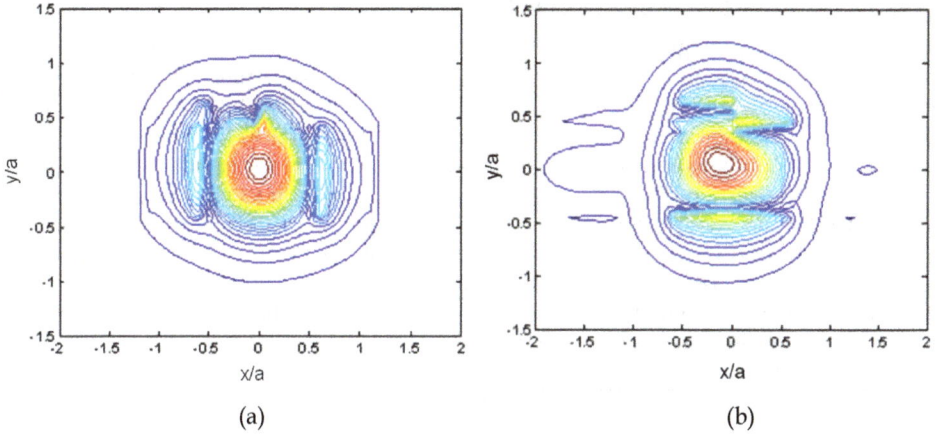

(a)   (b)

Fig. 10. The profile of (a) E_x and (b)E_y components of x-polarized and y-polarized modes of the structure shown in Fig. 3(b) obtained by semi-vectorial 3D BPM analysis ($t=0.8a$, $t_{up}=0.2a$, $w=0.6a$, $a=132.5$ μm, $n_{si}=3.48$ and $\lambda=500$ μm).

For $a/\lambda=0.265$ ($\lambda=0.5$ mm), 96% efficiency at $z=7.2$ mm (millimeter) was achieved. It is expected that by increasing or decreasing the normalized frequency, the power conversion efficiency reduce. To achieve high power conversion efficiency, the last silicon brick (top loaded layer) was no flipped around z-axis. In other word, the length of the last silicon brick was larger than 10a as it was predicted by normal mode analysis method, as well. The P.C.E. for $a/\lambda=0.275$ and 0.255 is larger than 75% at $z=7.2$ mm. Thus, it is expected to have a very high P.C.E. within the frequency band of the defect mode (0.258-0.267).

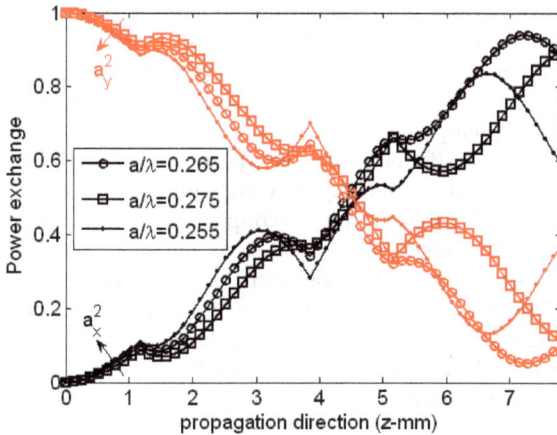

Fig. 11. Power exchange between the x-polarized and y-polarized wave versus the propagation length ($t=0.8a$, $t_{up}=0.2a$, $w=0.6a$, $a=132.5$ μm, $n_{si}=3.48$) for $a/\lambda=0.255$, 0.265 and 0.275 obtained by coupled-mode analysis.

3.4 3D-FDTD simulation of polarization rotator structure

Both Design methodologies suggest that high power exchange rate is expected to be observed within the frequency band of 0.258-0.267, the overlap frequency band between the fast and slow mode guiding. To verify the aforementioned results, 3D-FDTD was employed to simulate the polarization rotator. The simulated structure (Fig. 3(a)) consists of 70 rows of holes along the propagation direction (z-direction) and 11 rows of holes (including the defect row) in x-direction. The mesh sizes along the x, y and z-directions (Δx, Δy and Δz) are Δx=Δz=0.0331λ and Δy=0.0172λ. The perfectly matched layer (PML) boundary condition was applied for all three directions. Time waveforms in 3D_FDTD were chosen as a single frequency sinusoid. The spatial distribution of the incident field was Guassian. The frequency of the input signal lies within the frequency band of the defect mode (0.258-0.267 corresponding to 586-601 GHz). As the wave proceeds, the polarization of the input signal starts rotating. The power exchange between (E_x, E_y) and (H_x, H_y) components was observed. To achieve the maximum power conversion, the size of the last top silicon brick was 15a instead of 10a. Fig. 12 shows the contour plot of transverse field components, E_x, E_y, H_x and H_y at the input for a/λ=0.265. The input excitation is TE-like; E_y and H_x are the dominant components and have even parity as opposed to the non-dominant components E_x and H_y that have odd symmetry with respect to y=0 plane.

As the wave proceeds, the power exchange is observed between (E_y, E_x) and (H_x, H_y) components. The contour plot of E_x and H_y at a point close to the output are plotted in Fig. 13. It is seen that the parity of the E_x and H_y components have changed and become the dominant component. The amplitudes of E_y and H_x have been decreased more than an order of magnitude and reached to zero at the output plane. Thus, 90° rotation of polarization is realized at the output.

To show the power exchange between the two polarizations, the z-varying square amplitudes of E_x and E_y components were graphed. Fig. 14 shows $a_x^2(z)$ and $a_y^2(z)$ along the propagation direction for the normalized frequency of 0.265 corresponding to the free space wavelength of 500 μm. The two main elements contributing to the numerical noise are local reflections and imperfections of absorbing layer. Dots in the figure are the actual values of 3D-FDTD analysis. To have a smooth picture of $a_x^2(z)$ and $a_y^2(z)$ variations along the propagation direction, a polynomial fit to the data using least square method is also shown in the figure. Each plot consists of more than 100 data points. The FDTD "turn-on" transition of the input wave has also been included in the graph (first 0.5 mm). This portion is obviously a numerical artifact of the FDTD scheme. After almost 6 mm (12λ), the complete exchange, as can be seen in the Fig. 14, has taken place. Comparing this graph with coupled-mode (counterpart plot in Fig. 11), it is seen that the power exchange between the two polarizations takes place at smaller propagation distance; 6 mm in comparison with 7.2 mm. Moreover, the value of P.C.E obtained by 3D-FDTD is close to 100 %; whereas, P.C.E for the same wavelength for coupled-mode analysis is 96%. On the other hand, normal mode analysis method predicted that 100% polarization conversion could take place at less than 4.5L_π, 6 mm. Therefore, normal mode analysis design methodology provides more accurate results.

3D-FDTD simulations were repeated for other frequencies to obtain the frequency dependence of polarization conversion. The power exchange rate for both coupled-mode analysis and 3D-FDTD are graphed versus the normalized frequency in Fig. 15. Coupled-mode analysis shows that P.C.E of higher than 90% is achieved within the normalized frequency band of 0.258-0.267. 3D-FDTD simulation results show that the P.C.E higher than 90% is realized within the frequency band of 0.258-0.267; over which the defect mode lies.

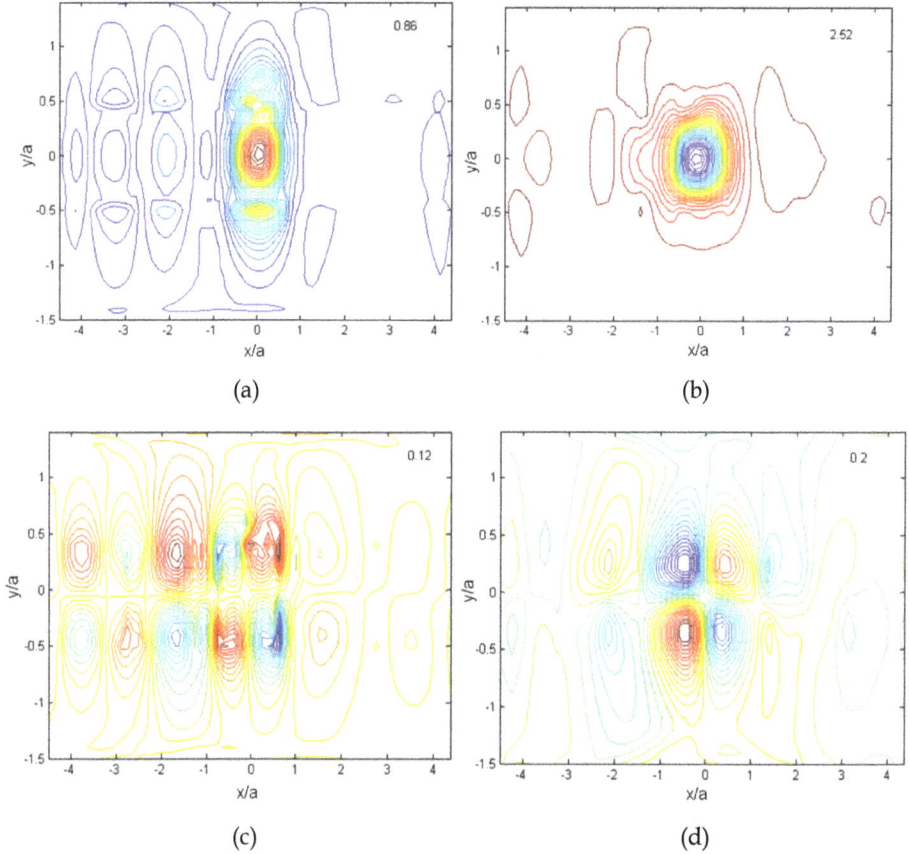

(a)

(b)

(c)

(d)

Fig. 12. The contour plot of the cross section of (a) E_y, (b) H_x, (c) E_x and (d) H_y components at the input plane (t=0.8a, t_{up}=0.2a, n_{si}=3.48, a/λ=0.265, λ=500 μm).

At normalized frequencies higher than 0.267, Ey starts leaking energy to the TE-like PC slab modes as it crosses the TE-like PC slab modes, Fig. 6(c). For example, FDTD simulation of the power exchange between x-polarized and y-polarized waves for a/λ=0.275 is graphed in Fig. 16. It is seen that for a/λ=0.275, the slope of the drop of $a_y^2(z)$ is much sharper than the slope of the rise of $a_x^2(z)$. More importantly, $a_y^2(z)$ is dropping much faster than that of a/λ=0.265, Fig. 14. This observation can be interpreted as if E_y is dissipating and leaking energy into the TE-like slab modes. Thus, a sudden drop on power exchange rate is observed at normalized frequencies higher than 0.267. Semi-vectorial BPM analysis utilized for modal analysis is not capable of including the PC modes; thus, in power exchange graph calculated by coupled-mode analysis for normalized frequency of a/λ=0.275 (Fig. 11), no power dissipation is observed as opposed to 3D-FDTD simulation (Fig. 16).

(a)

(b)

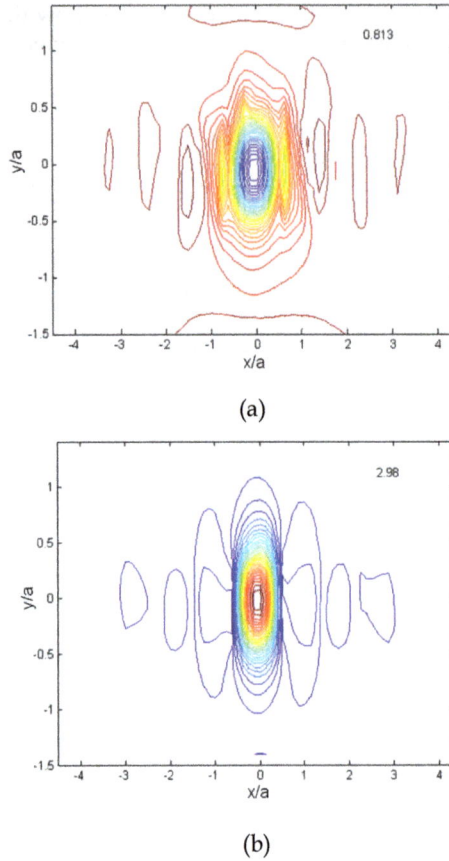

Fig. 13. Contour plot of the cross section of (a) E_x and (b) H_y at z=5.5 mm (t=0.8a, t_{up}=0.2a, n_{si}=3.48, a/λ=0.265, λ=500 µm).

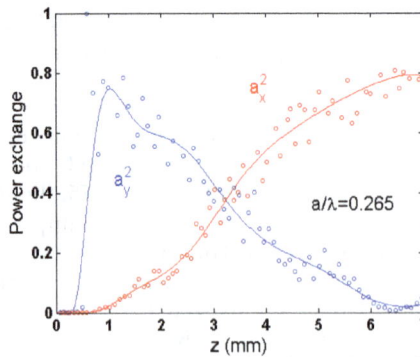

Fig. 14. Power exchange between the x-polarized and y-polarized wave versus the propagation length for a/λ=0.265 obtained by 3D-FDTD simulation (t=0.8a, t_{up}=0.2a, w=0.6a, a=132.5 µm, n_{si}=3.48, λ=500 µm).

At frequencies lower than $a/\lambda<0.255$, only y-polarized wave is guided; thus, no power exchange between the two polarization takes place. Our recommendation is to avoid this region for the design of the polarization rotator. Having compared FDTD and coupled-mode analyses, coupled-mode theory approach is effective within the frequency band where x-polarized and y-polarized guiding overlap. The other approach described in 3.3.1.2 can predict the frequency response of the PC based polarization rotator. In polarization rotation angel graph versus frequency obtained using normal mode analysis, a sudden jump in the polarization rotation angel at normalized frequencies larger than 0.268 and smaller than 0.257 is observed that is a sign of changing the behavior of the modes; hence the polarization rotator.

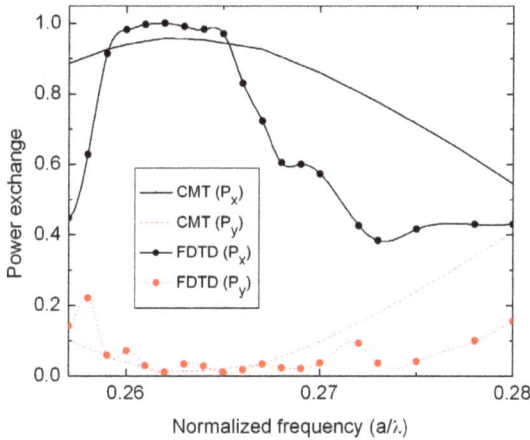

Fig. 15. Power exchange between the x-polarized and y-polarized waves versus frequency for both coupled-mode analysis and 3D-FDTD simulations ((t=0.8a, t_{up}=0.2a, w=0.6a, a=132.5 μm, n_{si}=3.48).

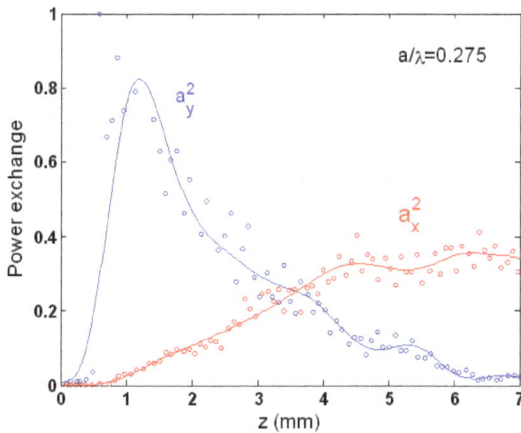

Fig. 16. Power exchange between the x-polarized and y-polarized wave versus the propagation length for a/λ=0.275 obtained by 3D-FDTD simulation (t=0.8a, t_{up}=0.2a, w=0.6a, a=132.5 μm, n_{si}=3.48, λ=500 μm).

4. Fabrication technique

In this section, the fabrication process of the Si membrane PC technology for THz application is presented. Highly resistive silicon (> 5-10 kΩ-cm) was employed to implement PC based membrane structures for THz applications. Here, a novel PC technology for the THz wave that is a potential candidate as a compact and low loss media for THz wave propagation was introduced. More importantly, this technology has a potential for integration with other optoelectronic and micro-electronic devices. Devices were fabricated in the frequency band of 200 GHz - 1 THz. The thickness of the device layer is chosen between 80 μm and 400 μm based on the design. The thickness of the buried oxide layer that separates the silicon device layer from the silicon handle layer could be varied between 0.5 μm-5 μm. Handle silicon layer is a low resistive silicon that attenuates the THz signal significantly and must be taken into consideration in the fabrication technology design.

The fabrication process consists of both front side and backside processing of the SOI wafer. The front side processing involves etching of deep holes into the silicon layer. Backside processing consists of opening window at the backside of the devices to prevent coupling of the THz wave to the lossy substrate modes and guarantee low loss propagation. To pattern the device layer, standard lithography is utilized as opposed to the optical PC structure that requires nanometer range lithography. In order to etch deep holes (> 100 μm), Deep Reactive Ion Etching (DRIE) is required. Thick photoresist must be utilized for photolithography to stand the long etching process. 4 μm thick photoresist (S1827) was employed as a soft mask for the DRIE process.

The etching of the holes was carried out using an optimized DRIE process to create holes with smooth vertical side walls and the desired aspect ratio. The Bosch process which alternated between two modes of nearly isotropic plasma etching using SF_6 for 7 sec and deposition of chemically inert passivation layer using C_2F_8 for 2 sec was employed. The SF_6 and C_2F_8 flow rates were 300 sccm and 150 sccm, respectively. The process temperature is kept at 20°C. The etching rate is around 5μm/min, and can vary slightly during the etching process. Buried oxide acts as an etch stop.

The second phase of the process is opening a window at the backsides of the SOI wafer to construct the membrane PC structure. Handle silicon is a thick silicon (525 μm), which was removed by wet chemical etching. 30% KOH etching at 90°C, which gave an etching rate of almost 50μm/hr, was used. The front side of the wafer must be protected from the KOH etching. The KOH etching was carried out by a custom made wet etching tool which only exposes the backside of the wafer to hot KOH solution. KOH mask that covers the unetched areas must withstand 10 hours of KOH at 90°C. A thick amorphous silicon nitride film (a-SiN) of 1μm, was deposited using PECVD technique to function as the hard mask at the backside of the SOI wafer. The second lithographic step was performed to pattern the SiN layer for the opening windows at the backside of the SOI wafer. Again buried silicon dioxide functions as the etching stop (Bayat et al., 2009).

The SEM image of a fabricated membrane PC slab waveguide is shown in Fig. 17 (a). It shows that the window under the active area (waveguide area) has been etched nicely. Fig. 17 (b), is SEM picture of the top side showing the air holes close-up. It can be seen that the walls are sharply etched. The backside etching is also very important and critical. Fig. 17 (c) shows the SEM picture of the backside. It can be seen that the back is etched uniformly; the oxide at the back can be easily removed by buffered HF (BHF).

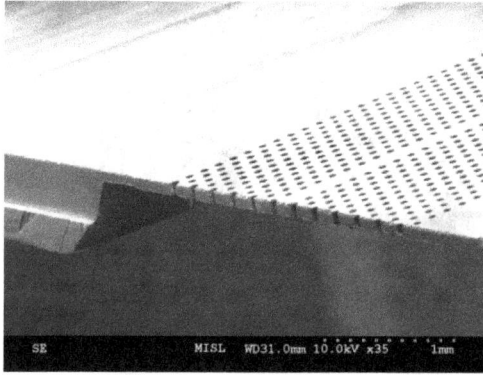

(a)

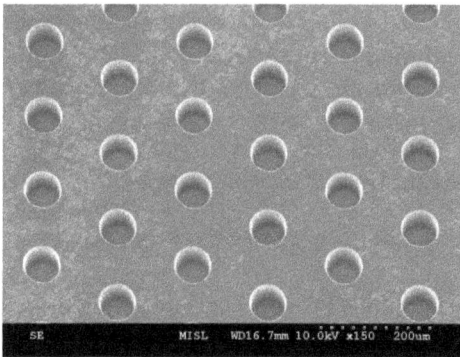

(b)

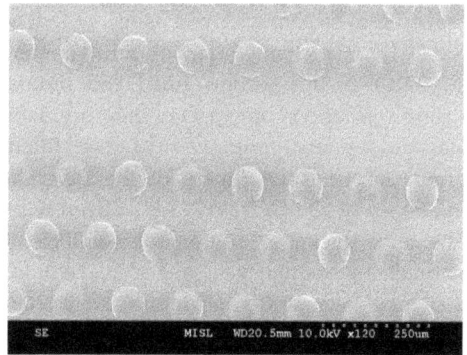

(c)

Fig. 17. SEM picture of fabricated (a) PC membrane slab waveguide (b) front side and backside of the PC structure.

Polarization rotation devices for potential applications in the THz frequency band (200 GHz – 1THz) were fabricated. The fabrication of this PC based polarization rotator is more complex in a sense that the front side processing requires two sets of masks. The first mask is employed to create the periodic loading layers. The second mask is for patterning of the PC slab waveguide. The third mask is used to open window at the back side of the structure. Fig. 18(a) shows the SEM picture of the periodic asymmetric loaded PC slab waveguide with square holes. The SEM picture shows that the walls are very sharp. In Fig. 18(b), the SEM picture of the periodic asymmetric loaded PC slab waveguide for circular air holes pattern is presented.

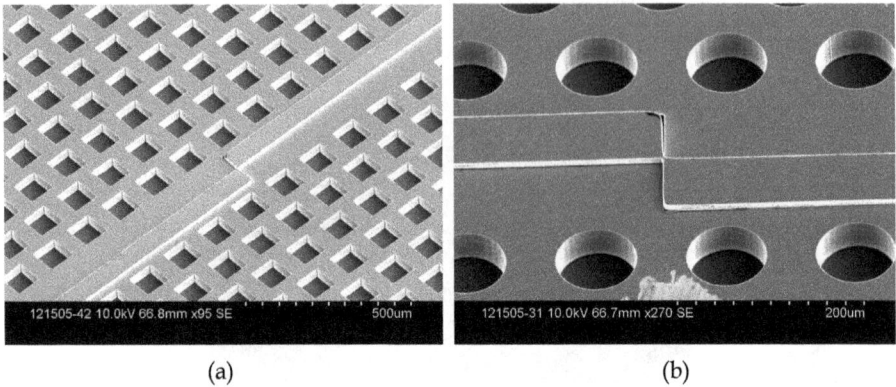

(a)   (b)

Fig. 18. SEM picture of (a) square hole PC polarization converter (b) circular hole PC and circular hole polarization converter.

5. Characterization setup

Series of devices in the frequency range of 0.2-1 THz are fabricated. The devices are being prepared to be characterized using Agilent Millimeter-wave PNA-X network analyzer (up to 500 GHz). The characterization setup is simulated using HFSS v.11 and SEMCAD (Bayat et al., 2010). In this section, full HFSS simulation results are presented that are considered strong validations of expected measurement outcomes. In the new setup, sub-mm metallic waveguides are employed as the interface between the PNA coaxial cables and input/output (I/O) tapers of the PC-slab waveguide devices. For example for central frequency of 200 GHz, WR-5 is utilized with size of 0.0510" × 0.0255" (1.295mm × 0.647mm). The total thickness of the polarization rotator assuming that 200 GHz corresponds to normalized frequency of $a/\lambda=0.263$ would be 0.0155" (0.395 mm); thus, there is a good match between WR-5 and input taper. A single-defect line PC slab waveguide is employed as the calibration reference. The waveguide is designed to guide both TE-like and TM-like wave (Bayat, et al., 2007).

Fig. 19(a) shows the sketch of the setup designed to couple electromagnetic wave in and out of the waveguide utilizing sub-mm metallic waveguides. To couple the electromagnetic wave to the defect line, a taper structure is utilized as shown in Fig. 19(a). The geometry of the I/O tapers must be designed properly to maximize the coupling efficiency to the defect line of PC slab waveguide.

In HFSS (v.11) simulations, the input wave is TE10 mode of the rectangular transition waveguide that has been polarized along y direction. The structural parameters of the PC slab waveguide are as follows: w=0.6a, t=0.8a, a=0.378 mm, n_{si}=3.48 and $\tan\delta=1\times10^{-4}$; where a, w, t, n_{si}, $\tan\delta$ are the unit cell, width of square holes, thickness of the PC slab waveguide, refractive index and loss tangent of silicon, respectively. The central frequency is set to 200 GHz corresponding to the normalized frequency of $a/\lambda=0.252$. The power transmission takes place through the PC defect line. The frequency response of the setup is plotted in Fig.

19(b). S_{11} (reflection) and S_{21} (transmission) are depicted by dashed and solid lines, respectively. The graphs show that the insertion loss is less than 2 dB in the entire band from 190 to 210 GHz. The return loss is higher than 20 dB. Thus, the waveguide can be employed as a wide band low loss transmission line.

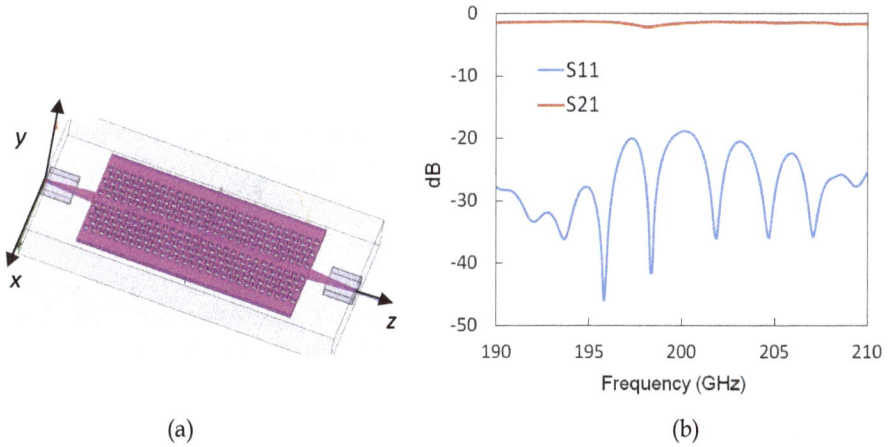

(a)   (b)

Fig. 19. (a) The schematic of the characterization setup consisting of PC slab waveguide, input/output tapers and rectangular waveguides (b) S_{11}(blue line)-S_{21}(red line) (return loss-insertion loss) plots of the PC slab waveguide

The same setup as in Fig. 19(a) has been used to characterize the polarization rotator. The input wave is TE10 mode with electric field pointing in y direction (E_y). For 90° polarization rotator, the input polarization rotates by 90° so that at the output plane the x-component of electric field is dominant. The output rectangular metallic waveguide is to be rotated by 90° to support E_x field. Fig. 20(a) shows the schematic of the polarization rotator with two alternating top loaded layers. Previously, it was shown that for this design the rotation angle for each top loaded layer is 6.5°; therefore, the polarization rotator with two top loaded layers rotates the input polarization by an angle of 2²*6.5°=26°. In this design, normalized frequency of a/λ=0.263 (where a and λ are the unit cell size and free space wavelength) is assigned to 200 GHz; thus, it is expected to see approximately 26° polarization rotation in the frequency band of 196-204 GHz corresponding to normalized frequency band of a/λ=0.258-0.267.

If the output taper shown in Fig. 19(a) was placed at the output, E_x component of the field would have been exposed to the geometry variation of the output taper imposing reversed rotation; the width of the taper in x-direction is decreasing along the propagation. To improve the polarization extinction ratio and enhance the coupling of E_x component to the rectangular metallic waveguide, the output taper was rotated by 90°, shown in Fig. 20 (a). Having rotated the output waveguide, it supports only E_x component and E_y component of the field reflects back. Thus, S_{21} and S_{11} parameters would provide a good measure of the polarization extinction ratio.

(a)   (b)

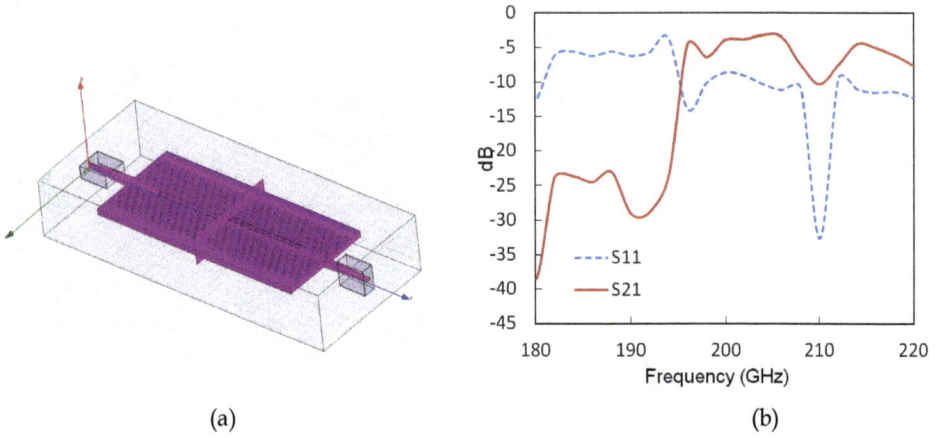

Fig. 20. (a) The schematic of the polarization rotator with the rotation angle of 26º with input/output tapers and waveguides (b) S_{11} (blue dashed line) and S_{21} (solid red line) plots of the structure.

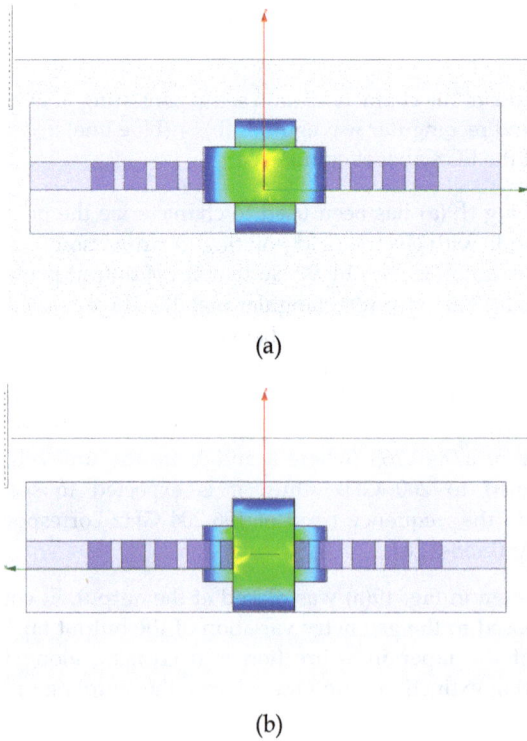

(a)

(b)

Fig. 21. Field distribution at the (a) input, and (b) output ports of the polarization rotator at 200 GHz.

Fig. 20(b) shows S_{11} and S_{21} plots from the HFSS simulations. It shows that within the frequency band of 196-206 GHz, S_{21} is higher than -4 dB and S_{11} is less than -10 dB. The bandwidth is in a good agreement with the design results presented in Sec. 3. The electric field distribution at the input and output ports are shown in Fig. 21. The input and output electric fields are laid out in x and y-direction, respectively. The electric field distribution clearly pictures the electric field rotation.

The structure shown in Fig. 22(a) is designed to rotate the input polarization by 90°. The 4.5 top loaded layers provide 90° rotation. The output taper has not been rotated 90° because of the fabrication limits and feasibility issues of such taper. Thus, we would expect to observe lower coupling efficiency of E_x component to the output rectangular waveguide. Due to the computational limits of HFSS, SEMCAD was employed to simulate the structure using 3D-FDTD method.

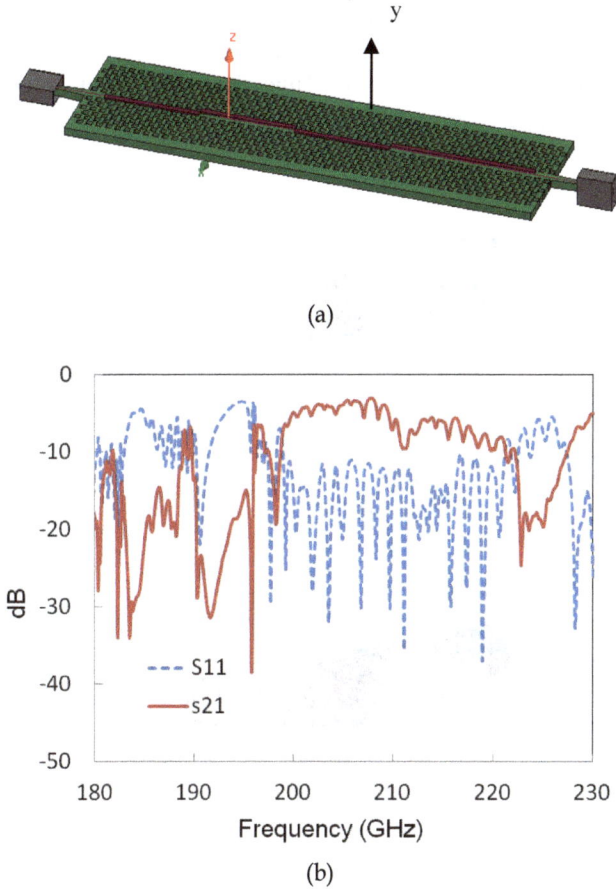

(a)

(b)

Fig. 22. (a) Schematic of the polarization rotator with an angle of rotation of 90° (b) S_{11} (blue dashed line) and S_{21} (solid red line) plots.

Fig. 22(b) shows the spectrum of S-parameters. It is very similar to S-parameters graph presented in Fig. 20(b). In Fig. 22(b), within the frequency band of 199-208 GHz, S_{21} and S_{11} are higher and less than -5 dB and -15 dB, respectively. Since the structure was designed for 90° rotation, higher polarization extinction ratio was expected in comparison with the polarization rotator with 26° angle of rotation. The S-parameter plot in Fig. 22(b) has large fluctuation and it is not as smooth as the S-parameter shown in Fig. 20(b). These fluctuations are due to the numerical noise of 3D-FDTD analysis.

Snap shots of Ex and Ey components at f=205 GHz is presented in Fig. 23. Fig. 23(a) shows that the TE10 mode has launched E_y –fields in the left side and it is well-confined inside the input taper and then couples into the defect line of the PC slab waveguide. On the other hand, E_x–field component in Fig. 23(b) is weak at the left (input) side of the defect line; the color bar shows that it is one order of magnitude smaller than E_y component. As E_y –field mode propagates inside the defect line of PC slab waveguide based polarization rotator, it gradually rotates and converts to E_x component. At the other end of the PC slab waveguide, E_x component seems to be one order of magnitude larger than E_y. At the output taper, E_x will expose to the geometry variation of the taper resulting in reverse polarization conversion. Thus, the polarization conversion efficiency would decrease.

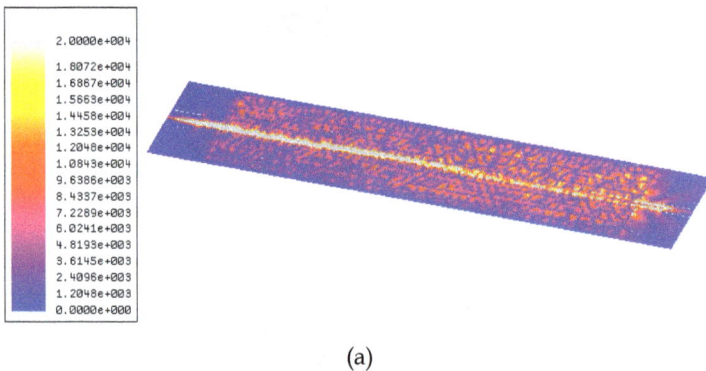

(a)

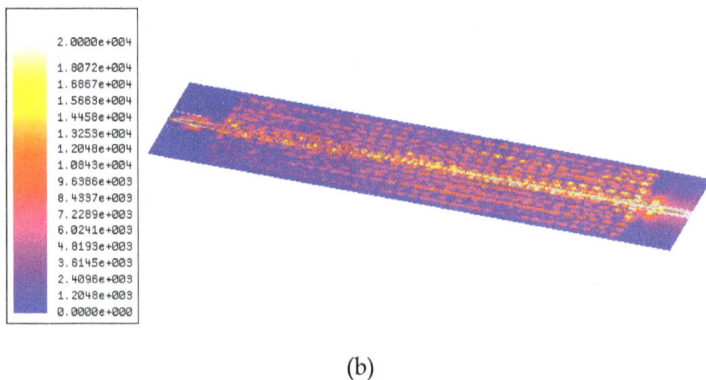

(b)

Fig. 23. Snap shots of (a) E_y and (b) E_x componnets obtained using 3D-FDTD analysis.

6. Conclusions

The focus of this chapter was on design and fabrication of PC slab waveguide based polarization processor devices. A summary of the key achievements are highlighted as following:

- A novel compact PC slab waveguide based polarization rotator was introduced and designed. PC based coupled-mode theory was developed to design the structure. Coupled-mode theory provides with a simple yet closed form method for initial design. Square-hole PC was preferred in order for simplicity of the closed-form formulations. The design was refined using rigorous electromagnetic numerical method (3D-FDTD). 3D-FDTD simulation results verified the robustness of coupled-mode theory for design and analysis of PC based polarization rotator.
- To extend the design to more general shape PC based polarization rotator, a design methodology based on vector-propagation characteristics of normal modes of asymmetric loaded PC slab waveguide was introduced. The vector-propagation characteristics of normal modes of the structure were calculated utilizing 3D-FDTD method combined with SFT. Profile and propagation constants of slow and fast modes of asymmetric loaded PC slab waveguide were extracted from 3D-FDTD simulation results. The half-beat length, which is the length of each loaded layer, and total number of the loaded layers are calculated using aforementioned data. This method provides with exact values of the parameters of the polarization rotator structure.
- SOI based PC membrane technology for THz application was developed. PC slab waveguide and polarization rotators were fabricated and characterized employing this technology.

7. References

Barwicz, T. & Watts, M.R. & Popović, M.A. & Rakich, P.T. & Socci, L. & Kärtner, F.X. & Ippen, E.P. & Smith, H.I. (2007) *Polarization-transparent microphotonic devices in the strong confinement limit,* Nature Photonics, Vol. 1, pp. 57 - 60.

Bayat, K. & Safavi-Naeini, S. & Chaudhuri, S.K. (2007). *Polarization and thickness dependent guiding in the photonic crystal slab waveguide,* ,Optics Express, Vol. 15, Issue 13, pp. 8391-8400.

Bayat, K. & Safavi-Naeini, S. & Chaudhuri, S.K. (2009). *Ultra-compact photonic crystal based polarization rotator* ,Optics Express, Vol. 17, Issue 9, pp. 7145-7158.

Bayat, K. & Safavi-Naeini, S. & Chaudhuri, S.K. & Barough, M.F. (2009). *Design and simulation of photonic crystal based polarization converter* ,J. Lightwave Technol., Vol. 27, Issue 23, pp. 5483-5491.

Bayat, K. & Rafi, G.Z. & Shaker, G.S.A. & Ranjkesh, N. & Chaudhuri, S.K. & Safavi-Naeini, S. (2010). *Photonic-crsytal based polarization converter for terahertz integrated circuit,* Microwave Theory and Tech., IEEE Trans., Vol. 58, Issue 7, pp. 1976-1984.

Bayat, K. & Safavi-Naeini, S. & Chaudhuri, S.K. (2007). *Polarization and thickness dependent guiding in the photonic crystal slab waveguide,* ,Optics Express, Vol. 15, Issue 13, pp. 8391-8400.

Chan, P.S. & Tsang, H.K. & Shu, C. (2003). *Mode conversion and birefringence adjustment by focused-ion-beam etching for slanted rib waveguide walls,* Opt. Lett., Vol. 28, No. 21, pp. 2109-2111.

Deng, H. (2005). *Design and characterization of silicon-on-insulator passive polarization converter with finite-element analysis,* Ph.D Thesis, University of Waterloo.

El-Refaei, H & Yevick, D. (2003) *An optimized InGaAsP/InP polarization converter employing asymmetric rib waveguides,* J. Lightwave Tech., Vol. 21, No. 6, pp. 1544-1548.

Genereux, F. & Leonard, S.W. & Van Driel, H.M. (2001) *Large birefringence in two-dimensional photonic crystals*, Phys. Rev. B., Vol. 63, pp. 161101.

Heidrich, H. & Albrecht, P. & Hamacherm, M. & Nolting, H.P. & Schroeter, H. & Weinert, C.M. (1992) *Passive mode converter with a periodically tilted InP/GaInAsP rib waveguide*, IEEE Photon. Tech. Lett., Vol. 4, No. 1, pp. 34-36.

Huang, J.Z. & Scarmozzino, R. & Nagy, G. & Steel, M.J. & Osgood, R.M. (2000) *Realization of a compact and single-mode optical passive polarization conveter*, IEEE Photon. Tech. Lett., Vol. 12, No. 3, pp. 317-319.

Huang, W. & Mao, Z.M. (1992) *Polarization rotation in periodic loaded rib waveguides*, IEEE J. Lightwave Tech., Vol. 10, pp. 1825-1831.

Haung, W.P. & Chu, S.T. & Chaudhuri, S.K. (1992) *Scalar coupled-mode theory with vector correction*, IEEE J. Quantum Electron, Vol. 28, pp. 184-193.

Lui, W.W. & Magari, K. & Yoshimoto, N. & Oku, S. & Hirono, T. & Yokoyama, K. & Huang W.P. (1997) *Modeling and design of bending waveguide based semiconductor polarization rotators*, IEEE Photon Tech. Lett., Vol. 9, No. 10, pp. 1379-1381.

Lui, W.W. & Xu, C.L. & Hirono, T. & Yokoyama, K & Huang, W.P. (1998) *Full-vectorial wave propagation in semiconductor optical bending waveguides and equivalent straight waveguides*, IEEE J. Lightwave Tech., Vol. 16, No. 5, pp. 910-914.

Lui, W.W. & Hirono, T. & Yokomaya, K. & Huang, W.P. (1998) *Polarization rotator in semiconductor bending waveguides: A coupled-mode theory formulation*, J. Lightwave Tech., Vol. 16, No. 5, pp. 929-936.

Johnson, S.G. & Fan, S. & Villeneuve, P.R. & Joannopoulos, J.D. (1999) *Guided modes in photonic crystal slabs*, Phys. Rev. B, Vol. 60, No. 8, pp.5751-5758.

Mertens, K & Scholl, B. & Schmitt, H.J. (1998) *Strong polarization conversion in periodically loaded strip waveguide*, IEEE Photon. Tech. Lett., Vol. 10, pp. 1133-1135.

Mrozowski, M. (1997) *Guided electromagnetic waves: properties and analysis*, England: Research Studies Press Ltd./ John Wiley & Sons Inc.

Obayya, S.S.A & Rahman, B.M.A & El-Mikati, H.A. (2000) *Vector beam propagation analysis of polarization conversion in periodically loaded waveguides*, IEEE Photon. Tech. Lett., Vol. 12, pp. 1346-1348.

Rahman, B.M.A & Obayya, S.S.A. & Somasiri, N. & Rajarajan, M. & Grattan, K.T.V. & El-Mikathi, H.A. (2001) *Design and characterization of compact single-section passive polarization rotator*, J. Lightwave Tech., Vol. 19, No. 4, pp. 512-519.

Shani, Y. & Alferness, R. & Koch, T. & Koren, U. & Oron, M. & Miller, B. I. & Young, M.G. (1991) *Polarization rotation in asymmetric periodic loaded rib waveguides*, Appl. Phys. Lett., Vol. 59, pp. 1278-1280.

Snyder, A.W. & Love, J.D. (1983), *Optical Waveguide Theory*, London :Chapman & Hall.

Tzolov, V.P. & Fontaine, M. (1996) *A passive polarization converter free of longitudinally-periodic structures*, Optics Communications, Vol. 27, pp. 7-13.

Van Dam, C. & Spickman, L.H. & Van Ham, F.P.G.M. & Groen, F.H. & Van Der Tol, J.J.G.M. & Moerman, I. & Pascher, W.W. & Hamacher, M. & Heidrich, H. & Weinert, C.M. & Smit, M.K. (1996) *Novel compact polarization converters based on ultra short bend*, IEEE Photon. Tech., Vol. 8, No. 10, pp. 1346-1348.

Wang, Z. & Dai, D. (2008) *Ultrasmall Si-nanowire-based polarization rotator*, J. Opt. Soc. Am. B, Vol. 25, No. 5, pp. 747-753.

Watts, M.R. & Haus, H.A. (2005) *Integrated mode-evolution-based polarization rotators*, Opt. Lett., Vol. 30, Issue 2, pp. 138-140.

14

Photonic Crystal Coupled to N-V Center in Diamond

Luca Marseglia
*Centre for Quantum Photonics, H. H. Wills Physics Laboratory & Department of Electrical
and Electronic Engineering, University of Bristol, BS8 1UB
United Kingdom*

1. Introduction

In this work we aim to exploit one of the most studied defect color centers in diamond , the negatively charged nitrogen vacancy (NV^-) color center, a three level system which emits a single photon at a wavelength of $637nm$ providing a possible deterministic single photon emitter very useful for quantum computing applications. Moreover the possibility of placing a NV^- in a photonic crystal cavity will enhance the coupling between photons and NV^- center. This could also allow us to address the ground state of the NV^- center, whose spin, could be used as qubit. It is also remarkable to notice that for quantum computing purposes it is very useful to increase the light collection from the NV^- centers, and in order to do that we performed a study of another structure, the solid immersion lens, which consists of an hemisphere whose center is at the position of an emitter, in this case the NV^- center, increasing the collection of the light from it. In order to create these structures we used a method called focused ion beam which allowed us to etch directly into the diamond many different kinds of structures. In order to allow an interaction between these structures and the NV^- centers we need to have a method to locate the NV^- center precisely under the etched structures. We developed a new technique (Marseglia et al. (2011)) where we show how to mark a single NV^- center and how to etch a desired structure over it on demand. This technique gave very good results allowing us to etch a solid immersion lens onto a NV^- previously located and characterized, increasing the light collection from the NV^- of a factor of $8\times$.

2. Introduction to Nitrogen Vacancy center in diamond

Diamond has emerged in recent years as a promising platform for quantum communication and spin qubit operations as shown by Gabel et al. (2006), as well as for "quantum imaging" based on single spin magnetic resonance or nanoscopy. Impressive demonstrations in all these areas have mostly been based on the negatively-charged nitrogen vacancy center, NV^-, which consists of a substitutional nitrogen atom adjacent to a carbon vacancy. Due to its useful optical and magnetic spin selection properties, the NV^- center has been used by Kurtsiefer et al. (2000) to demonstrate a stable single photon source and single spin manipulations (Hanson et al. (2006)) at room temperature. A single-photon source based on NV^- in nano-diamond is already commercially available, and a ground state spin coherence time of $15ms$ has been observed in ultra-pure diamond at room temperature. At present, one of the biggest issues preventing diamond from taking the lead among competing technologies

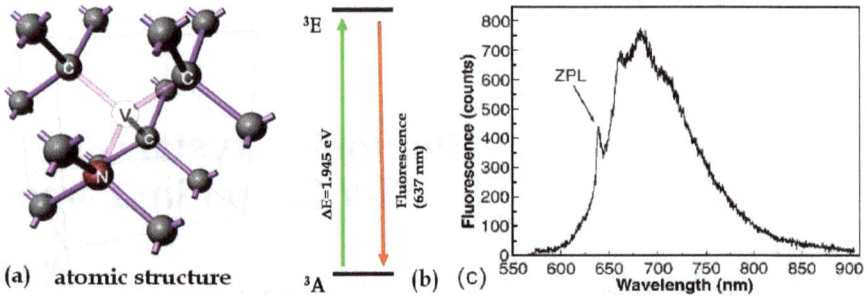

(a) atomic structure **(b)** **(c)**

Fig. 1. a)Atomic structure of NV⁻center in diamond(N=nitrogen V=Vacancy, C=Carbon) b) Energy level scheme of NV⁻center. c) Fluorescence spectrum of a single NV⁻defect center. The wavelength of the zero phonon line (ZPL) is $637nm$ ($1.945eV$). Excitation was at $514nm$ (Image taken from Gruber et al. (1997))

is the difficulty in fabricating photonic devices to couple and guide light. For the realization of large-scale quantum information processing protocols (e.g. via photonic module approaches) or for quantum repeater systems, it will be necessary to connect NV⁻centers through "flying" qubits such as photons. To achieve this, micro-cavities and waveguides are needed to enable the transfer of quantum information between the electron spin of the NV⁻center and a photon. In this work I will show some applications of diamond useful for quantum computing. Synthetic diamonds can be doped in order to create implanted NV⁻center which interacts with light, as described further. From its discovery, it has not been very clear if the NV⁻were a proper two level system. Recently it has been shown that it has properties more typical of a three level system with a metastable level. In its ground state it has spin $s = 1$ and different emission rates for transitions to the ground states, so NV⁻center can be also exploited in order to achieve spin readout.

3. Interaction of N-V center with light

The NV⁻center in diamond occurs naturally or is produced after radiation damage and annealing in vacuum. As described earlier is made by substitutional nitrogen atom adjacent to a vacancy in carbon lattice in the diamond as depicted in Fig.1a. The NV⁻center has attracted a lot of interest because it can be optically addressed as a single quantum system as discussed by van Oortt et al. (1988). The NV⁻center behaves as a two level system with a transition from the excited state to the ground state providing a single photon of $637nm$, as shown in Fig.1b. This is a very useful characteristic for quantum information purposes because it can be used as single photon source. Let us remember that a characteristic of the NV⁻center is a zero-phonon line (ZPL), in the spectrum at room temperature, at $637nm$ as shown in Fig.1c, the zero-phonon line constitutes the line shape of individual light absorbing and emitting molecules embedded into the crystal lattice. The state of NV⁻center ground state spin strongly modulates the rate of spontaneous emission from the $^3E \leftrightarrow ^3A$ sub-levels providing a mechanism for spin read out as discussed by Hanson et al. (2006). We have recently shown theoretically (Young et al. (2009)) that spin readout with a small number of photons could be achieved by placing the NV⁻centre in a subwavelength scale micro-cavity with a moderate Q-factor($Q \sim 3000$). So one of our aims is to optimize the output coupling of photons from diamond color centers into waveguides and free space to

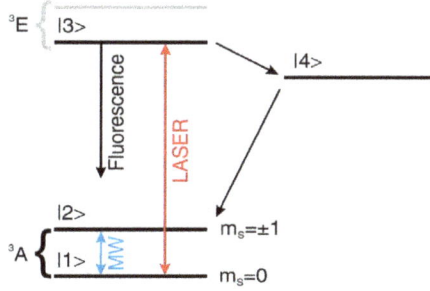

Fig. 2. Energy level scheme of the nitrogen vacancy defect center in diamond. The greyed out lines correspond to the $m_s = \pm 1$ sublevels (Image taken from Jelezko et al. (2004b))

increase the efficiency of single photon sources and to enable faster single spin read-out. In order to do that we want to study resonant structures. These structures confine the light close to the emitter allowing cavity-QED effects to be exploited to direct an emitted photon into a particular spatial mode and will allow us to enhance the ZPL. An improvement of the photon emission rate and photon indistinguishability for NV⁻ can be achieved due to the (coherent) interaction with the highly localized photon field of the cavity. In principle a high-Q micro-cavity can be realized directly in diamond but the first experimental demonstrations with micro-disk resonators and photonic crystal cavities, made for example by Wang et al. (2007), suffered from large scattering losses due to the poly-crystalline nature of the diamond material used. The fabrication of high-Q cavities in single crystal diamond is very challenging because vertical optical confinement within diamond requires either a $3D$ etching process or a method for fabricating thin single crystal diamond films. We want analyze photonic crystal structures in diamond and fabrication methods to achieve efficient spin read-out in low-Q cavities. Electronic spin resonance (ESR) experiments performed by Jelezko et al. (2004a) has shown that the electronic ground state of NV⁻ center (^3A) is paramagnetic. Indeed the electronic ground state of the NV⁻ center is a spin triplet that exhibits a 2.87GHz zero-field splitting defining the z axis of the electron spin. An application of a small magnetic field splits the magnetic sublevel $m_s = \pm 1$ energy level structure of the NV⁻ center, as we can see in Fig.2. Electron spin relaxation times (T_1) of defect centers in diamond range from millisecond at room temperature to seconds at low temperature. Several experiments have shown the manipulation of the ground state spin of a NV⁻ center using optically detected magnetic resonance (ODMR) techniques, the main problem in using ODMR is that detection step involves observing fluorescence cycles from the NV⁻ center which has a probability of destroying the spin. Another characteristic of NV⁻ center useful for quantum information storage is the capability of transferring its electronic spin state to nuclear spins. Experiments performed by van Oortt et al. (1988) have shown the possibility of manipulating nuclear spins of NV⁻. Nuclear spins are of fundamental importance for storage and processing of quantum information, their excellent coherence properties make them a superior qubit candidate even at room temperature.

4. Beyond the two level system model

In order to study the dynamics of the NV⁻ center, remembering that $m_e \ll m_C$ where m_e is the value of the mass of the electrons and m_C is the value of the mass of carbon atom,

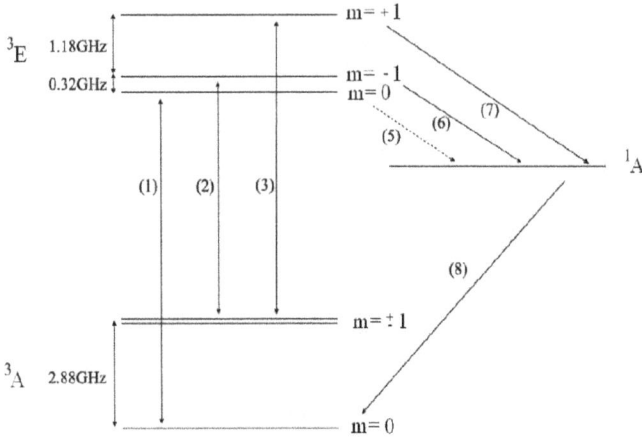

Fig. 3. Experimentally measured energy level diagram of the NV center in diamond showing the experimentally determined ground and excited state splitting. The defect has zero phonon line at 637 nm, with width of order MHz at low temperatures (image taken from Young et al. (2009)).

we make the so called Born-Oppenheimer approximation, in which we consider the nuclei fixed in a crystal geometry and the coordinates of the electrons are considered with respect to them. When a defect is present it breaks down the crystal symmetry and, regarding the NV$^-$center, we have a contribution of one electron from each carbon atom, the nitrogen contributes two electrons, and an extra electron comes from the environment as described by Gali et al. (2008), possibly given by substitutional nitrogen, so ending with a total number of six electrons. The excited level of the NV$^-$ is not yet very well understood explicitly but there are many theoretical descriptions using group theory and partially confirmed by experimental results. The joint use of group theory and numerical calculations has led to predictions of the ordering of the levels of the ground state and excited states of the NV$^-$. Taking account the coulomb interaction, spin-orbit effect and spin-spin coupling it explains the splitting of the levels initially degenerate giving rise to different transition between them. Without entering in to detailed group theory calculations we can summarize by stating that the hamiltonian of the system is composed of three elements, the coulomb interaction H_C, the spin orbit interaction H_{SO} and the spin-spin interaction H_{SS}.

$$H = H_C + H_{SO} + H_{SS} \qquad (1)$$

The dynamics of the system is resolved by solving the hamiltonian in free space with coulomb interaction as potential and the spin-orbit and spin-spin interaction were eventually added as perturbations. We show the detailed energy level structure of the NV$^-$center in Fig.3. The overall effect can be summarized as follows, the coulomb interaction, splits the degeneracy between singlet and triplets of the ground state and the first excited state, the spin orbit interaction splits the states which have $M_{s=0}$ and $M_{s=\pm1}$ and finally the spin-spin splits the A levels. The end result is that the optical transitions between ground and excited states $(1-3)$ occur at different energies. In the absence of external fields the ground state is a spin triplet split by 2.88 GHz due to spin–spin interactions. Manson et al. (2006) showed that the excited state is a triplet split by spin-spin interactions, but with the further addition of spin-orbit coupling . Recent experimental evidence performed by Tamarat et al. (2008), has uncovered this excited state structure (Fig.3). The net effect of spin-spin and spin-orbit interactions is to create a detuning $\approx$ 1.4 GHz ($6\mu eV$) between the transitions $^3A_{(m=0)} \rightarrow ^3E_{(m=0)}$ (transitions 1)and $^3A_{(m=+1)} \rightarrow ^3E_{(m=+1)}$ (transition 3). A similar detuning of $\approx$ 2.5 GHz ($10\mu eV$) exists for the $^3A_{(m=-1)} \rightarrow ^3E_{(m=-1)}$ (transition 2), the rates for these three transition is $k_1 = k_2 = k_3 = 77MHz$ which gives a spontaneous emission (SE) lifetime $\tau \approx 13ns$. The energy level structure is not simply a ground and excited triplet state, there also exists an intermediate singlet state ^1A arising from Coulomb interactions. There is a probability of the transition $^3E \rightarrow ^1A$, with different rates depending on the spin. For the $^3E_{m=\pm 1}$ states (transitions 6, 7) both theoretical predictions and experimental results suggest that the decay rate is around $k_6 = k_7 = 30MHz$ giving a spontaneous emission (SE) lifetime $\tau \approx 30ns$. For the $^3E_{m=0}$ state (transition 5) theoretically the rate of decay to the singlet should be zero, however, experimental observations made by Jelezko & Wrachtrup (2004) have shown the rate to be $\approx 10^{-4} \times 1/\tau$. Since the ^1A singlet state decays preferentially to the $^3E_{m=0}$ state (transition 8), then it is clear from the rates above that broadband excitation leads to spin polarization in the spin zero ground state. Since transition 8 is non-radiative then there will be a dark period in the fluorescence when ^1A becomes populated, and as the decay rate from $^3E_{m=\pm 1}$, $k_8 = 3.3MHz$, to the singlet state is much larger than from $^3E_{m=0}$, the change in intensity measures the spin state. Clearly using fluorescence intensity to detect the spin state has a probability to flip the spin, therefore it would seem necessary for a scheme to suppress this. However, spin-flip transitions are essential to initialize the system. Thus a compromise is required between the perfectly cyclic spin preserving transitions required for readout and the spin flip transitions needed for reset.

5. Photonic crystals

To take advantage of atom-photon coupling using NV^-, as required by many quantum protocols, cavity structures are required. Again, concentrating on monolithic diamond solutions, photonic crystal cavities are the most natural structures to explore. A photonic crystal structure modulates the propagation of light in a way that is analogous to the way a semiconductor crystal modulates the motion of electrons. In both cases a periodic structure gives rise to 'band-gap' behavior, with a photon (electron) being allowed or not allowed to propagate depending on its wave vector. In photonic crystals the periodicity is comprised of regions of higher and lower dielectric constants. The basic physical phenomenon is based on diffraction, the period needs to be of the order of a half-wavelength of the light to be confined. For visible light the wavelength goes from $200nm$ (blue) to $650nm$ (red), leading to a real challenge in order to make the fabrication of optical photonic crystals because of the small dimensions. Breaking the periodicity in a controlled way creates nanocavities that confine light to extremely small volumes in which the lightmatter interaction is dominated by cavity quantum electrodynamic. We have previously described the characteristics of the NV^- center, a three level system which is promising as an efficient room temperature source of single photons at a wavelength of $637nm$. We pointed out that the NV^- center looks very promising for performing quantum spin readout, which is also useful for quantum computing purposes. Zero-phonon emission, at $637nm$, accounts for only a small fraction ($\sim$ 4%) of NV^-fluorescence, with the majority of emitted photons falling in the very broad ($\sim 200nm$) phonon-assisted sideband. By coupling the NV^-center to a photonic crystal cavity, spontaneous emission in the phonon sideband can be suppressed and emission in the zero-phonon line can be enhanced (Su et al. (2008)) so the photonic crystals offers a controllable electromagnetic environment, ideal for the compact integration and isolation of the fragile quantum system. The challenges of engineering the parameters of the photonic crystal in diamond at this scale are not trivial, as described further where we will show how to tune a cavity to increase the efficiency of light collection from an emitter placed in it. Indeed, a single photon emitted by a NV^- could then interact with another NV^- allowing entanglement between both qubits represented by the spin of the NV^- centers ase described by Neumann et al. (2008). High-Q resonators of different kinds have been fabricated in non-diamond materials and coupled to NV^- emission from nano-diamonds. Since we are concerned here with developing monolithic photonics, it is necessary to fabricate cavities in the diamond itself. It should be noted that photonic crystal cavities have been fabricated in diamond films and an un-coupled Q-factor as high as 585 at $637nm$ has been measured by Wang et al. (2007). The polycrystal nature of the material used in those demonstrations makes it unsuitable for our purposes due to enhanced scattering and background fluorescence. We aim to fabricate photonic crystal cavities in ultra-high-purity type IIa single-crystal diamond (Element Six) grown by chemical vapor deposition. This material has extremely low levels of nitrogen (less than $1ppb$), and very few native NV^- centers, making it the ideal material for creating NV^- centers in a controlled fashion by implantation and annealing. In order to have strong coupling we need to have a cavity with high Q factor and small modal volume, but a cavity with a more moderate Q would still be useful. In particular, a scheme for reading out the ground state spin of an NV^- center has been described by our group (Young et al. (2009)), that requires a Q (before coupling) of only ~ 3000. This scheme exploits the zero-field splitting in the NV^- center ground state and uses narrow band resonant excitation to achieve high-fidelity read-out of the ground state spin with just a few excitation cycles.

6. Crystals geometry

A rigid body is called symmetric if it remains identical to itself after a translation, a rotation or a reflection, which are called the symmetrical operations of the body. A composition of two symmetrical operations is still a symmetrical operation and every symmetrical operation has got its own inverse which is still a symmetrical operation, so all the operations which satisfies these requirements form a set. The general element of this set can be described by a matrix transformation in cartesian coordinates as follows

$$\begin{pmatrix} x_1' \\ x_2' \\ x_3' \end{pmatrix} = \begin{pmatrix} \alpha_{11} & \alpha_{12} & \alpha_{13} \\ \alpha_{21} & \alpha_{22} & \alpha_{23} \\ \alpha_{31} & \alpha_{32} & \alpha_{33} \end{pmatrix} \begin{pmatrix} x_1 \\ x_2 \\ x_3 \end{pmatrix} + \begin{pmatrix} a_1 \\ a_2 \\ a_3 \end{pmatrix} \tag{2}$$

the choice of the parameters lead to interesting subsets as for example if we choose all the value $a_i = 0$ we have the so called point sets. if we choose the α matrix as unitary

$$\alpha = \begin{pmatrix} 1 & 0 & 0 \\ 0 & 1 & 0 \\ 0 & 0 & 1 \end{pmatrix} \tag{3}$$

we have the translations set which we will focus on. The crystal contains a lattice of equivalent points known as *Bravais Lattice*, a mathematical entity defined as the set of points which looks

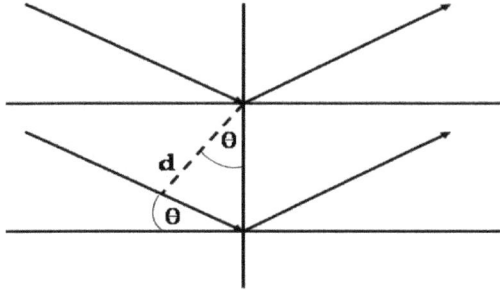

Fig. 4. Bragg Hypothesis

always the same after any translations between the points on the Bravais lattice itself so, for example if a translational vector $\overrightarrow{R}$ connects two points of the set, an integer application of the the same vector can connect any point of the Bravais lattice to the appropriate point of itself. A *Crystal Lattice* is obtained if a simple structure, called the base, is repeated in space in order to create a Bravais lattice, which can be seen as a way of locating identical copies of the base in the space. In a physical crystal the base is made by set of atoms or ions called the unit cell. It is possible to generate the whole set of points of a Bravais lattice with three primitive vectors. All the translations are linear combinations of these three vectors

$$\overrightarrow{R} = n\,\overrightarrow{a}_1 + m\,\overrightarrow{a}_2 + p\,\overrightarrow{a}_3 \tag{4}$$

with n, m, p integers. The solid made by the three primitive vectors is called a primitive cell, which has the lattice points at its vertices. The repetition of the primitive cell gives rise to the crystal lattice.

7. Bragg scattering

If we want to study the scattering of the light incident on a crystal lattice we shall start from the Bragg approximation, in which the crystal is considered to be made from lattice planes. The light is incident on these planes with angle of incidence θ. The underlying hypothesis is that each plane partially reflects the light. If **d** (Fig.4) is the distance between a pair of planes the constructive interference condition between the reflected rays is

$$2d \sin \theta = n\lambda \tag{5}$$

where λ is the wavelength and n is an integer. We now generalize this principle to the scattering of light from a three dimensional crystal following. Let's suppose a monochromatic wave of wave vector $\overrightarrow{k}$ and angular frequency ω hits the crystal with amplitude

$$\mathbf{F}(\overrightarrow{x}) = \mathbf{F}_0 exp(i(\overrightarrow{k} \cdot \overrightarrow{x} - \omega t)) \tag{6}$$

If $\overrightarrow{R}$ is the position of an atom in the lattice the amplitude of the wavelength incident on it at $t = 0$ is

$$\mathbf{F}(\overrightarrow{R}) = \mathbf{F}_0 exp(i\overrightarrow{k} \cdot \overrightarrow{R}) \tag{7}$$

So the wave spread by the atom is a spherical wave which has amplitude at a point $\vec{\rho}$ given by $exp(ikr)/r$, where $\vec{r} = \vec{\rho} - \vec{R}$. The whole amplitude is given by

$$exp(i\vec{k} \cdot \vec{R}) \cdot exp(ikr)/r \simeq exp(i\vec{k} \cdot \vec{R} + ikr)/\rho \qquad (8)$$

If we choose the origin of the crystal and suppose that the position $\vec{\rho}$ in which we are looking the wave is far from the the the crystal so that $r \simeq \rho$ and if θ is the angle between $\vec{\rho}$ and $\vec{R}$. Then we can say that:

$$r^2 = (\vec{\rho} - \vec{R})^2 = \rho^2 + R^2 - 2\rho R cos\theta \qquad (9)$$

$$r = \rho\sqrt{1 - 2\frac{R}{\rho}cos\theta + \frac{R^2}{\rho^2}} \simeq \rho - Rcos\theta \qquad (10)$$

and the amplitude of the wave is determined by the phase factor.

$$exp(i\vec{k} \cdot \vec{R} + ik\rho - ikRcos\theta) = exp(ik\rho)exp(i\vec{k}\vec{R} - ikRcos\theta) \qquad (11)$$

At great distance from the source the spherical wave can be considered to be a plane wave with $\vec{k'} = \vec{k}\,cos\theta$ as wave vector with the same modulus of the of the incident wave, propagating in a direction which forms an angle θ respect to $\vec{R}$ so with value ρ and with the amplitude depending by the position of the atom as

$$exp(-i\Delta\vec{k} \cdot \vec{R}) \qquad (12)$$

with

$$\Delta\vec{k} = \vec{k'} - \vec{k} \qquad (13)$$

If $\vec{a}$, $\vec{b}$ and $\vec{c}$ are the primitive vectors of the Bravais lattice so we can say

$$R = m\vec{a} + n\vec{b} + p\vec{c} \qquad (14)$$

where m,n and p are integers. Let's suppose these three numbers can vary between 0 and $M - 1$ so the crystals contains M^3 primitive cells. The total amplitude in the position ρ will be proportional to

$$A \equiv \sum_{\vec{R}} exp(-i\Delta\vec{k} \cdot \vec{R}) = \sum_{nmp} exp[-i(m\vec{a} + n\vec{b} + p\vec{c}) \cdot \Delta\vec{k}] =$$

$$(\sum_{m} exp[-im(\vec{a} \cdot \Delta\vec{k})])(\sum_{n} exp[-in(\vec{b} \cdot \Delta\vec{k})])(\sum_{p} exp[-ip(\vec{c} \cdot \Delta\vec{k})]) \qquad (15)$$

so the intensity is given by

$$|A|^2 = |\sum_{m} exp[-im(\vec{a} \cdot \Delta\vec{k})]|^2 ... |^2 ... |^2 \qquad (16)$$

any of these value can be computed using the geometrical series and multiplying by the complex conjugate we can show that

$$|\sum_{m} exp(-im(\vec{a} \cdot \Delta\vec{k}))|^2 = \frac{sin^2\frac{1}{2}M(\vec{a} \cdot \Delta\vec{k})}{sin^2\frac{1}{2}(\vec{a} \cdot \Delta\vec{k})} \qquad (17)$$

This function of $\vec{a} \cdot \Delta \vec{k}$ is a peak function which has the absolute maximum values at

$$\vec{a} \cdot \Delta \vec{k} = 2\pi q \tag{18}$$

where q is an integer number, which gives the equation 17 the value of $\mathbf{M}^2$. The width of these maximum values is measured from the consequent zeros which are at

$$\vec{a} \cdot \Delta \vec{k} = 2\pi q + \epsilon \tag{19}$$

where ϵ is the smallest number different from zero so that

$$\sin(\frac{1}{2}M\epsilon) = 0 \tag{20}$$

giving us

$$\frac{1}{2}\mathbf{M}\epsilon = \pi \tag{21}$$

$$\epsilon = \frac{2\pi}{\mathbf{M}} \tag{22}$$

So the width of the maximum is proportional to $\frac{1}{M}$ and the area of the peak is given by the height ($\propto M^2$) times the width ($\propto \frac{1}{M}$) so it's proportional to the number $\mathbf{M}$ of atoms present in the line where the $\vec{a}$ vector lies. So taking all of equation 16 we see that when the crystal contains M^3 atoms the peaks of the scattered intensity will be proportional to M^3. They appear in the direction which satisfies simultaneously the three **Laue equations** as we see

$$\vec{a} \cdot \Delta \vec{k} = 2\pi q; \ \vec{b} \cdot \Delta \vec{k} = 2\pi r; \ \vec{c} \cdot \Delta \vec{k} = 2\pi s \tag{23}$$

where q,r and s are three integer numbers. The Laue Equation can be easily solved rewriting $\Delta \vec{k}$ as linear combination of multiple integers of three vectors $\vec{\mathbf{A}}$, $\vec{\mathbf{B}}$ and $\vec{\mathbf{C}}$

$$\Delta \vec{k} = q\vec{\mathbf{A}} + r\vec{\mathbf{B}} + s\vec{\mathbf{C}} \tag{24}$$

so that

$$\begin{aligned}
\vec{\mathbf{A}} \cdot \vec{a} = 2\pi, \quad \vec{\mathbf{B}} \cdot \vec{a} = 0, \quad \vec{\mathbf{C}} \cdot \vec{a} = 0, \\
\vec{\mathbf{A}} \cdot \vec{b} = 0, \quad \vec{\mathbf{B}} \cdot \vec{b} = 2\pi, \quad \vec{\mathbf{C}} \cdot \vec{b} = 0, \\
\vec{\mathbf{A}} \cdot \vec{c} = 0, \quad \vec{\mathbf{B}} \cdot \vec{c} = 0, \quad \vec{\mathbf{C}} \cdot \vec{c} = 2\pi,
\end{aligned} \tag{25}$$

where the $\vec{\mathbf{A}}$ vector is perpendicular to the plane made by $\vec{b}, \vec{c}$, the vector $\vec{\mathbf{B}}$ is perpendicular to the plane made by $\vec{a}, \vec{c}$ and the $\vec{\mathbf{C}}$ vector is perpendicular to the plane made by $\vec{a}, \vec{b}$. The equations in 25 can be all satisfied choosing

$$\vec{\mathbf{A}} = 2\pi\frac{\vec{b} \times \vec{c}}{\vec{a} \cdot b \times \vec{c}}; \ \vec{\mathbf{B}} = 2\pi\frac{\vec{c} \times \vec{a}}{\vec{a} \cdot b \times \vec{c}}; \ \vec{\mathbf{C}} = 2\pi\frac{\vec{a} \times \vec{b}}{\vec{a} \cdot b \times \vec{c}} \tag{26}$$

these are the primitive vectors of a new lattice called the **reciprocal lattice**. For any crystal lattice in the real world, which will be called the direct lattice, we can consider a corresponding mathematical entity called the reciprocal lattice. The translation vectors of the crystal lattice have the dimension of a length [L], the ones of the reciprocal lattice have the dimension of a $[L]^{-1}$. If we consider a function which has the spatial periodicity of the lattice

$$f(\vec{r}) = f(\vec{r} + \vec{R}) \ with \ \ \vec{R} = n\vec{a} + m\vec{b} + p\vec{c} \tag{27}$$

it can be represented by the Fourier Series

$$f(\vec{r}) = \sum_{\vec{k}} \tilde{f}(\vec{k}) exp(i\vec{k} \cdot \vec{r}) \tag{28}$$

and the only vectors in the sum are the translational vectors of the reciprocal lattice

$$\vec{G} = h\vec{A} + k\vec{B} + l\vec{C} \tag{29}$$

with

$$e^{i\vec{G} \cdot \vec{R}} = 1 \tag{30}$$

because

$$\vec{G} \cdot \vec{R} = 2\pi(hm + kn + lp) = 2n\pi \tag{31}$$

where n is an integer, so the reciprocal lattice is the Fourier transform of the direct lattice. It is very important to note that every translational vector is perpendicular to a lattice plane of the direct lattice. So let's consider a translational vector of the reciprocal lattice,

$$\vec{G} = h\vec{A} + k\vec{B} + l\vec{C} \tag{32}$$

and let m,n and p be three integers obtained considering the greatest common divisor $\mathcal{D}$ of $h,k,$and l and dividing by $h,k,$and l. So the three numbers m,n and p define the lattice plane which intersects the line $\vec{a}$ in the point $m\vec{a}$, the line $\vec{b}$ in the point $n\vec{b}$ and the the line $\vec{c}$ in the point $p\vec{c}$. The vectors $m\vec{a} - n\vec{b}$, $m\vec{a} - p\vec{c}$ and $n\vec{b} - p\vec{c}$ lie on the same plane and are perpendicular to $\vec{G}$. Indeed

$$\vec{G} \cdot (m\vec{a} - n\vec{b}) = 2\pi(mh - nk) = 0$$
$$\vec{G} \cdot (m\vec{a} - p\vec{c}) = 2\pi(mh - pl) = 0 \tag{33}$$
$$\vec{G} \cdot (n\vec{b} - p\vec{c}) = 2\pi(nk - pl) = 0$$

because by construction

$$m = \frac{\mathcal{D}}{h}; n = \frac{\mathcal{D}}{k}; p = \frac{\mathcal{D}}{l}; \tag{34}$$

and so

$$mh - nk = \mathcal{D} - \mathcal{D} = 0$$
$$mh - pl = \mathcal{D} - \mathcal{D} = 0 \tag{35}$$
$$nk - pl = \mathcal{D} - \mathcal{D} = 0$$

The plane is uniquely fixed by the integers h,k and l called the **Miller indexes** of the plane. It is remarkable to note that the Laue equations can be rewritten in the explicit form as

$$\Delta \vec{k} = \vec{G} \tag{36}$$

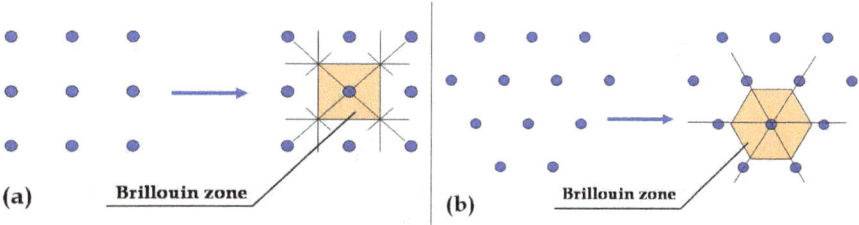

Fig. 5. a) First Brillouin Zone of a Square Lattice. b) First Brillouin Zone of a Hexagonal Lattice

the equation 36 is equivalent to the Bragg relation because the distance d from the plane with Miller indexes from the origin is equal to $\frac{2\pi}{|\vec{G}|}$. So if $\vec{R}$ ia a generic point of the lattice plane (hkl) we can say that

$$d = \frac{\vec{R} \cdot \vec{G}}{|\vec{G}|} = \frac{2\pi}{|\vec{G}|} \tag{37}$$

it follows that

$$|\Delta \vec{k}| = \frac{2\pi}{\lambda} 2 \sin\theta = \frac{2\pi}{d}$$

$$2d \sin\theta = \lambda \tag{38}$$

where 2θ is the angle between the incident direction $\vec{k}$ and the diffracted direction $\vec{k'}$ and remembering that $\Delta \vec{k}$ is orthogonal to the plane, θ is also the incident angle of the plane itself. It is evident that the vector of the reciprocal lattice $n\vec{G}$ shows the plane which is far from the origin of the value nd. For each lattice plane corresponds a constructive interference to the diffracted wave. It is more convenient to consider the *proximity cell* instead of the primitive cell. The Proximity Cell is the polyhedron whose faces lie on the plane which are orthogonal to the middle points of the lines who have the lattice point and its next one as extremes. The proximity cell of the reciprocal lattice is known as the First Brillouin Zone. By repeating the procedure with the further lattice point we can construct the second Brillouin Zone, the third and so on. Two of the most common lattice structures are the square lattice, whose first Brillouin zone we can see in Fig.5a and the hexagonal lattice whose first Brillouin zone is depicted in Fig.5b. Some points of high symmetry of these structures are very useful for the computation of the bandgap, as we will se further, and they are called the critical points.

8. Electromagnetic waves in the lattice and Bloch wave representation

We have showed a situation in which we used the so called Born (or single scattering) approximation where the lattice is weak point scatterer in a constant index. Background which is suitable to describe x-ray scattering. Now we move on to look at structures that are not point scatterers but high contrast dielectrics where bandgaps can occur. The field associated with the propagation of the electromagnetic waves is proportional to a vector $\vec{R}$ of the direct lattice and to a variable σ which can vary on the degree of freedom of the field inside the primitive cell made by the translational vector. For a wave function, σ must be specified in the whole set of the points of the primitive cell and the field of the lattice vibrations has a number of degree of freedom for each cell which is three times the number of the atom of each cell. The normal modes of the field $\xi(\sigma; \vec{R})$ can be obtained as solutions of an eigenvalue problem as

$$\Omega\xi(\sigma; \vec{R}) = \lambda\xi(\sigma; \vec{R}) \tag{39}$$

where Ω is a linear operator, like the Hamiltonian. The allowed values for λ with the constraints are the eigenvalues of Ω with eigenfunctions $\xi(\sigma; \vec{R})$. The translational symmetry can be described via a translational operator $\mathbb{T}(\vec{R})$ defined as

$$\mathbb{T}(\vec{R_i})\xi(\sigma; \vec{R}) = \xi(\sigma; \vec{R} + \vec{R_i}) \tag{40}$$

It is important to remark that the vector $\vec{R} + \vec{R_i}$ is still a lattice vector. By applying twice the $\mathbb{T}$ operator we have

$$\mathbb{T}(\vec{R_2})\mathbb{T}(\vec{R_1})\xi(\sigma; \vec{R}) = \mathbb{T}(\vec{R_2})\xi(\sigma; \vec{R} + \vec{R_1}) = \tag{41}$$

$$\xi(\sigma; \vec{R} + \vec{R_1} + \vec{R_2}) = \mathbb{T}(\vec{R_1})\mathbb{T}(\vec{R_2})\xi(\sigma; \vec{R}) \tag{42}$$

so two translational operators commute. Let's consider the dynamic operator Ω. By the periodicity of the lattice every point of the lattice should be the same for the operator Ω, it means that

$$\mathbb{T}(\vec{R_i})\Omega\xi(\sigma; \vec{R}) = \Omega\mathbb{T}(\vec{R_i})\xi(\sigma; \vec{R}) \tag{43}$$

so for any ξ and for any $\mathbb{T}$, Ω commutes with all the $\mathbb{T}$ of the lattice. So we can find the eigenfunctions of 39 which are simultaneously eigenfunctions of the translational operators of the lattice. If $\xi(\sigma; \vec{R})$ is an eigenfunction of Ω and $\mathbb{T}$ and $C(\vec{R_i})$ is the eigenvalue of $\mathbb{T}(\vec{R_i})$ so

$$\mathbb{T}(\vec{R_1})\xi(\sigma; \vec{R}) = \xi(\sigma; \vec{R} + \vec{R_1}) = C(\vec{R_1})\xi(\sigma; \vec{R}) \tag{44}$$

if we apply another translational operator $\mathbb{T}(\vec{R_2})$

$$\mathbb{T}(\vec{R_2})\mathbb{T}(\vec{R_1})\xi(\sigma; \vec{R}) = C(\vec{R_2})C(\vec{R_1})\xi(\sigma; \vec{R}) \tag{45}$$

but the product $\mathbb{T}(\vec{R_2})\mathbb{T}(\vec{R_1})$ is equivalent to the translational operator $\mathbb{T}(\vec{R_2} + \vec{R_1})$

$$\mathbb{T}(\vec{R_2} + \vec{R_1})\xi(\sigma; \vec{R}) = C(\vec{R_2} + \vec{R_1})\xi(\sigma; \vec{R}) \tag{46}$$

and so we can deduce

$$C(\vec{R_2} + \vec{R_1}) = C(\vec{R_2})C(\vec{R_1}) \tag{47}$$

for any choice of the vector $\vec{R_1}$ and $\vec{R_2}$. Hence the eigenvalue C must have an exponential behavior in $\vec{R}$:

$$C(\vec{R}) = e^{\vec{\mu} \cdot \vec{R}}\xi(\sigma, 0) \tag{48}$$

if the vector $\vec{\mu}$ would have real part different from zero, there should be directions in the infinite crystal in which the amplitude of the field would increase exponentially. In order to avoid that $\vec{\mu}$ must be of the form $i\vec{k}$ where $\vec{k}$ is a real vector. From this moment we will call $\xi(\sigma, 0)$ simply $\xi(\sigma)$, so the eigenfunctions of the dynamic problem must have the form of the well known Bloch Theorem:

$$\xi(\sigma; \vec{R}) = \xi_{\vec{k}}(\sigma)e^{i\vec{k} \cdot \vec{R}} \tag{49}$$

and the eigenfunctions of Ω are obtained by varying the vector $\vec{k}$ of the eigenvalue $exp(i\vec{k} \cdot \vec{R})$ of the translational operator $\mathbb{T}(\vec{R})$. Now the reciprocal lattice becomes useful, if the vector $\vec{k'}$ and $\vec{k}$ differ by a value of a translational vector of the reciprocal lattice $\vec{G}$

$$\vec{k'} = \vec{k} + \vec{G} \tag{50}$$

so remembering that $\vec{G} \cdot \vec{R} = 2n\pi$ where n is an integer, we can say

$$exp(i\vec{k'} \cdot \vec{R}) = exp(\vec{G} \cdot \vec{R})exp(i\vec{k} \cdot \vec{R}) = exp(i\vec{k} \cdot \vec{R}) \tag{51}$$

so $\vec{k'}$ and $\vec{k}$ identify the same eigenvalue of the operator $\mathbb{T}(\vec{R})$. The set of the solutions $\zeta_{\vec{k}}$ of the dynamic problem is the same as the set of the solutions of $\zeta_{\vec{k}+\vec{G}}$. So without loss of generality we can focus our attention just on the primitive cell of the reciprocal lattice hence the **First Brillouin Zone**

9. Photonic band structures in the solid

A lot of results can be provided with the study of the simplest case; a periodic potential in one dimension V(x) so that

$$V(x) = V(x+a) \tag{52}$$

where a is the lattice constant. The Schrödinger Equation

$$-\frac{\hbar^2}{2m}\frac{d^2\psi}{dx^2} + V(x)\psi = E\psi \tag{53}$$

has two real, linear, independent solutions $u(x)$ and $v(x)$. Eq.52 implies that $u(x+a)$ and $v(x+a)$ are still solutions of the Schrödinger Equation, and so have to be linear combinations of $u(x)$ and $v(x)$;

$$u(x+a) = \alpha u(x) + \beta v(x)$$

$$v(x+a) = \gamma u(x) + \delta v(x) \tag{54}$$

with α, β, γ and δ real numbers. So it's useful to consider that in this case the Wronskian

$$\mathbf{W}(x) = v\frac{du}{dx} - u\frac{dv}{dx} = constant \tag{55}$$

independent from x, and this implies

$$\mathbf{W}(x+a) = v(x+a)\frac{du(x+a)}{dx} - u(x+a)\frac{dv(x+a)}{dx} =$$
$$= (\alpha\delta - \beta\gamma)\{v\frac{du}{dx} - u\frac{dv}{dx}\} = \tag{56}$$
$$= (\alpha\delta - \beta\gamma)\mathbf{W}(x)$$

so it means that

$$\alpha\delta - \beta\gamma = 1 \tag{57}$$

Let's consider now a generic solution of the Schrödinger Equation

$$\psi(x) = pu(x) + qv(x) \tag{58}$$

and let's call λ the value of ψ under translation operation a

$$\psi(x+a) = \lambda\psi(x) \tag{59}$$

so we can rewrite

$$pu(x+a) + qv(x+a) = \lambda(pu(x) + qv(x)) \tag{60}$$

the equation (54) implies that

$$\alpha p + \gamma q = \lambda p$$

$$\beta p + \delta q = \lambda q \tag{61}$$

which has a solution different from zero when

$$(\alpha - \lambda)(\delta - \lambda) - \beta\gamma = 0$$

$$\lambda^2 - (\alpha + \delta)\lambda + \alpha\delta - \beta\gamma = 0 \tag{62}$$

so finally just for the value of the equation

$$\lambda^2 - (\alpha + \delta)\lambda + 1 = 0 \tag{63}$$

which brings out two different cases: in the first one the two solutions λ_1 and λ_2 are real and different, and by $\lambda_1\lambda_2 = 1$ we have $\lambda_1 > 1$ and $\lambda_2 < 1$. If we translate ψ by a number of times n we have

$$\psi(x+na) = \lambda_1^n\psi(x) = \lambda_2^n\psi(x) \tag{64}$$

both solutions diverge, $\lambda_1^n\psi(x)$ for $x \to \infty$ and $\lambda_2^n\psi(x)$ for $x \to -\infty$. In the second case the two solutions are the complex conjugate of the other, and both module=1 so we can write

$$\lambda_1 = e^{ika}; \lambda_2 = e^{-ika} \tag{65}$$

in this case there is a degeneration of the eigenvalue of the energy. By the way it is useful to note the particular case in which both solutions are equal to 1 or -1. Because of equation 65 we can write

$$\psi(x+na) = e^{ikna}\lambda(x) \tag{66}$$

which is the Bloch theorem, and implies also that if we write $\psi(x)$ in the form

$$\psi(x) = u(k;x)e^{ikx} \tag{67}$$

so u is periodic both in the direct lattice

$$u(k;x+a) = u(k;x) \tag{68}$$

and the reciprocal one

$$u(k + \frac{2\pi}{a};x) = u(k;x) \tag{69}$$

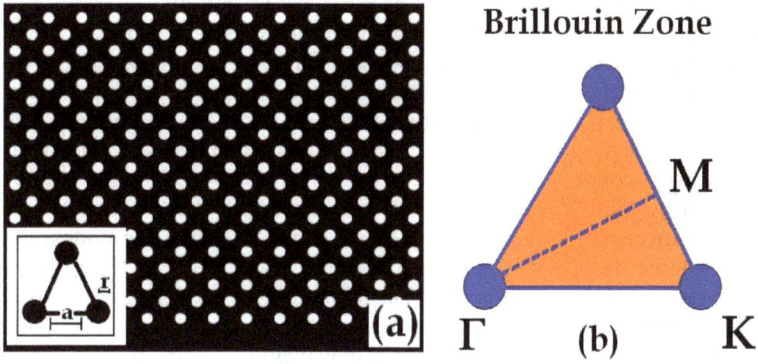

Fig. 6. a) two dimensional hexagonal array: (inset Primitive cell of the two dimensional hexagonal array showing lattice spacing a and radius r). b) First Brillouin Zone showing the critical points Γ, K and M

the physical solution allowed are the Bloch waves, so the plane waves modulated by a function which has the periodicity of the lattice. It is convenient to shrink k just to the First Brillouin Zone which is in this case the interval $[-\frac{\pi}{a}, \frac{\pi}{a}]$. The eigenvalue E($k$) and E(-$k$) are degenerate and so E(k) is an even function of k which has also the periodicity of the reciprocal lattice.

$$E(k + \frac{2\pi}{a}) = E(k) \tag{70}$$

So if we want to perform a study of how the whole crystal propagates electromagnetic waves, it will be sufficient just to focus the study on the reciprocal lattice.

10. Two-dimensional hexagonal photonic crystal structure

Our aim is to fabricate a structure which will behave as resonant cavity for the single photon emission of the NV^- center. The best choice to pursue this goal would have been a 3D photonic crystal structure with a NV^- placed in its center, but unfortunately the fabrication of this kind of structure is challenging. So we decide to follow a different path using a quasi$-3D$ structure. In fact combining the photonic crystal feature and the total internal reflection (TIR), we obtain a structure which confines the light in the three directions XYZ. Indeed the light is confined by distributed Bragg reflection in the plane of periodicity (XY) and by total internal reflection in the perpendicular plane (Z), so we aim to fabricate a photonic crystal in a thin membrane. Let us start to consider the photonic crystal structure, as we have seen before, performing different choices of primitive cells gives rise to many different kind of structures, we have chosen a primitive cell which consists of an equilateral triangle of air holes, as seen in the inset of Fig.6(a). The repetition of the primitive cell gives rise a two dimensional hexagonal array, as shown in Fig.6(a). In Fig.6(b) we have shown the first brillouin zone for the two dimensional hexagonal lattice with its critical points. We decided to study this kind of structure because it is suitable for fabrication requirements and at the same time this photonic crystal structure has a bandgap tunable in the region of the desired wavelength with current fabrication technology. As previously discussed, in order to study the behavior of the whole crystal, we can just focus our attention on the primitive cell. The parameters of the primitive cell to take account of, are the radius of the holes **r**, and the lattice constant **a**, namely the distance between holes centers. These two parameters allow to compute the whole behavior of the primitive cell and so of the photonic crystal which it makes. The first stage of the simulation process is to calculate the photonic band-gap of this structure, which gives a starting point for optimizing the lattice constants and hole radii. In order to compute the bandgap of the structure, we used a simulation software known as MIT Photonic-Bands (MPB) package, which is a free program for computing the band structures (dispersion relations) and electromagnetic modes of periodic dielectric structures, on both serial and parallel computers. It was developed by Johnson & Joannopoulos (2001) at MIT along with the Joannopoulos group. This program computes definite-frequency eigenstates (harmonic modes) of Maxwell's equations in periodic dielectric structures for arbitrary wavevectors, using fully-vectorial and three-dimensional methods. It is especially designed for the study of photonic crystals, but is also applicable to many other problems in optics, such as waveguides and resonator systems. (For example, it can solve for the modes of waveguides with arbitrary cross-sections). Remembering the first Brillouin zone depicted in Fig.6(b), if we imagine to wave vector k "moving" in a path assuming all the values from the critical point Γ through K and M and finally coming back to Γ, as also shown in the inset of Fig.7, we visualize the horizontal axis in the bandgap picture depicted in Fig.7. In that diagram we can appreciate how the behavior of the electromagnetic field inside the primitive cell responds to the variation of the frequency. So we can appreciate the first guided TE mode, the second guided TE mode and the bandgap between them, the yellow line, the values of the normalized frequency in that region gives us the range values for the lattice constant, in order to have a confinement for the desired wavelength. We performed many different simulations varying the ratio of the radius over the lattice constant, namely $\frac{r}{a}$ of the primitive cell, so we finally found the best choice, $\mathbf{r} = 0.30\mathbf{a}$ which gave us the widest bandgap for the triangular lattice. Now we take the middle value of the bandgap which gave us the value for $\frac{a}{\lambda}$=**0.375** giving the ratio for the computation of the lattice constant $a = 0.375\lambda$, where λ is the frequency of the light. So if we want to have a two-dimensional hexagonal photonic crystal structure which has a bandgap centered to wavelength of 637 nm, namely a PC structure resonant with the NV^- center emission, we have to use the values for lattice constant $a = 238.875$ nm and radius $r = 71.6625$ nm. Once we have a range for the values of the relevant parameters, we started to simulate the photonic crystal cavity made by a hexagonal array of air holes in diamond with three missing holes in the middle, also known as $L3$ cavity as shown in Fig.7b. In order to perform the simulations of the behavior of the cavity we used a Finite difference time domain (FDTD) software developed at University of Bristol by Professor Railton. In the FDTD method the time-dependent Maxwell's equations are discretized using central-difference approximations to the space and time partial derivatives the electric field vector components in a volume of space are solved at a given instant in time the magnetic field vector components in the same spatial volume are solved at the next instant in time the process is repeated over and over again until the desired transient or steady-state electromagnetic field behavior is fully evolved. In brief, the method involves dividing three-dimensional space into a grid of unit cells. Each cell is assigned six nodes, where the components of the electric and magnetic fields are stored. These values are updated at each time step by calculating the response to an incident field or excitation. The excitation takes the form of a Gaussian modulated sinusoid, as is appropriate for a dipole. A Fourier transform is taken on data sampled over time. The electric and magnetic fields ($\vec{E}$ and $\vec{H}$) at specific frequencies can be calculated at all points on the grid. The parameters that can be varied to optimize the performance of the cavity are the lattice constant, $\mathbf{a}$, the radius of the air holes $\mathbf{r}$. We performed many simulations changing the values of the lattice constant and

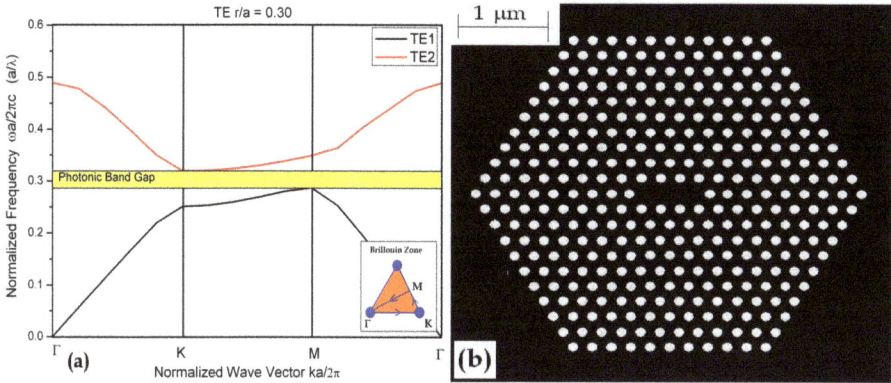

Fig. 7. a) Bandgap of the Hexagonal PC structure with the ratio r/a=0.3: Horizontal axis plots the normalized wavelength around the Brillouin zone going from $\Gamma - K - M - \Gamma$, Vertical axis shows normalized frequency $\frac{\omega a}{2\pi c} = \frac{a}{\lambda}$. b) L3 photonic crystal cavity structure

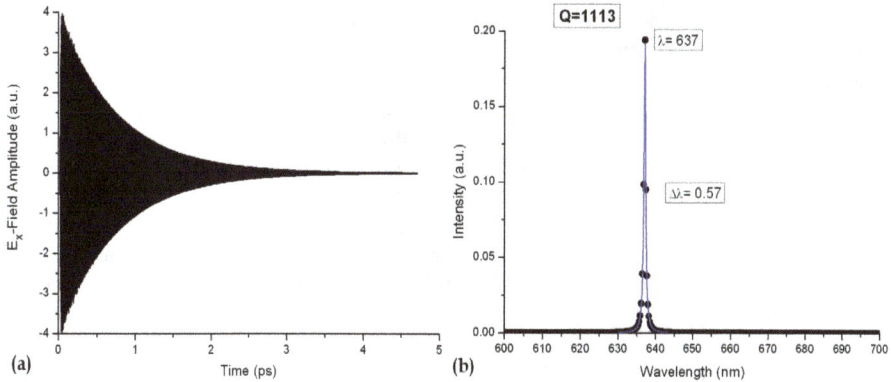

Fig. 8. a) Ring-down of the Ex component of the electromagnetic field inside the cavity. b) Fourier Transform of the Ex plot with Lorentzian Fit leading to an estimation of Q

the radius in the range gave by the bandgap calculation, in order to have a cavity resonant frequency of the NV^- emission. So we ended as results of the simulations with a L3 cavity resonant to a single mode frequency of $637nm$ with a small quality factor $Q = 1113$. In Fig.8(a) we can see the simulated ring down of the E_x component of the electromagnetic field inside the L3 cavity, in Fig.8(b) is shown its Fourier Transform with a Lorentzian fit which allowed us to calculate the resonant frequency, $\lambda = 637nm$ and the Full Half Width Maximum (FHWM) $\Delta\lambda = 0.57$. We checked in a large range surrounding the resonant frequency considering many different values but this left the value of quality factor unchanged, so we decided for clarity purposes to show a small range surrounding the resonant frequency. In this early we chose a cavity thickness of 500 nm which was not optimum The second step consisted of the calculation of the thickness of the membrane, in the direction perpendicular to the photonic crystal structure the field is confined by total internal reflection (TIR) by creating a suspended membrane. To ensure single-mode operation at wavelength $\lambda = 637nm$, a slab thickness of $\sim \lambda/4n$ is required as discussed by Joannopoulous et al. (1995). So we simulated the behavior

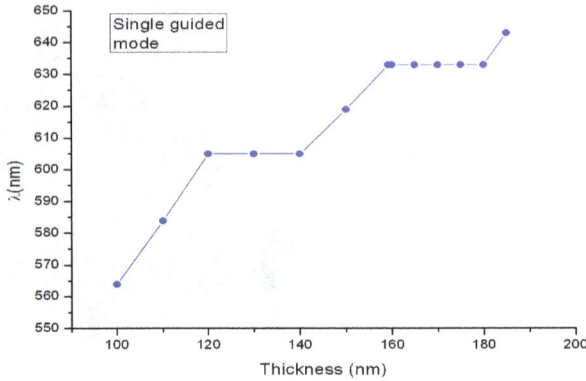

Fig. 9. Wavelength of the confined single mode emission in the membrane respect to the variation of its thickness

of the light from a source placed in the center of a membrane with a photonic crystal L3 cavity for different thickness of the membrane. As we can see in in Fig.9, where it is plot the change of the wavelength of the confined single mode with respect to the thickness of the membrane. The plot shows the expected blue-shift as we thin the membrane. We can correct for this shift by scaling all dimensions after optimizing the Q-Factor. In order to increase Q-factor of the light inside the cavity having a small modal volume, Akahane et al. (2003) demonstrated that the light should be confined gently in order to reduce scattering into leaky modes. The strategy to obtain gentler confinement is to change the condition for Bragg reflection at the cavity edge. We changed the radii of the holes at the side of to the cavity and eventually we shifted the holes at the end to the cavity. So first we varied the size of the holes in the first lines next to the cavity, as shown in the inset of Fig.10, modifying the geometry of the lattice structure surrounding the defect, following the idea proposed by Zhang & Qiu (2004). We improved so far the value of the simulated quality factor, as shown in Fig.10, where it is plot the behavior of simulated quality factor as function of the radius of the holes in the first lines next to the cavity. We note that a value for the small radius $R_1 = 0.198a$ gives the highest value for the quality factor. After that modification we performed a further modification in which we aim to gently confine the mode at the edge of the activity by shifting the nearest holes as we can see in the inset of Fig.11. So if we shift the position of the air holes at the edge we change the reflection conditions, but on the other way round the light penetrates more inside the mirror and is reflected perfectly. In Fig.11 we have plot the behavior of the simulated quality factor as function of the shift of the nearest holes, and we can clearly see that there is a peak for $d = 0.11a$, which is the value of the distance at which we shift the holes. Choosing a thickness of 185 nm we now optimize the Q-factor as shown in Fig.12a where we plot the behavior of the Quality factor respect to change of the thickness. We can see clearly that it is highest for *thickness* = 185*nm* which is the value we have chosen for the next steps. The quality factor, Q, can be separated into the in-plane value, $Q_{\parallel}$, and a vertical value, $Q_{\perp}$. $Q_{\parallel}$ can, in principle, be made arbitrarily high by increasing the number of periods. The cavity is surrounded by 14 periods in all directions. Simulation results showed that increasing the number of periods to 25 changes the quality factor by less than 2%. This means that the total Q approaches $Q_{\perp}$ as described also by Tomljenovic-Hanic et al. (2006). Finally in order to increase the quality factor with a small modal volume we decided to keep 14 periods surrounding the

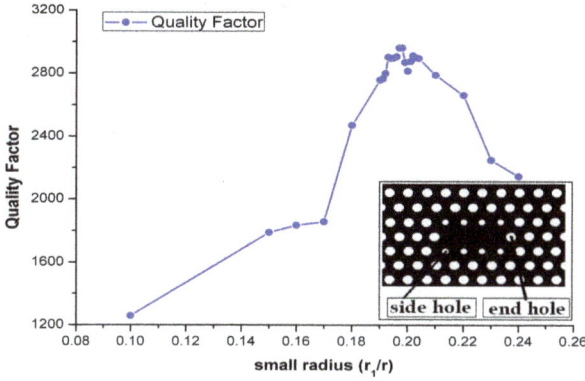

Fig. 10. Quality factor as function of the radius of the nearest holes of the L3 cavity. Inset:schematic diagram of the L3 cavity modified with different value for the holes in the first lines next to the cavity

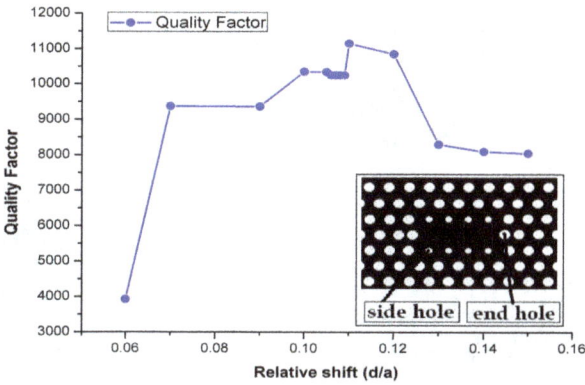

Fig. 11. Quality factor as function of the shift of the nearest holes of the L3 cavity. Inset:schematic diagram of the L3 cavity modified with different value for the holes in the first lines next to the cavity

cavity, because we did not want to have too many holes for fabrication purposes, the final structure perfectly optimized is shown in Fig.12b. In Fig.13 we plot the intensity ($|E^2|$) of the electromagnetic field, higher in red, lower in blue, from top view ($x - y$ plane)Fig.13a and side view ($z - y$ plane)Fig.13b respectively. Clearly we see that the light is confined by distributed Bragg reflection in the plane of periodicity (xy) and by total internal reflection in the perpendicular plane (z). The result of increasing the total size of the photonic crystal is that we achieve a highest quality factor $Q \simeq 32.000$. This value we simulated agrees with the best values reached by other groups like Tomljenovic-Hanic et al. (2009). We show the time decay and the fourier transform in Fig.14. In order to verify that the photonic crystal cavity produce an enhancement of the Purcell factor of the NV$^-$ center we need also to calculate the modal volume of the cavity. The modal volume, defined by Coccioli et al. (1998) is

$$V_{eff} = \frac{\int \epsilon(r)|E(r)|^2 d^3 r}{[\epsilon(r)|E(r)|^2]_{max}} \qquad (71)$$

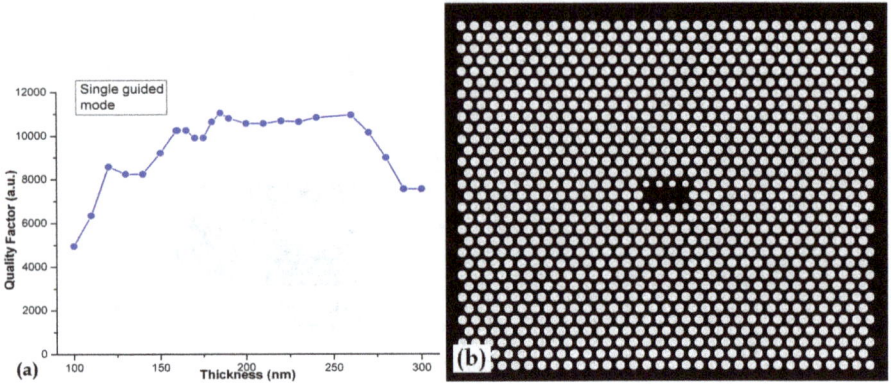

Fig. 12. a)Quality factor for the confined single mode emission in the membrane according to the variation of its thickness, b)L3 photonic crystal cavity structure modified with smaller radius and nearest holes shifted

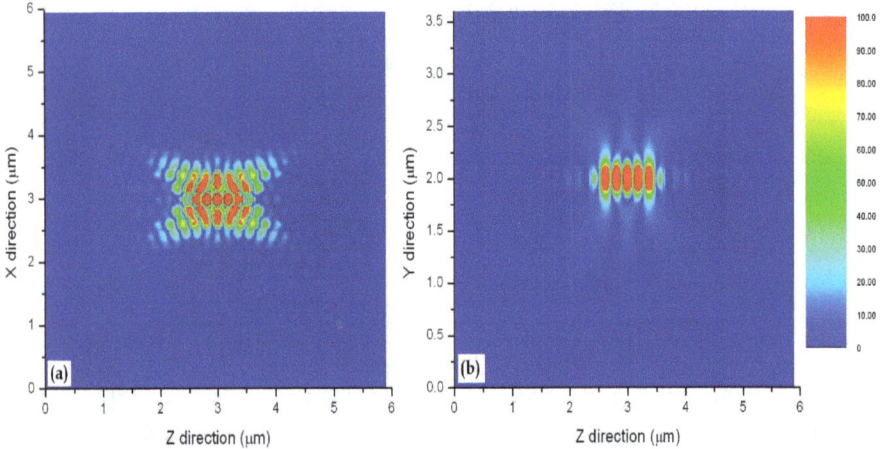

Fig. 13. a) Plot of the intensity of electromagnetic field confined by the photonic crystal bandgap cavity,(top view of $x - z$ plane). b) Plot of the intensity of electromagnetic field confined by total internal reflection, (top view of $y - z$ plane). higher intensity in red lower in blue)

where $\epsilon(r)$ is the dielectric constant and $|E(r)|^2$ is the electric field intensity at position r and the integral is normalized by the maximum intensity in the cavity. In order to estimate V_{eff} we used a procedure shown in Ho et al. (2011) which creates different frequency snapshots at different position in the computational grid during the simulation, recording all the information of the electromagnetic field in each slice. Using this method and the structure depicted in Fig.12 we estimate the modal volume of the field inside the photonic crystal to be $V_{eff} = 0.0162\mu m^3$ which is a value close to the one we assumed on our work on non-demolition measurement, described in Young et al. (2009), and is also consistent with results obtained by other groups as for example Tomljenovic-Hanic et al. (2009). The modal volume can be normalized to the cubic wavelength of the resonant mode $(\frac{\lambda}{n})^3$ in a medium

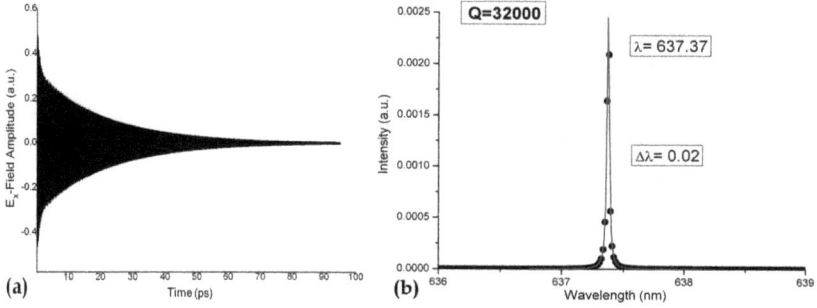

Fig. 14. a) Ringdown of the E_x component of the electromagnetic field inside the optimized cavity. b) Fourier Transform of the E_x plot with Lorentzian Fit leading to an estimation of Q=32000

of refractive index (n) defined as:

$$V_{opt} = \frac{V_{eff}}{(\lambda/n)^3} \tag{72}$$

So for our case the value of the normalized volume for a the wavelength of the NV$^-$ center ($\lambda = 637$ nm) in the diamond which has refractive index $n = 2.4$ corresponds to

$$V_{opt} = 0.8665(\frac{\lambda}{n})^3. \tag{73}$$

Remembering the Purcell factor, as described in Fox (2006),

$$F_p = \frac{3Q\lambda^3}{4\pi^2 n^3 V_{eff}} \tag{74}$$

we can estimate the enhancement of spontaneous emission rate of the NV$^-$ center inside the photonic crystal cavity

$$F_p = 2.8 \times 10^3. \tag{75}$$

Remembering the definition of the coupling rate and radiative decay (Fox (2006)) we can see that in this case $g_0/2\pi = 11.7028$ GHZ, with a radiative decay rate $\kappa/2\pi = 14.7174$ GHZ. The NV$^-$ center usually has a typical emission lifetime $\tau = 12$ ns which gives us a rate $\gamma/2\pi = 1/\tau = 0.0833$ GHZ. Finally we can compare these three main parameters, g_0, κ and γ, in order to reach the strong-coupling regime the coherent coupling g_0 between the transition and the cavity field has to exceed the decay rates of the color center γ and the cavity κ, obeying $4g_0 > (\kappa + \gamma)$. In this particular case we can estimate $4g_0/(\kappa + \gamma) > 3.16$, which tells us that we would be in the strong coupling regime.

11. Fabricating photonic crystals using focus ion beam etching

Having simulated photonic crystal structure cavities we have begun fabrication via focused ion beam etching (FIB). Our aim is to create a suspended membrane with the "Noda" cavity previously described. In the first fabrication step, the diamond crystal is undercut by turning side-on and etching to obtain a 200nm thick slab attached to the bulk (a suspended slab). We needed to etch the membrane first, because if we have done the photonic crystal structure first, at the stage in which we etch the membrane some sputtering could have filled the

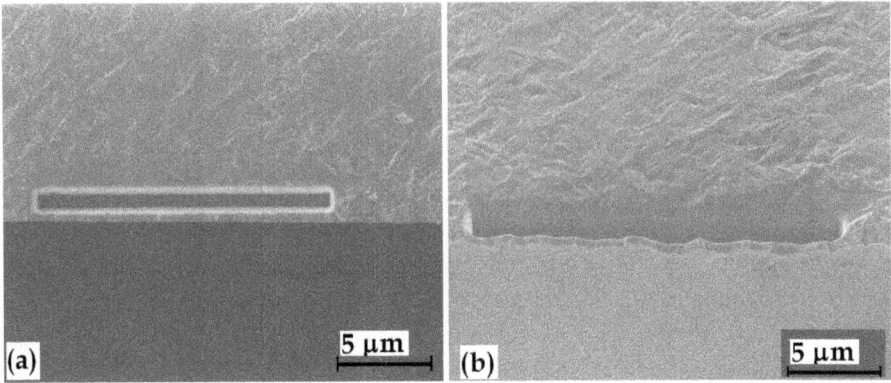

Fig. 15. secondary electron image of a etched membrane in the diamond sample. a) a top view of the etched membrane, b) 45° Tilted view of the etched membrane

Fig. 16. a) Tilted view of the L3 cavity taken with FIB at different tilt and magnitude. b) larger image of the membrane and the cavities

holes. In order to etch the membrane we mounted the sample on a stage, and then titled it to 90° and after we covered the implanted NV⁻ center array zone with silver, in order to protect the implanted NV⁻s, we etched a thin membrane of $200nm$ according to the results of the simulation previously shown. In Fig.15(a) we can see a top view of the membrane and in Fig.15(b) we can see an image of the same membrane tilted by 45°. After we made the membrane we repositioned the sample horizontally and finally we etched the hexagonal air hole array with cavity formed from three filled holes. Fig.16 shows two views, tilted of 45° at different magnifications of the resulting structure. Both were secondary electron images taken with FIB after the etching. In Fig.16(a), we can see the photonic crystal cavity and etched in the membrane which is more evident in Fig.16(b) where we have a scan over a larger area which shows the size of the cavities compared to the suspended membrane. In the top view, shown in Fig.17(a), we can observe the cavity and notice some imperfections in it due to the FIB technique which creates deposit of etched material during the scanning. In Fig.17(b) we can see an image taken with a confocal microscope,in which blue color means low intensity and red color means high intensity. Fig.17(b) is remarkable because we can clearly

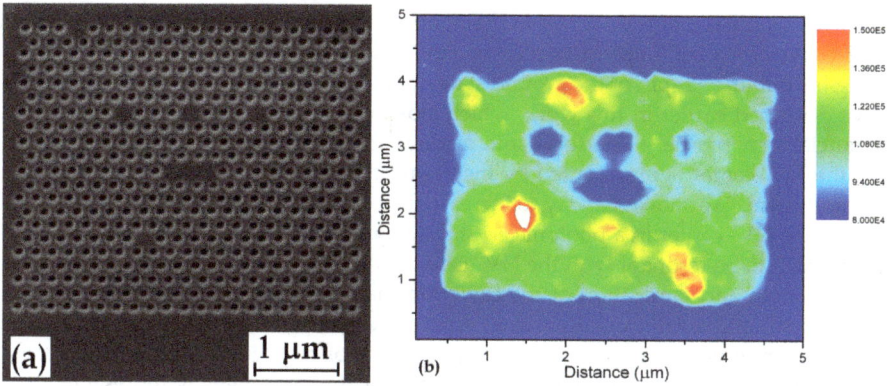

Fig. 17. a) Secondary electron emission image of the top view of the L3 cavity. b) Fluorescence image taken with the confocal microscope (color red: high intensity, color blue:low intensity).

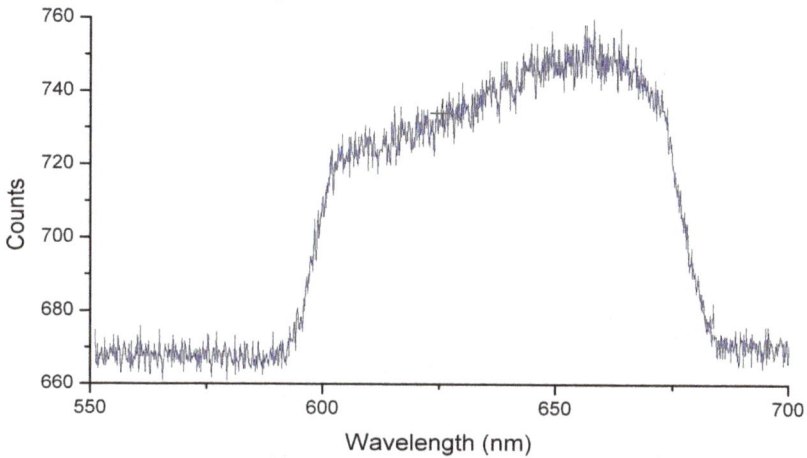

Fig. 18. Image of the emission spectrum taken at the center of the photonic crystal cavity. No narrow peaks are seen as would be expected from cavity resonance effects.

see reduced fluorescence in the un-etched zone, forming the cavity. Because there is no (or less) etch damage in these regions. This is an encouraging result because it means that if there were a NV^- center in the cavity we might be able to see it. We performed some measurement of the spectrum of the light emitted from the cavity region. Unfortunately we were not able to see any enhancement of the signal as we might expect from a cavity resonance, but just a broad emission as shown in Fig.18. At this stage we decided to make a step back and to perform a preliminary study about the real possibility of coupling a single NV^- center to a larger structure etched in the diamond with FIB.

12. Conclusion

In this work we discussed about the feasibility of NV^- centers as single photon emitters and how to use its spin as qubit for quantum computing applications, remarking the many advantages that the use of NV⁻ center in diamond would produce. One of the key challenges in order to perform a real implementation of a quantum computer concerns the possibility of handling the qubit. The spin of the ground state of the NV⁻ center shows some characteristics we have described earlier in this work, which looks very promising for this purposes. One of the crucial step in order to perform spin readout and non demolition measurement of the spin of the NV⁻ center is represented by increasing the coupling between the light and the solid state system. We showed a way to increase the coupling between the NV⁻ and the light by placing the NV⁻ center in a photonic crystal cavity. We characterized the photonic crystal cavity tuning it to be resonant with the NV⁻ center emission, having had encouraging results in the simulation and fabrication of the cavity, in that we reached a reasonable high value of quality factor and small modal volume. Another very important aspect in order to build a quantum computer is represented by the possibility of handling a single photon source. In order to use NV⁻ as single photon emitter, one of the challenges is represented by the light collection. We developed in order to increase the light collection from the NV⁻ center by etching a solid immersion lens around it, proposing a technique (Marseglia et al. (2011)) in order to locate NV⁻ centers with accuracy of $10nm$, and fabricate structure around them. In future works we will use the technique we have recently developed, in order to create a photonic crystal around a single NV⁻ center. Another important path we want to follow consist in exploiting another useful color center in the diamond, the chromium center. This center acts as a single photon emitter as well but with a narrower spectrum. It has a resonant wavelength of $755nm$ which is in the wavelength range for the si photon counting detector, allowing us to detect them with high efficiency. We are very interested in using the technique we have developed in order to etch photonic crystal around single chromium center and coupling with it, this would allow us to increase the light collection from the chromium center permitting it to be used as an ultra bright single photon emitter. As it emits at $755nm$ it is compatible with integrated photonic circuits being developed in our group by Politi et al. (2008). Similarly the nickel-nitrogen complex (NE8) center in diamonds, studied by Gaebel et al. (2004); Rabeau et al. (2005), has narrow emission bandwidth of $1.2nm$ at room temperature with emission wavelength around $800nm$, again suitable for Si detectors and quantum photonic circuits. In addition, in this spectral region little background light from the diamond bulk material is detected, which made it an interesting possible candidate for single photon source. Once we are able to locate an NE8 (or other suitable narrowband) center we will extend the registration procedure developed to allow fabrication of photonic crystal structure around individual defects, ending in the measurement of Q-factors and Purcell enhanced emission. In order to handle and guide the light emitted from the source a detailed study of parameters of a photonic crystal waveguide in the diamond will be required as demonstarted by Song et al. (2007). We will simulate the behavior of the electromagnetic field inside the cavity and how it will couple with the waveguide. A good response will lead us to fabricate and then measure the effective coupling. We will also explore different etching techniques such as Reactive-Ion Etching (RIE) which will allow us to create membranes and very thin structures in diamond with a high precision. This will be useful in order to create different structures around registered NV⁻ centers, for instance with photonic crystal nanobeam cavities studied by Deotare et al. (2009). This kind of structures as the remarkable advantage to be very easy to fabricate offering a huge quality factor and a very small modal volume.

13. References

Akahane, Y., Asano, T., Song, B.-S. & Noda, S. (2003). High-q photonic nanocavity in a two-dimensional photonic crystal, *Nature* 425: 944.

Coccioli, R., Borodsky, M., Kim, K. W., Rahmat-Samii, Y. & Yablonovitch, E. (1998). Smallest possible electromagnetic mode volume in a dielectric cavity, *IEE Proc.-Optoelectron.* 145: 391–397.

Deotare, P. B., McCutcheon, M. W., Frank, I. W., Khan, M. & Loncar, M. (2009). Coupled photonic crystal nanobeam cavities, *Appl. Phys. Lett.* 95: 031102.

Fox, M. (2006). *Quantum Optics*, 2 edn, Oxford University Press.

Gabel, T., Dohman, M., Popa, I., Wittmann, C., Neumann, P., Jelezko, F., Rabeau, J. R., Stavrias, N., Greentree, A. D., Prawer, S., Meijer, J., Twamley, J., Hemmer, P. R. & Wrachtrup, J. (2006). Room-temperature coherent coupling of single spins in diamond, *Nature Physics* 2: 408 – 413.

Gaebel, T., Popa, I., Gruber, A., Domhan, M., Jelezko, F. & Wrachtrup, J. (2004). Stable single-photon source in the near infrared, *New Journal of Physics* 6: 98.

Gali, A., Fyta, M. & Kaxiras, E. (2008). Ab initio supercell calculations on nitrogen-vacancy center in diamond: Electronic structure and hyperfine tensors, *Physical Review B* 77: 155206.

Gruber, A., Dräbenstedt, A., Tietz, C., Fleury, L., Wrachtrup, J. & von Borczyskowski, C. (1997). Scanning confocal optical microscopy and magnetic resonance on single defect centers, *Science* 276.

Hanson, R., Mendoza, F. M., Epstein, R. J. & Awschalom, D. D. (2006). Polarization and readout of coupled single spins in diamond, *Phys. Rev. Lett.* 97: 087601.

Ho, Y.-L. D., Ivanov, P. S., Engin, E., Nicol, M., Taverne, M. P. C., HU, C., Cryan, M. J., Craddock, I. J., Railton, C. J. & Rarity, J. G. (2011). Three-dimensional fdtd simulation of inverse three-dimensional face-centered cubic photonic crystal cavities, *IEEE J. Quantum Electron.* in press.

Jelezko, F., Gaebel, T., Popa, I., Gruber, A. & Wrachtrup, J. (2004a). Observation of coherent oscillation of a single nuclear spin and realization of a two-qubit conditional quantum gate, *Phys. Rev. Lett.* 93: 7.

Jelezko, F., Gaebel, T., Popa, I., Gruber, A. & Wrachtrup, J. (2004b). Observation of coherent oscillations in a single electron spin, *Phys. Rev. Lett.* 92: 7.

Jelezko, F. & Wrachtrup, J. (2004). Read-out of single spins by optical spectroscopy, *J. Phys.: Condens. Matter* 16: 104.

Joannopoulous, J. D., Meade, R. D. & Winn, J. N. (1995). *Photonic crystals: Molding the flow of light*, Princeton University Press.

Johnson, S. G. & Joannopoulos, J. D. (2001). Block-iterative frequency-domain methods for maxwell's equations in a planewave basis, *Optics Express* 8: 173.

Kurtsiefer, C., Mayer, S., Zarda, P. & Weinfurter, H. (2000). Stable solid-state source of single photons, *Phys. Rev. Lett.* 85: 290.

Manson, N. B., Harrison, J. P. & Sellars, M. J. (2006). Nitrogen-vacancy center in diamond: Model of the electronic structure and associated dynamics, *Phys. Rev. B* 74: 104303.

Marseglia, L., Hadden, J. P., Stanley-Clarke, A. C., Harrison, J. P., Patton, B., Ho, Y.-L. D., Naydenov, B., Jelezko, F., Meijer, J., Dolan, P. R., Smith, J. M., Rarity, J. G. & O'Brien, J. L. (2011). Nano-fabricated solid immersion lenses registered to single emitters in diamond, *Appl. Phys. Lett.* 98: 133107.

Neumann, P., Mizuochi, N., Rempp, F., Hemmer, P., Watanabe, H., Yamasaki, S., Jacques, V., Gaebel, T., Jelezko, F. & Wrachtrup, J. (2008). Multipartite entanglement among single spins in diamond, *Science* 320: 1326.

Politi, A., Cryan, M. J., Rarity, J. G., Yu, S. & O'Brien, J. L. (2008). Silica-on-silicon waveguide quantum circuits, *Science* 320: 646–649.

Rabeau, J. R., Chin, Y. L., Prawer, S., Jelezko, F., Gaebel, T., & Wrachtrup, J. (2005). Fabrication of single nickel-nitrogen defects in diamond by chemical vapor deposition, *Appl. Phys. Lett.* 86: 131926.

Song, B.-S., Noda, S. & Asano, T. (2007). Photonic devices based on in-plane hetero photonic crystals, *Science* 300: 1537.

Su, C. H., Greentree, A. D. & Hollenberg, L. C. L. (2008). Towards a picosecond transform-limited nitrogen-vacancy based single photon source, *Optics Express* 16: 6240.

Tamarat, P., Manson, N. B., Harrison, J. P., McMurtrie, R. L., Nizovtsev, A., Santori, C., Beausoleil, R. G., Neumann, P., Gaebel, T., Jelezko, F., Hemmer, P. & Wrachtrup, J. (2008). Spin-flip and spin-conserving optical transitions of the nitrogen-vacancy centre in diamond, *New Journal of Physics* 10: 045004.

Tomljenovic-Hanic, S., Greentree, A. D., de Sterke, C. M. & Prawer, S. (2009). Flexible design of ultrahigh-q microcavities in diamond-based photonic crystal slabs, *Optics Express* 17: 6465.

Tomljenovic-Hanic, S., Steel, M. J., de Sterke, C. M. & Salzman, J. (2006). Diamond based photonic crystal microcavities, *Optics Express* 14: 3556.

van Oortt, E., Manson, N. B. & Glasbeekt, M. (1988). Optically detected spin coherence of the diamond n-v centre in its triplet ground state, *J. Phys. C: Solid State Phys.* 21: 4385–4391.

Wang, C. F., Hanson, R., Awschalom, D. D., Hu, E. L., Feygelson, T., Yang, J. & Butler, J. E. (2007). Fabrication and charcterization of two-dimensional photonic crystal microcavities in nanocrystalline diamond, *Appl.Phys.Lett.* 91: 201112.

Young, A., Hu, C. Y., Marseglia, L., Harrison, J. P., O'Brien, J. L. & Rarity, J. G. (2009). Cavity enhanced spin measurement of the ground state spin of an nv center in diamond, *New Journal of Physics* 11: 013007.

Zhang, Z. & Qiu, M. (2004). Small-volume waveguide-section high q microcavities in 2d photonic crystal slabs, *Optics Express* 12: 3988–3995.

15

Photonic Crystals for Plasmonics: From Fundamentals to Superhydrophobic Devices

Remo Proietti Zaccaria et al.[*]
Nanobiotech Facility, Italian Institute of Technology, Genova
Italy

1. Introduction

In the last couple of decades we have been witnessing an enormous technological advancement in the field of micro-technology to the extent that nowadays we talk about nanotechnology. Faster computers, LCD based mobiles, nanoparticles for UV absorption in suntan lotions are just few of many examples where nanotechnology plays a fundamental role. The merit of this is mainly in the advance of the fabrication methods. Present techniques such as Focused Ion Beam (FIB) lithography guarantee a resolution of less than 10 nanometers which is about five times more precise than ten years before. Also Photonic Crystals (PhCs), among the others, take advantage from this extremely high resolution level allowing a downscale that permits the realization of structures which in principle can work at vey high energy. Historically PhCs were known as Bragg mirrors and only in 1987 (Yablonovitch, 1987; Sajeev, 1987) with the works of Yablonovitch and Sajeev the term Photonic Crystals was introduced. Nowadays, besides their natural application as filters in particular under full band gap conditions, PhCs see a number of applications: optical fibers (Birks et al., 1997; Zhao et al., 2010), vertical cavity surface emitting lasers (Yokouchi et al., 2003), high reflection coatings, temperature sensors (Song et al., 2006), high efficiency solar cells (Bermel et al., 2007), electric field detectors (Song & Proietti Zaccaria, 2007), non-linear analysis (Malvezzi et al., 2002; Malvezzi et al., 2003), just to name a few. Many are the techniques for the fabrication of PhCs, for example by means of focused-ion beam (Cabrini et al., 2005), two-photon fabrication (Deubel et al., 2004), laser-interference (Proietti Zaccaria et al., 2008a) or waver-fusion techniques (Takahashi et al., 2006). Here we shall focus on the role that PhCs can play for another exciting discipline known as *Plasmonics*. It refers to the capability of some devices of sustaining an *optical surface* mode, namely an electromagnetic wave travelling at the interface between two different materials such as a dielectric and a metal. Such a wave originates from the coupling of incident photons on the interface with

---

[*] Anisha Gopalakrishnan[1], Gobind Das[1], Francesco Gentile[1,2], Ali Haddadpour[3], Andrea Toma[1], Francesco De Angelis[1], Carlo Liberale[1], Federico Mecarini[1], Luca Razzari[1], Andrea Giugni[1], Roman Krahne[1] and Enzo Di Fabrizio[1,2]
[1]*Nanobiotech Facility, Italian Institute of Technology, Genova ,Italy*
[2]*BIONEM lab., Departement of Clinical and Experimental Medicine, Magna Graecia University, viale Europa, Catanzaro, Italy*
[3]*Department of Electrical and Computer Engineering, University of Tabriz, Iran*

the existing free electrons. This kind of mode is known as Surface Plasmon Polariton (SPP). Photonic crystals, as a translational modulation of the refractive index, have already been playing a very crucial role in plasmonics. In fact, they can provide the missing wave vector for the coupling between photons and free electrons of a metal layer (Raether, 1988). Here we have chosen to face three situations relating PhCs and Plasmonics:

a. two dimensional (2D) metallic photonic crystal for maximizing the optical band gap;
b. metallic photonic crystal structures for Surface Enhanced Raman Spectroscopy (SERS);
c. few molecules detection through super-hydrophobic crystals (De Angelis et al., 2011).

The first topic concerns the fundamentals of SPP and PhCs. No particular application will be suggested, but mostly we will focus on the theory behind the generation of SPP inside a metallic photonic crystal. The next two topics are, on the other hand, strictly related to applications. We will concentrate our attention on Raman spectroscopy, as a very important tool for the investigation of the optical properties of many kinds of samples, such as semiconductors or proteins. In particular, metallic ordered (periodic) structures will be used either as artificial SERS substrates or as combiners of SERS and super-hydrophobic effect for few molecules detection. This description will offer a general overview of the important functions that PhCs can hold in Plasmonics and how we could start thinking more intensively of PhCs realized with metallic materials.

2. Two dimensional metallic PhCs for maximum full band gap

One of the main issues when dealing with two dimensional PhCs is the maximum full band gap that the crystal can provide. Specifically, in 2D the crystal manifests two non degenerative polarizations known as TE and TM. The former refers to solutions with the electric field *in* the plane of symmetry of the crystal, the latter with the magnetic field in the same plane. In order to produce a *full* band gap, both TE and TM must provide an optical gap in the *same* spectral region. This is not an easy task and different solutions have been proposed (Joannopoulos et al., 1995), however mostly referring to dielectric PhCs. Here we shall examine how metallic PhCs can improve the chance of a common band gap between TE and TM with respect to a dielectric structure.

2.1 Dielectric photonic crystals

It has been demonstrated that the geometry of PhCs strongly affect whether TE or TM band gap will be generated. In particular, when the crystal is formed by isolated regions of high dielectric, the TM gap is favored; when the crystal is formed by connected regions of high dielectric, the TE gap is favored. Hence, the geometry plays a crucial role. In regards, when 2D photonic crystals are considered, only five possible geometrical arrangements exist. These are called Bravais lattices. Considering the complementary conditions for the realization of a TE/TM gaps, the best Bravais lattice to maximize the probability of obtaining a full band gap is the hexagonal one. This kind of Bravais lattice posses the highest possible rotational symmetry in two dimensions. In view of this information, we have instead chosen the square photonic crystal (Galli et al., 2002), namely a configuration not supporting a full band gap for isotropic dielectric materials (Wang et al., 2005; Proietti Zaccaria, 2008b). In this manner we expect to enhance the different results in terms of full band gap rising either from a dielectric crystal or a metallic one.

2.1.1 Full band gap in dielectric 2D photonic crystals

The chosen dielectric PhC is made of circular columns of high refractive index material, namely silicon (n=3.6) surrounded by air (n=1). The radius of the columns is r=300nm and lattice period P=1μm. As expected, only TM polarization shows zero transmission regions, in particular three band gaps below 1μm-1 are shown in Fig. 2.1. On the other hand, TE does not sustain any band gap. These results remain true even increasing the columns dielectric value or changing the columns radius. No full band gap is then found for this kind of structure.

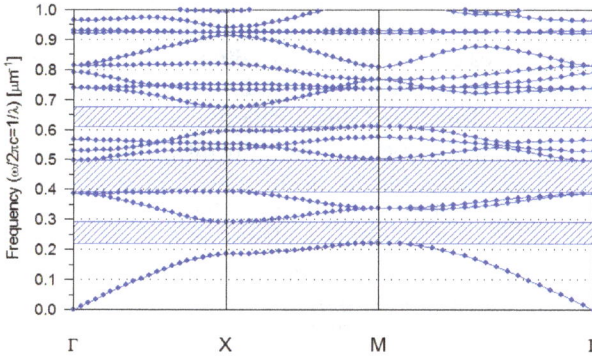

Fig. 2.1. TM band structure for a 2D square silicon columns photonic crystal. Three band gaps are shown.

2.2 Metallic photonic crystals

The behavior of 2D metallic photonic crystals is fundamentally different from what expected by 2D dielectric photonic crystals (Zhao et al., 2009; Sakoda et al., 2001; Ito & Sakoda, 2001). In fact, the use of metallic dispersive materials strongly modifies the light behavior in periodic structures, both for TM and TE polarizations. In particular, TM polarization shows a cut-off frequency ω_c and no modes are found below it. This implies the existence of a TM gap below ω_c. From a physical point of view this is related to the existence of free electrons in metallic materials. Similarly, TE polarization shows a behavior which is absent in dielectric 2D photonic crystals. In fact, in specific range of frequencies, metallic photonic crystals show TE polaritonic band gap close to the plasma frequency. Physically it is associated to the creation of surface plasmon polaritons on the metallic columns of the crystal. These peculiarities of metallic PhCs have the merit of increasing the chance of a full band gap also for square crystals. Hence, maintaining the same geometrical configuration as in the previous section, namely a square structure of columns in air, we have numerically analyzed the band gap formation when metallic columns are considered instead of dielectric ones.

2.2.1 Full band gap in metallic 2D photonic crystals: Drude vs. Lorentz model

We start by considering the Drude model (Rakic' et al., 1998) to describe the metallic parts of the crystal:

$$\varepsilon(\omega) = 1 - \frac{f_0 \omega_p^2}{\omega^2 - i\omega\,\Gamma_0} \tag{2.1}$$

where ω_p is the plasma frequency, Γ_0 is the damping constant and f_0 is the oscillator strength. We recall that this model is a simplified version of the realistic Drude-Lorentz model, in fact it does not describe the material resonances (absorption) of the metal. Nevertheless, the Drude model is commonly used ought to the reduced calculation complexity. Furthermore, it is often assumed that for noble metals such as Ag or Au, no difference is expected between the Drude and the Drude-Lorentz description in the visible range. We will show that this assumption is not always justified.

By choosing $\omega_p = 2\pi c / P$, $\Gamma_0 = 0.01\omega_p$ and $f_0 = 1$ in Eq. (1), a square PhC with columns of radius $r = 0.472P$ and TM polarization, the transmission spectrum of Fig. 2.2 was obtained. The figure shows different curves each associated to a different incident angle. In fact, in order to determine a band gap, two possible ways can be chosen: i) band structure calculation; b) transmission spectrum calculation. Considering the dispersive properties of the metal, the band structure calculation presents some difficulties which make the transmission spectra method the simplest way for the calculation of band gaps. However, when this method is considered, the light must impinge on the PhC under a number of angles, namely the angles covering the irreducible Brillouin zone (Zhao et al., 2009). In our case it means to look at the range [0-45] degrees. Zero transmission regions which are common to all the simulations can be identified as band gaps. For clarity we have chosen only 0 and 45 degrees (however, for the present case, this choice will not affect the identification of optical gaps).

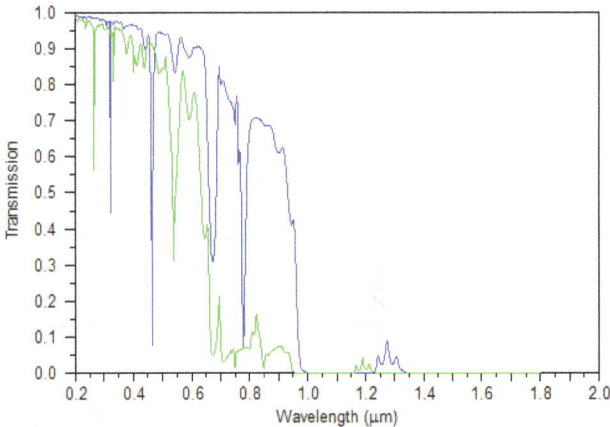

Fig. 2.2. TM transmission spectra for a Drude square PhC. Different lines (colors) correspond to different incident angles (blue: 0°; green: 45°). Two band gaps are found.

Fig. 2.2 clearly shows two band gaps. In particular the lowest frequency one (above 1.35μm) is associated to the cut-off frequency ω_c as confirmed by the band structure calculation in (Sakoda et al., 2001).

Interesting, when TE polarization is considered, a wide band gap is obtained as shown in Fig. 2.3. This result stresses the difference with similar geometries having only dielectric parts where TE does not provide any gap. In fact, if the TE and TM transmission spectra are overlapped, a full band gap is found.

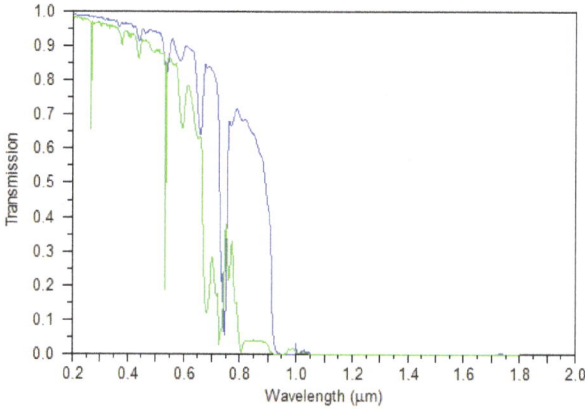

Fig. 2.3. TE transmission spectra for a Drude square PhC. Different lines (colors) correspond to different incident angles (blue: 0°; green: 45°).

These results confirm that the use of metallic PhCs increases the chance of a full band gap even for square-like Bravais lattices. However we have to keep in mind that the modeling leading to the previous results is based on a simplified Drude description. Hence, it is now important to asset the role that possible Lorentz contributes could play in the overall gap calculation. By moving to the Drude-Lorentz description from the Drude equation (1):

$$\varepsilon(\omega) = 1 - \frac{f_0 \omega_p^2}{\omega^2 - i\omega \, \Gamma_0} + \sum_{j=1}^{k} \frac{f_j \omega_p^2}{\left(\omega_j^2 - \omega^2\right) + i\omega \Gamma_j} \tag{2.2}$$

and assuming k=1, f_1=1, ω_1=800nm and Γ_1=0.1ω_p a different transmission spectrum is obtained as shown in Fig. 2.4.

We can see that considering also the Lorentz term strongly modifies the transmission spectrum to the extent that for the present case the gap around 1.1μm is suppressed. It is then of fundamental importance to compare the Drude and the Drude-Lorentz models when real materials, such as Ag, Au or Al, are considered. In fact, their description implies taking into account a number of Lorentz peaks which, as we have seen, might strongly modify the transmission spectrum compared to a simple Drude description.

2.2.2 Photonic crystal with realistic metallic materials

We will consider two materials: Ag and Al. The former is usually considered for plasmonic applications in the visible range whereas the latter can play a very interesting role in the UV range. We shall calculate their transmission spectra both for TM and TE in order to identify possible full band gaps.

First we will assume a Drude description. In case of Al, the experimental parameters to build up the model are ω_p= 14.98eV, Γ_0=0.047eV and f_0=0.523 (Rakić et al., 1998). Under these conditions the transmission/absorption spectra, both for TM and TE polarization, are shown in Fig. 2.5. We have considered two incident angles, 0° and 45° which, in the present case, will be enough to determine the existence of a full band gap.

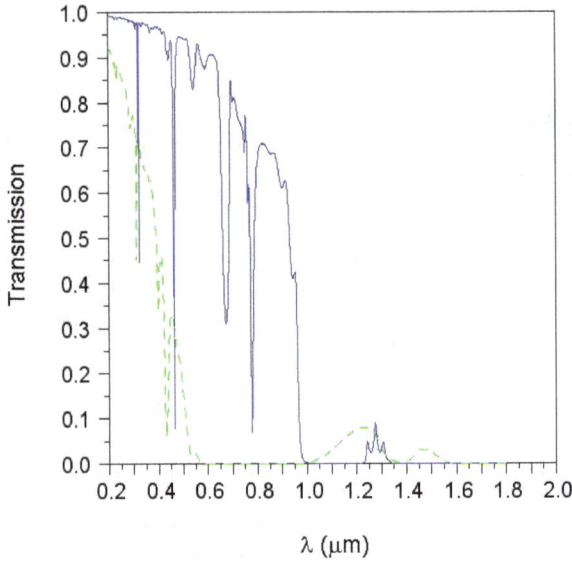

Fig. 2.4. TM transmission spectra of a square metallic PhC with radius r=0.472·P and material properties defined by (continue line – Drude model) $\omega p = 2\pi c/P$ and damping constant $\Gamma_0 = 0.01 \cdot \omega_p$ (0° incident angle of Fig. 2.2); (dashed line – Drude-Lorentz model) $\omega p = 2\pi c/P$ and damping constant $\Gamma_1 = 10 \cdot \Gamma_0$ for the resonance at 800nm. Incident angle of 0° was assumed.

The first important result to be noticed is the existence of a full band gap in the range 700-750nm. This achievement is consistent with the full band gaps obtained with the arbitrary Drude model introduced in Figs. 2.2 and 2.3. Furthermore, it is interesting to observe the behavior of the absorption both for TM and TE. In fact, the former shows no peaks in any band gap region whereas the latter has a smooth behavior, inside the gap regions, only for high wavelengths. It is explained recalling that only TE polarization shows a peculiar band gap which originates from the creation of surface plasmon polaritons on the metallic cylinders of the crystal, namely absorption peaks have to be observed in the gaps. The frequency region supporting the surface plasmon polariton modes starts roughly just below the plasma frequency of the metal, which in case of Al is 14.98eV=83nm, and its width depends on the geometry of the crystal.

When the Drude-Lorentz model is considered the transmission for both TM and TE is shown in Fig. 2.6. By comparison with Fig. 2.5 not substantial differences in terms of band gaps can be noticed, neither for TM nor for TE. In fact, the 700-750nm full band gap is found also in the Drude-Lorentz description. The only noticeable change is the increase of the absorption with respect to the transmission mainly below 1.0μm ought to the differences in the refractive index profile between the Drude and the Drude-Lorentz model of Al. In fig. 2.7 the two models are plotted. A strong absorptive peak around 800nm is shown in the Drude-Lorentz model.

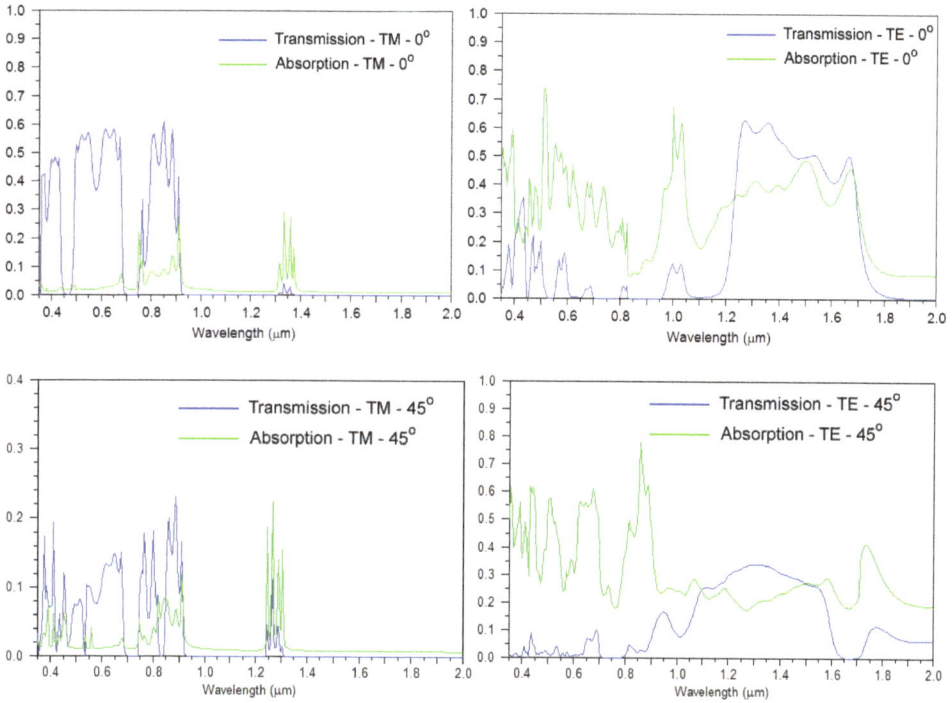

Fig. 2.5. TM/TE absorption/transmission spectra for a Drude square PhC. Aluminium is considered. Lattice constant P=1μm, cylinder radius=300nm. The source incident angle was 0º and 45º.

Moving now to PhCs based on Ag, the results for the Drude and Drude-Lorentz models are shown in Figs. 2.8 and 2.9, respectively. The optical parameters are ω_p=9.01eV, Γ_0=0.048eV, f_0=0.845, ω_1=0.816eV, Γ_1=3.866eV and f_1=0.065 (Rakić et al., 1998). In this case the spectrum range was chosen from 50nm to 2μm in order to exploit the band gap behavior from the UV to the IR region.

The Drude model for Ag shows a full band gap between 710nm and 750nm in a way similar to the Al structure (we recall that for both the materials we have used the same geometry). Once again, differently from square dielectric periodic structures, metallic square PhCs can provide full band gaps. Similarly to the Al case, the Drude-Lorentz model registers an increase in the absorption in the frequency region below 1.0μm. In particular, the three lowest wavelengths absorption peaks shown in the Drude-Lorentz model of Fig. 2.9 correspond to three well defined resonant peaks of silver, as confirmed by Fig. 2.10.

Furthermore, by observing the TE spectra for the silver PhC of Fig. 2.9, it can be noticed that absorptive peaks inside zero transmission regions are shown below 400nm, namely close to the silver plasma frequency of 137nm. This is the fingerprint of polaritonic gaps created by surface plasmon polaritons on the surface of the silver columns. On the other hand, TM spectra show a zero transmission region above 1.4μm which can be associated to the cut-off frequency ω_c.

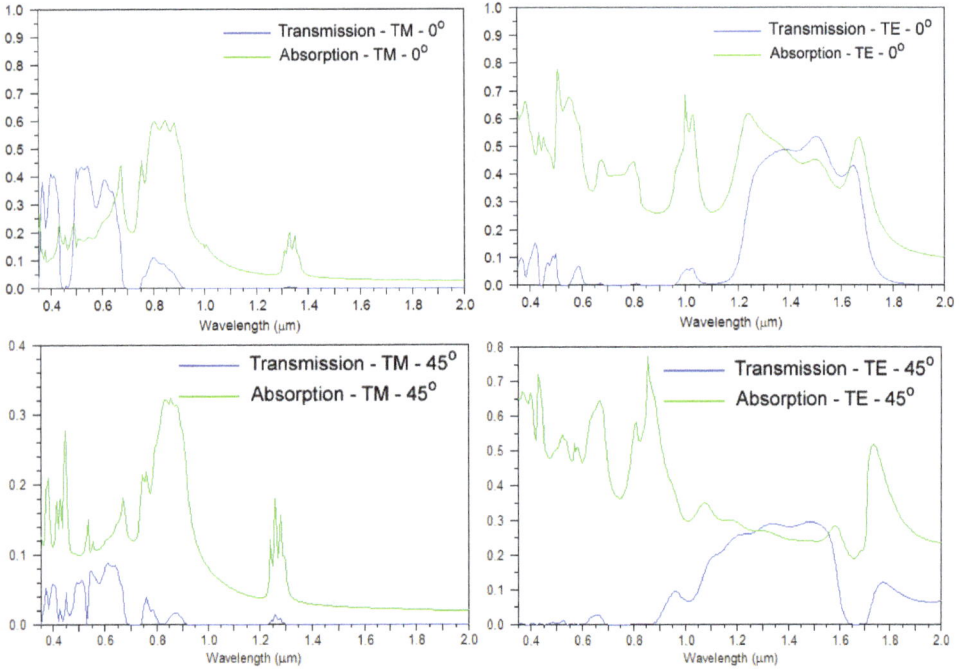

Fig. 2.6. TM/TE absorption/transmission spectra for a Drude-Lorentz square PhC. Aluminium is considered. Lattice constant P=1μm, cylinder radius=300nm, incident angles 0° and 45°.

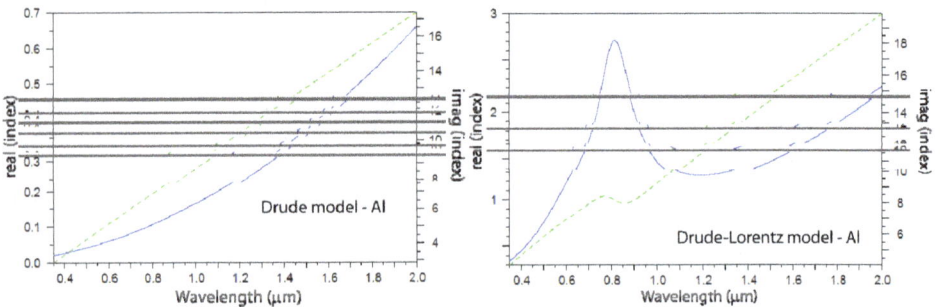

Fig. 2.7. Real (continuous/blue line) and imaginary (dashed/green line) parts of Al refractive index for the Drude and the Drude-Lorentz models.

Finally a consideration about the Drude and the Drude-Lorentz models for the Ag PhC. The simulations have shown that when moving to the Drude-Lorentz model the transmission spectrum remains similar to the Drude counterpart for energy far away from ω_p. When this value is approached, differences between the two models become appreciable. This is related to the typical absorption peaks of Ag located before 400nm.

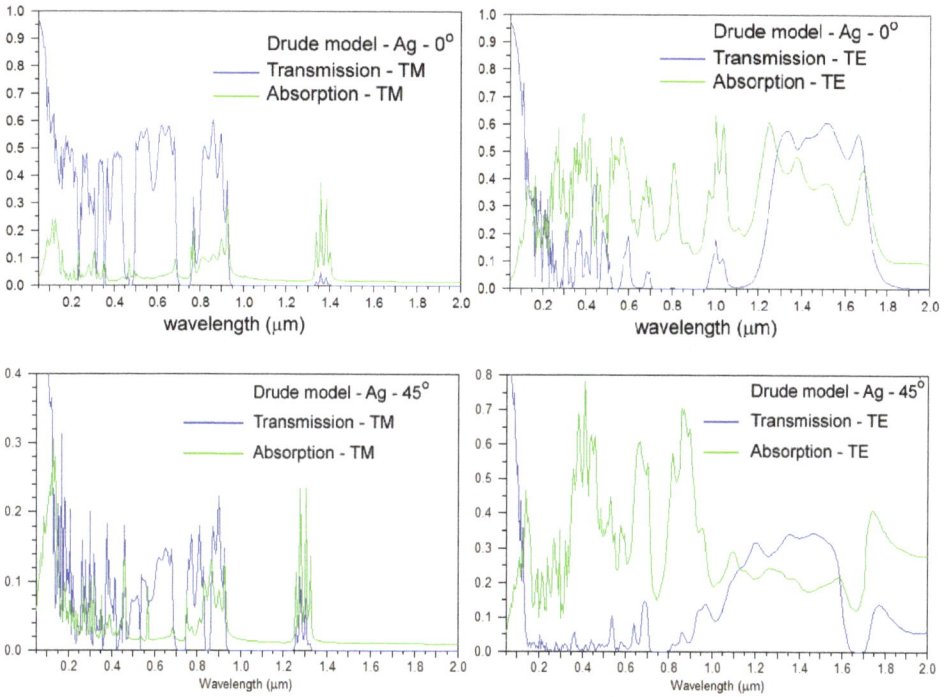

Fig. 2.8. TM/TE absorption/transmission spectra for the Drude square PhC. Silver is considered. Lattice constant P=1μm, cylinder radius=300nm, incident angles 0° and 45°.

In conclusion, we have shown that metallic PhCs can provide a full band gap not possible for analogues dielectric PhCs. This characteristic is related to the dispersive properties of the metallic parts of the crystal. Furthermore, we have shown that the Drude and Drude-Lorentz model provide the same results as long as the frequency range is far away from the metal plasma frequency. In fact, when approaching ω_p, discrepancies between the two models start rising that implies the failure of the Drude model at low frequencies.

3. Metallic PhCs for surface enhanced Raman spectroscopy

In recent years, plasmonics based sensor device such as surface enhanced Raman spectroscopy (SERS) has attracted lots of attention to the scientific community worldwide. SERS is a technique using which an increase in optical signal of the molecule situated in the vicinity of nano-metallic surface can be observed when electromagnetic light is being impinged on it (Haynes et al., 2005). There are many techniques such as electron beam lithography (EBL) (Das et al., 2009; Kahl et al., 1998), colloidal technique (Kneipp et al., 1997; Nie & Emory, 1997; Coluccio et al., 2009), metal island film (Constantino et al., 2001), etc. by which SERS substrate can be fabricated. There is always a competition regarding the quality of the nanostructure, fabrication area, time and cost. EBL is the most efficient technique to

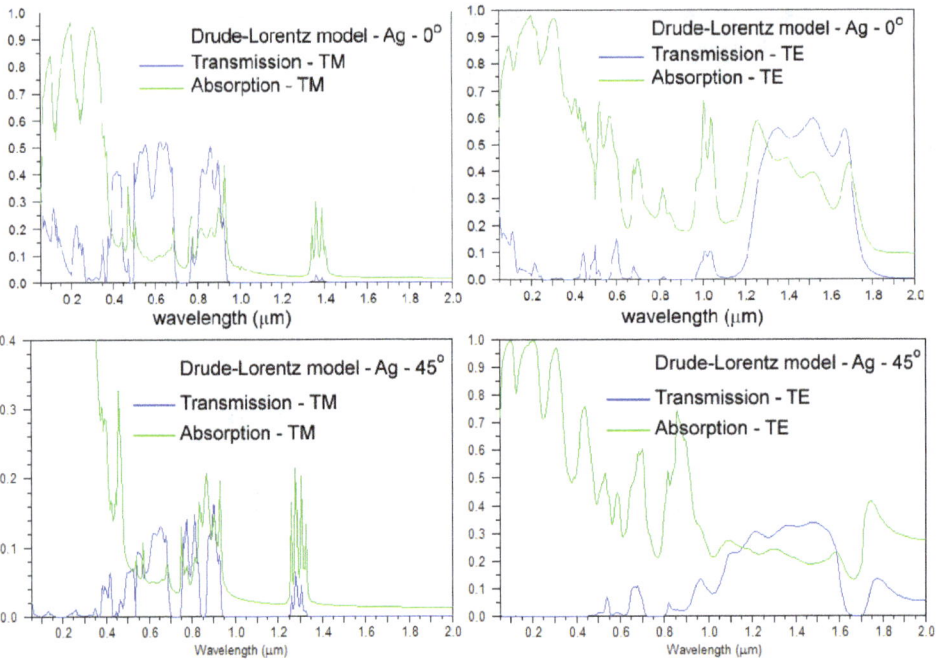

Fig. 2.9. TM/TE absorption/transmission spectra for the Drude-Lorentz square PhC. Silver is considered. Lattice constant P=1μm, cylinder radius=300nm, incident angles 0º and 45º.

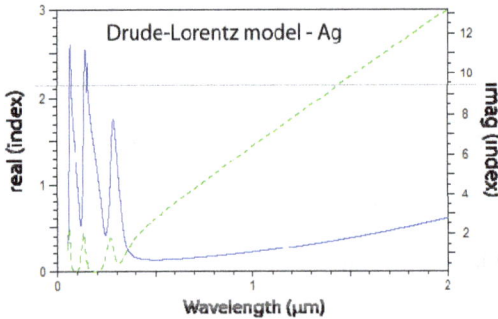

Fig. 2.10. Real (continuous/blue line) and imaginary (dashed/green line) parts of Ag refractive index for the Drude-Lorentz model. Three resonant peaks are shown in the UV region. Silver plasma frequency ω_p=9.04eV~137nm.

reproduce the nano-patterned SERS substrates but faces a critical limitations when it is about large area fabrication whereas colloidal technique is well known for large area metal deposition but showing minimum reproducibility. In the past, Zhang et al. (2006) reported a new technique called nanosphere lithography in which shadow evaporation through self-assembled arrays of polystyrene nanosphere was used to fabricate a large area SERS substrate. Herein, we propose a new way to fabricate a SERS substrate based on anodic porous alumina (APA) which could be a trade-off between the two above limits such as large fabrication area and SERS enhancement factor.

In the present work, APA substrate, having hexagonal periodicity, has been fabricated. The fabricated APA substrates were used as templates for the preparation of nanopatterned gold surfaces, obtained by gold film deposition of ~25 nm thickness covering the APA features. Using this technique, we are able to achieve reproducible SERS substrates with the wall thickness and pore diameter down to 40 and 60 nm, respectively. It is noticeable that the substrate shows very efficient SERS signal even in presence of intrinsically fluorescent molecules. The substrate surface morphology was characterized by both atomic force microscopy (AFM) and scanning electron microscopy (SEM), and the performance as a SERS substrate was tested with cresyl violet (CV). These substances fluoresce at distinct wavelengths in the visible spectrum from red to violet region. Even if it is known that nano-metallic surface acts as a fluorescence quenching substrate (Dulkeith et al., 2002), it is the first time, in our knowledge, that the large area SERS substrate on APA template was employed for SERS analysis in such detail.

In order to provide the surfaces with plasmon functionality, gold was thermally evaporated from a tungsten boat onto the APA substrates (APA), starting the deposition at a base chamber pressure of 2.0×10^{-6} mbar and proceeding at a 0.5 Å/s until a final total thickness of ~25 nm was reached. During deposition the sample holder was rotated at 1 RPM in order to improve the uniformity and the morphological quality of the gold layer. The substance of interest was deposited over resulting gold-coated APA substrate (termed 'AuAPA') using chemisorption technique. In this process, the substrate was dipped in a solution containing the molecule of interest. After incubation, the substrate was removed from the solution, gently rinsed to remove excess molecules those are not attached directly to the metal surface. Thereafter, the samples were dried in N_2 flow, and finally stored in a desiccator before SERS measurements. Cresyl violet (CV) dyes was tested for SERS probes with the concentration in the order of 10^{-6} M.

Fig. 3.1 shows the schematic picture of honey-comb structures, the SEM and AFM image of AuAPA SERS substrate. AFM and SEM images show the clear formation of SERS honey-comb structures. The gold-coating over APA template is fixed to the 25 nm.

CV (Fig. 3.2a for optical absorption spectrum and molecular structure) is an organic dye molecule intensively used in biology and medicine for histological stain. It is an effective stain applied for highlighting acidic components of tissues and is commonly used for nerve tissue sections. SERS measurements were carried out for CV deposited on AuAPA SERS substrate at different positions in the range of 300-1400 cm^{-1}. Raman measurements were also performed after deposition of a 2 μl drop of CV on a flat non-patterned silicon wafer substrate (see Fig. 3.2a) which can be considered as a positive control sample. The Raman spectrum of CV on Si surface is relatively featureless, with an exponentially increasing fluorescence background and a single characteristic peak of CV at ~591 cm^{-1} with low intensity. In the inset of Fig. 3.2a, the zoomed Raman spectrum of CV in the range of 400-650 cm^{-1} is also shown.

In Fig. 3.2a, SERS substrate background measurement (no dye on AuAPA pattern) was carried out and shown in the inset of Fig. 3.2a, which shows a flat background response without any Raman band at all. Various measurements were carried out on this SERS AuAPA substrate at different points, giving identical Raman vibrational frequency of CV.

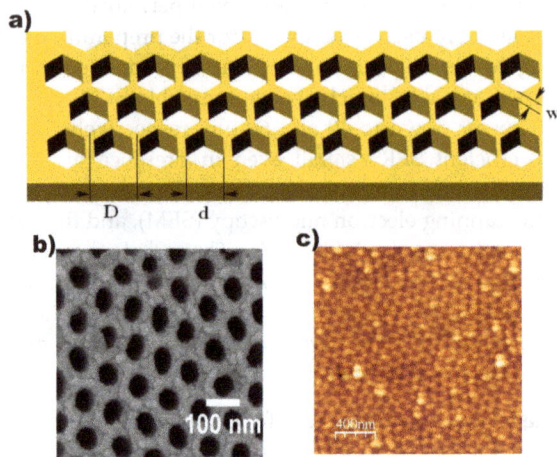

Fig. 3.1. a) Illustration of Honey-comb structure; SEM and AFM images after gold-coating of 25 nm are shown in Fig. b) and c).

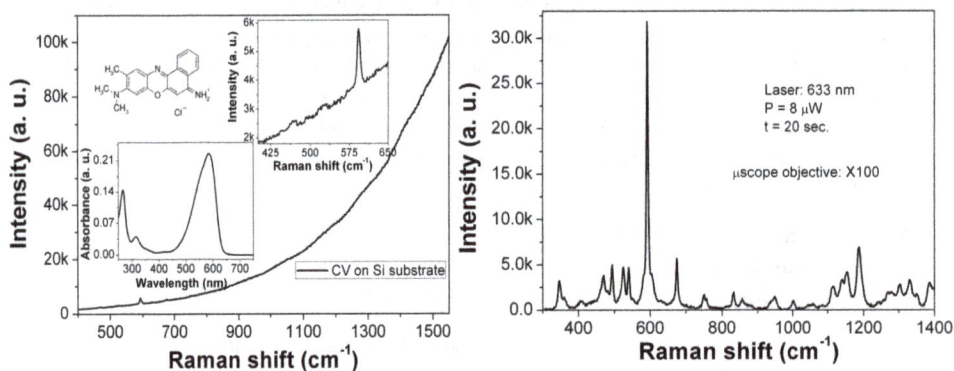

Fig. 3.2. a) Cresyl violet deposited on Si substrate, showing high fluorescence background. Molecular structure, optical absorption spectrum and zoomed Raman spectrum of CV in the range of 400-650 cm-1 are shown in the inset of the same Fig.; b) SERS spectrum of CV is also illustrated, showing the CV spectrum due to an efficient fluorescence quenching by honey-comb structure.

SERS spectra of the CV molecule, performed on the AuAPA substrates with pore diameter (~60 nm) and wall thickness (~40 nm) (Fig. 3.2b), the characteristic vibrational bands of CV are observed (Vogel et al., 2000). Intense Raman bands centred at around 348, 591, 675 and 1186 cm-1 can be attributed to the out of plane sceleton deformation, combination of in-plane N-H$_2$ and ring bending, ring deformation and combination of N-H$_2$ rocking and C-H$_x$ rocking, respectively (Vogel et al., 2000; Kudelski, 2005). From the vibrational bands observed in the Fig. 3.2b it is most probable that the CV is oriented in such a way that the – N-H$_2$ group is closer to the gold film, leading to the strong Raman scattering and, consequently, higher Raman signal. Additionally, in Fig. 3.2b, the high exponential background appeared in Fig. 3.2a for the CV on silicon substrate has disappeared. Hence, significant fluorescence quenching is also illustrated, as already observed in the past for gold nanoparticles (Dulkeith et al., 2002). SERS substrate background measurement (no dye on AuAPA pattern) was carried out and found a flat background response without any Raman band at all.

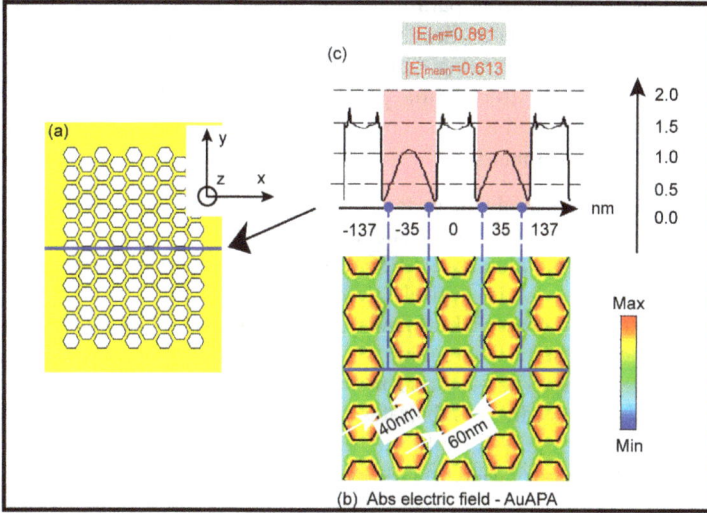

Fig. 3.3. a) The overall honey-comb structure. b) The overall electric field distribution in x-y plane, in this case, the polarization of incident light is along x-direction; c) absolute value of the electric field profile along the highlighted horizontal line. The excitation light is λ= 633 nm on gold surface with dielectric function ε$_{real}$= 9.83 and ε$_{imaginary}$= 1.97.

In order to demonstrate the enhanced electric field due to honey-comb nanostructure, we have also performed a series of simulations with the intent of calculating the near field electromagnetic distribution on AuAPA substrates. This kind of analysis can, in fact, provide information on the level of excitation of any sample deposited on a SERS substrate. In Fig. 3.3, simulations of the electric field due to the LSP generation on SERS substrates at z=0 nm (i.e. on the front surface of substrate) using CST software are presented. In particular, a 25 nm thick golden slab showing honeycomb pattern with periodicity of 100 nm, and sub-wavelength air holes (Lezec et al., 2004) with diameter of 60 nm, respectively, was simulated. The source light was fixed to 633 nm with linear polarization along the x-axis. The overall simulated structure is shown in Fig. 3.3a. For all the periodicity patterns, the results show an electric field strongly affected by the material discontinuity. In fact, the absolute value of the total electric field shows some nodes at the gold-air interface. The x-polarization creates anti-phase hot spots along the x-axis which determines the abrupt change of the electric field as in Fig. 3.3b. The profile, shown in Fig. 3.3c, gives even a better image of this behavior. The sharp doubled horns existing inside the holes are due to the hexagonal geometry of the holes. Important features emerging from the simulations are both the mean (E_{mean}) and the effective (E_{eff}) value of the electric field outside the holes (shaded regions in Fig. 3.3c). The former is simply defined as the average field whereas the latter is given by the difference between the maximum and minimum values in the hole.

In conclusion, home-built AuAPA substrates with pore size of 60 nm and wall thickness of 40 nm was utilized for large area SERS substrates. CV was deposited using chemisorption technique, using which a monolayer of molecules can be achieved. CST simulations of the electric field distribution on these ordered and reproducible SERS substrates were also performed, keeping the structural parameters as close as possible to the experimental honey-comb nanostructures. Theoretical results follow the same trend of experimental findings. The major advantage of using nanoporous alumina substrates, as compared to the traditional colloidal coating or lithographic processing, is a good trade-off between the high enhancement factor 'G' obtained and the large surface area produced. The respective SERS enhancement factor 'G' is estimated to be ~4×10^6.

Further research should be made in order to optimize 'G' on the basis of the substrate parameters of pore size, wall thickness, and thickness of the gold coating. The next step towards reproducible micro-fabricated SERS devices would be the transfer of APA to technological substrates such as optical glass or silicon wafer, and combining APA patterning with optical lithography. In this way, large-area SERS bio- and chemical sensors assays for e.g. parallel biomedical screening of different body liquids or even tissues could be carried out, conveniently. The easy and inexpensive processing required for APA SERS fabrication would also make these substrates disposable, opening the way to their large scale applications. Such a high SERS enhancement as demonstrated here can provide single molecule sensitivity in a sensor based on labeling using fluorescence dyes.

4. Integration of biophotonics and superhydrophobic devices for the detection of few bio-molecules

In biomedical disciplines including medicine, biology or biotechnology, there is a huge interest in understanding the pathogenesis of diseases, and to develop new biomarkers for their diagnosis. In particular, early detection can lead to significant benefits in terms of efficient and timely treatment. In these regards, the blood contains a multitude of unstudied and unknown biomarkers that could reflect the ongoing physiologic state of tissues and organs. The low-molecular weight region of the blood proteome is an important source of diagnostic markers (Posadas et al., 2001). Nevertheless, this 'mine' of data for diagnosis is often practically inaccessible by conventional approaches, in that the solutions at study are extremely diluted, and contain a multitude of substances that would blur the signal of interest. To exploit this 'mine' of data for diagnosis, new devices are required with very low detection limits. *Nanotechnology*, as a whole, offers unprecedented opportunities to realize such devices, thus increasing the efficacy of diagnosis (Liotta et al., 2003). In the paragraphs above, the theory of plasmons and the use of plasmonic devices have been largely described. Here, the integration of this technology with superhydrophobic surfaces (SHSs) is introduced.

The aim of this paragraph is to expound on the contributions that this combination can possibly offer to clinical medicine, specifically in the discipline of early diagnostics. In particular, it revolves around the study and the micro/nano fabrication of SHSs which incorporate biophotonic devices. Well assessed mechanisms such as (i) superhydrophobicity and (ii) SERS (Surface Enhanced Raman Scattering) are recapitulated and integrated into a single micro- nano- system. The combination of these two would represent a boost towards the detection and the analysis of few (or single) molecules. In the following, SHSs and nano optics based photonic devices are treated separately. After an introduction of the most important features of these, their combination and the effects thereof are discussed. Notice that, differently from plasmonic devices, SHSs are exposed to a good detail in that they are introduced here for the first time in the text.

In short, the major novelty, here, is the simultaneous use of wetting mechanisms (that arise due to the superhydrophobic surfaces, SHSs) and sensitive materials (that include, randomly distributed silver nanograin aggregates, regular arrays of metallic nano dots, adiabatic nanofocusing cones) to increase the response of nano optics based spectroscopy devices (see the cartoon representation of Fig. 4.1). By doing so, otherwise inaccessible information about the biological moieties at study is disclosed.

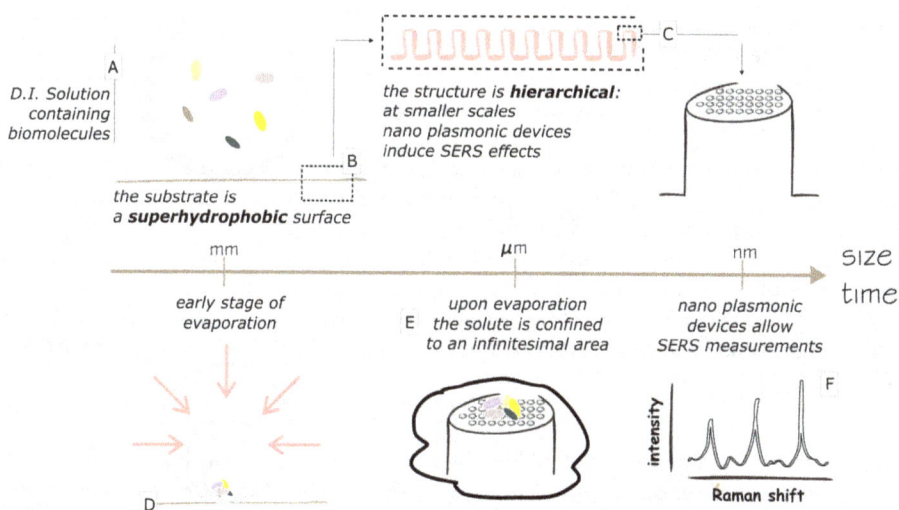

Fig. 4.1. Raman spectroscopy through super-hydrophobicity.

4.1 Superhydrophobic surfaces

It is well known that a drop post upon a solid surface develops a contact with the solid described by the sole parameter θ_e (Fig. 4.2A) that is the equilibrium contact angle at the interface between the liquid and the solid. θ_e obeys the celebrated Young equation (Young, 1805):

$$\cos\theta_e = \frac{\gamma_{SV} - \gamma_{SL}}{\gamma_{LV}} \tag{4.1}$$

where γ_{ij} is the surface tension between the phase i and j, and the letters S, L, V stand for the solid, liquid and vapour, and thus equation (1) may be regarded as a simple balance of forces. For sufficiently small droplets the dominant force becomes the liquid-vapor surface tension and gravitation may be neglected. The dimensionless Bond number can be consequently introduced as $Bo = \rho \times g \times R^2 / \gamma_{LV}$, where ρ is the density of the liquid, R is the radius of the spherical drop prior the deposition upon the surface, and g is the acceleration due to gravity. When $Bo \ll 1$ gravitational effects vanish and the shape of the droplet may be assumed spherical everywhere. For a drop of water with $\gamma_{LV} = 72.9$ mJ/m^2, $\rho = 1000$ kg/m^3, and diameter $d = 2 \times R = 1$ mm, it follows that $Bo \sim 0.035$, and thus the physics of micrometric or submillimetric drops is correctly governed by surface tension solely.

Superhydrophobicity (Fig. 4.2B) is a phenomenon whereby a drop post upon a surface would preserve its original spherical shape rather than spreading or wetting indefinitely the plane of contact (Lafuma & Quéré, 2003; Blossey, 2003). The theoretical conundrum explaining this mechanism dates back to the mid forties of the last century and is very well assessed. In the celebrated model of Cassie, a surface would be superhydrophobic on account of the pockets of air that remain trapped between the liquid and the substrate (Fig. 4.2C), and the smaller the fraction of solid in contact with the drop (ϕ) the larger the apparent contact angle. In particular, the increased contact angle $\theta_e{}^c > \theta_e$ the drop experiences may be theoretically written as (Cassie & Baxter, 1944)

$$\cos \theta_e^c = -1 + \phi_s \left(\cos \theta_e + 1 \right) \tag{4.2}$$

where ϕ_s is the fraction of solid in contact with the drop. The less ϕ_s the larger the apparent contact angle. At the limit of $\phi_s \to 0$, the drop would paradoxically float in air. The Cassie model is intuitive in that predicts that a drop upon a patterned surface 'sees' a contact angle that is proportional to the fraction of air in contact with the drop. A flat surface would have $\phi_s = 1$ and accordingly $\theta_e^c = \theta_e$. Another mechanism (Wenzel model) explains the increased contact angle on account of the roughness, and thus relies upon geometric effects solely (Fig. 4.2D). The modified angle $\theta_e{}^w$ is related to the unmodified contact angle θ_e as (Lafuma & Quéré, 2003):

$$\cos \theta_e^w = r \cos \theta_e \tag{4.3}$$

Where r is the solid roughness defined as the ratio between the real the projected surface, and thus the effect of surface roughness is to amplify the wetting. Noticeably, when $\theta_e < \pi/2$, r increases the hydrophilicity of the surface, whereas for $\theta_e > \pi/2$ roughness promotes hydrophobicity. In spite of the fact that they do both induce hydrophobicity, these situations are very different when considering their adhesive properties. In Wenzel state drops are found to be highly pinned. On the contrary, in the Cassie state the drop sits mainly upon air, and this increases the contact angle.

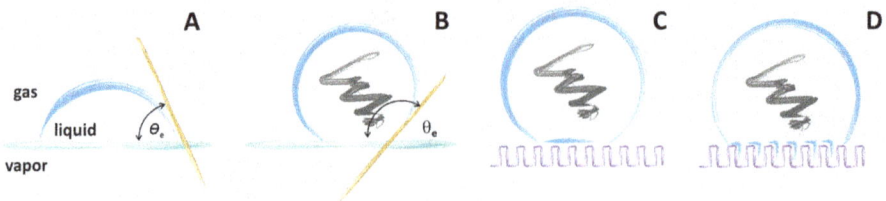

Fig. 4.2. Superhydrophobicity and contact angle.

In the Cassie state the drop would roll upon the surface and it would progressively reduce its contact area during a process of evaporation. The drop would then maintain the Cassie state over time, thus avoiding collapse and any irreversible transition to Wenzel until a critical radius of impalement is achieved. The mechanisms inducing collapse in a slowly evaporating droplet are two and, namely, (i) the drop could either touch the surface below the posts, or (ii) the surface free energy gained as the drop collapses wins over the surface free energy lost by increased contact with the hydrophobic posts. While the first mechanism regards surfaces decorated with short posts or pillars, the second relies upon an energetic argument and is independent on the pillars' height. In many practical situations one should consider the latter criterion solely (De Angelis et al., 2011). In particular, both analytical calculations and numerical simulations show that the critical radius of impalement depends upon the distance between the pillars δ and θ_e as

$$r_{\min} = \frac{\delta}{\cos \theta_e} \qquad (4.4)$$

and thus the closer the pillars the smaller the final area of contact. From the analysis above it stems out that surfaces with large ϕ (that is, with sufficiently dense pillars), would guarantee a stable Cassie configuration. Notice though that large ϕ would induce small contact angles, and thus the choice for the best design parameters is not trivial, and it is indeed a matter of optimization.

SHSs retain unique properties in terms of wettability that can be reviewed as follows: (i) SHSs have superior adhesive properties, in the sense that they exhibit vanishing friction coefficients; (ii) a droplet, post upon these surfaces, would accordingly preserve a quasi-spherical shape while evaporates, and the contact area at the interface would thus progressively reduce; (iii) SHSs can be artificially reproduced using micro and nano fabrication techniques. Using the properties above, micro textured surfaces may be successfully exploited to concentrate tiny amounts of moities over micrometric areas, and consequently measure these moieties with unprecedented accuracy (De Angelis et al., 2011; Accardo et al., 2011; Gentile et al., 2011a; Gentile et al., 2011b; Di Fabrizio et al., 2008).

4.1.1 Fabrication of SHSs

SHSs typically comprise a *periodic* hexagonal lattice of cylindrical Si micro pillars with a certain diameter and pitch (Fig. 4.3A). Due to the surface patterning the drop experiences an increased contact angle that can be theoretically predicted (Fig. 4.3B). The diameter d of the pillars and the distance δ between the pillars (gap) may be arbitrarily imposed, and can be chosen in accordance to a criterion of optimal design. This criterion would guarantee the best trade-off between sufficiently dense forests of pillars, that prevent the early collapse of the drop, and diluted structures, that instead assure large contact angles (De Angelis et al., 2011). The height of the pillars h is chosen to be as large as, at least, two times d, and this would avoid the spontaneous impalement of the drop. For the present configuration, d=10 μm, δ=20 μm, and h=20 μm. The pillars are arranged to recover a honeycomb lattice; notice that when these are combined with adiabatic nanofocusing cones the symmetry of the pattern is broken and, on account of this, the biomolecules are enforced to deposit upon the cones. The micro pillars are typically realized combining optical lithography, electroless growth and Bosch Reactive Ion Etching (RIE) techniques. Non conventional biophotonic nanostructures as those described below, conveniently positioned upon the pillars, would complete a hierarchical structure thus permitting the identification of proteins in the single molecule regime (Fig. 4.3C). These 'two-stages' micro nano structures function as SHSs with an increased contact angle ranging from 155° to about 175°.

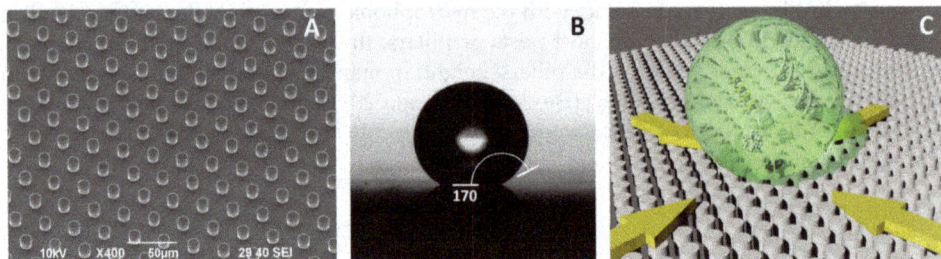

Fig. 4.3. SEM image of a photonic crystal for superhydrophobicity applications.

4.1.2 Current applications of SHSs

Many applications exist revolving around the use of superhydrophobic surfaces and mainly founding upon two key features of these, that are the limited contact area and the ability of a drop to slide or slip upon such surfaces (Li et al., 2007). Water repellent coatings for radar domes, satellite dishes and glass are widely reported and currently in use. These coatings provide self-cleaning under the action of rain in that they mimic the self-cleaning action of the Lotus leaf. Other utilizations include antibiofouling paints for boats, antisticking of snow for antennas and windows, self-cleaning windshields for automobiles, the separation of water and oil, and the manufacture of water-proof fireretardant clothes. The latter would represent examples of a target market that is the huge household-commodity sector, nevertheless new analytical potentials also do exist which carry the promise of unparalleled opportunities in life sciences and in the high tech field of biotechnology. In Biotechnology many efforts are devoted to manipulate, control and analyse relevant molecules as DNA and proteins, and artificially (nanotech) fabricated superhydrophobic surfaces could be an effective advance towards this objective. Let's consider, for instance, the case of microarrays. These are libraries of biological or chemical entities (probes) immobilised in a grid on a solid surface, the probes would be then interacting with targets (i.e., geneses, proteins, cDNA...) to provide deep insight into DNA sequences. It is desirable that the drop containing the targets would evaporate avoiding pinning and the formation of noxious ring like structures (it is in fact well known that pinning during evaporations causes final solute formations disposed in doughnut shape, and this is commonly known as the coffee-ring effect, and here suitable superhydrophobic substrates come into play.

The transport, separation and mixing of moieties dispersed in droplets has an enormous interest likewise. Superhydrophobic surfaces are slippy, and thus a droplet would move upon the application of an infinitesimal external force field: tilting the surface (and thus gravitation) or electrostatic or surface waves induced force fields would move the droplet in a controllable way (see also below in materials and methods). On the other hand, variations in the patterning or texture of the surface would generate regions where the adhesive properties are also different, thus creating well defined and distinguished tracks or patterns, and the drop could be exteriorly positioned into a precise area or point of the substrate. This argument is the ground for novel droplet-based microfluidic systems as reported in a number or publications and patents, and these surface would be in fact lab-on-chip systems which deliver the ability of performing bio-chemical detections and reactions with incomparable accuracy and efficiency. Recently, the use of superhydrophobic surfaces to concentrate and localize a solute has been proposed.

4.2 BioPhotonic devices

Biophotonic nanostructures can be integrated to SHSs to obtain devices with advanced sensing capabilities. Here we shall recapitulate in short specific types of nano-geometry based plasmonic device, and namely (i) electroless grown random assemblies of silver nanograins; (ii) regular arrays of metallic nano dots; (iii) adiabatic nanofocusing cones. Nevertheless, the method can be extended to a number of different plasmonic nanostructures, as those broadly described in the paragraphs above (De Angelis et al., 2011; Accardo et al., 2011; Gentile et al., 2011a; Gentile et al., 2011b; Di Fabrizio et al., 2008; Gentile et al., 2010).

4.2.1 Electroless grown silver nanograins

The electroless growth is a process whereby ionic silver is reduced and deposited as metallic silver upon silicon, via a redox reaction, and according to a mechanism that may be adequately described by a diffusion limited aggregation model. A satisfactory comprehension of the method may be found in (Coluccio et al., 2009). Here it is briefly recalled that, conveniently employing this process, and depending upon the growth parameters, silver nanograins may be obtained with an average size as small as few tens of nanometers, which compose efficient SERS substrates (Fig. 4.4A).

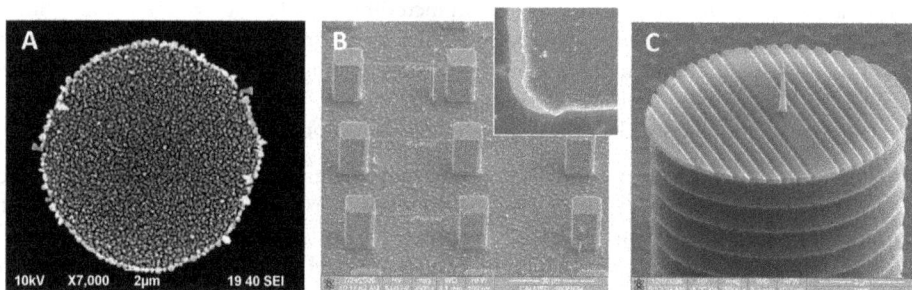

Fig. 4.4. SEM images of a metallic photonic crystal and plasmonic device.

4.2.2 Regular arrays of metallic nano dots

These are gold or silver hemisphere positioned upon the substrate as to reproduce a regular square lattice. Optical lithography, electron beam lithography, Reactive Ion Etching, evaporation techniques are routinely used for the fabrication of these devices. Differently from the case above, the shape, dimension, position of the dots is deterministic in the limit of the fabrication process. Thereby SERS effects arise from the rational design of the nano structure, founding upon the theory of surface plasmon polariton scattering (Fig. 4.4B) (De Angelis et al., 2011).

4.2.3 Adiabatic nanoficusing cones

These are extremely small conical geometries whereby crystal together with a plasmonic waveguide focuses the excitation laser to the apex of the waveguide, enabling a photon confinement equivalent to the radius of curvature of the nanofabricated tip (De Angelis et al., 2008; De Angelis et al., 2010). The fabrication process is accomplished on the basis of three steps. The grating is milled on the surface of the silicon micropillar by focused ion beam milling. The nanocones are growth on the top of the silicon tapered pillar by employing electron beam induced deposition (EBID) from a Platinum-based gas precursor. A thin layer of silver (40 nm) is finally deposited upon the device by means of thermal evaporation. These devices exploit the surface plasmon polariton adiabatic compression whereby the electro-magnetic field is locally enhanced (Fig. 4.4C) (De Angelis et al., 2011).

4.3 The device as a whole

The above analysis of SHSs and of the related properties thereof may be summarized as follows: (i) SHSs retain unique properties in terms of wettability, in particular a certain mass of water, in shape of a drop, would be repelled by such surfaces; (ii) SHSs have superior adhesive properties, in the sense that they exhibit vanishing friction coefficients; (iii) a droplet, post upon these surfaces, would accordingly preserve a quasi-spherical shape while evaporates, and the contact area at the interface would thus progressively reduce. Node (iii), above, is the key feature for such surfaces, in that it would enable to concentrate tiny amounts of agents (biomolecules) over micrometric areas. Imagine to deposit a drop of an extremely diluted solution upon a textured, superhydrophobic substrate. The drop would evaporate over time and thus the solution would get more and more concentrated. At the late stage of evaporation, the residual solute would be confined within an incredibly small region of the plane. With an appropriate design, few molecules may be conveniently enforced to confine into the smallest area conceivable, at the limit upon a sole pillar. Nano geometry based biophotonic devices, conveniently tiling these surfaces, would probe/detect the moieties with heretofore unattainable resolution limits (the process, as a whole, is recapitulated in Fig. 4.1).

The devices introduced would perform SERS measurements as well as (and definitely not better than) conventional SERS substrates. Nevertheless, here, the beneficial effects of super hydrophobicity and nanogeometry based spectroscopy are combined and conveyed into a unique platform, and from the combination of the two novel properties arise permitting the identification of proteins or analytes in the single molecule regime.

4.4 Measurements

Here, we report briefly on some experiments that would demonstrate the potentials of the method. Small drops of D.I. water containing Rhodamine molecules were gently positioned upon the substrates as in Fig. 4.4A. The evaporation process was followed over time until an irreversible transition to a pinning (Wenzel) state occurred. Few molecules were conveniently enforced to confine into a small area. Solutions were investigated with concentration as small as 10^{-18} M, that is, in the atto molar range. Figs. from 4.5A to C are SEM images of the residual solute of Rhodamine at the end of the process of evaporation. The magnification of the images is different: Fig. 4.5A was captured at the low magnification factor of 150⊚, whereby the initial footprint of the drop is clearly visible; Fig. 4.5B and C were acquired setting higher magnification factors as 800⊚ and 1500⊚, respectively. Notice, from these, the solute extremely concentrated to the extent that the Rhodamine is accumulated into a small area clearly bridging the pillars. In any event, the residual solute is smaller than few tens of microns. Considering that the initial diameter of deposition is about 1200 μm, while the final deposit is as large as 40 μm, the concentration capability of the substrate is, at the very least, in the order of, roughly, $(1200/40)^2{\sim}1000$, that is, the problem is scaled down by three orders of magnitude. Notice that this analysis relies upon geometric effects solely.

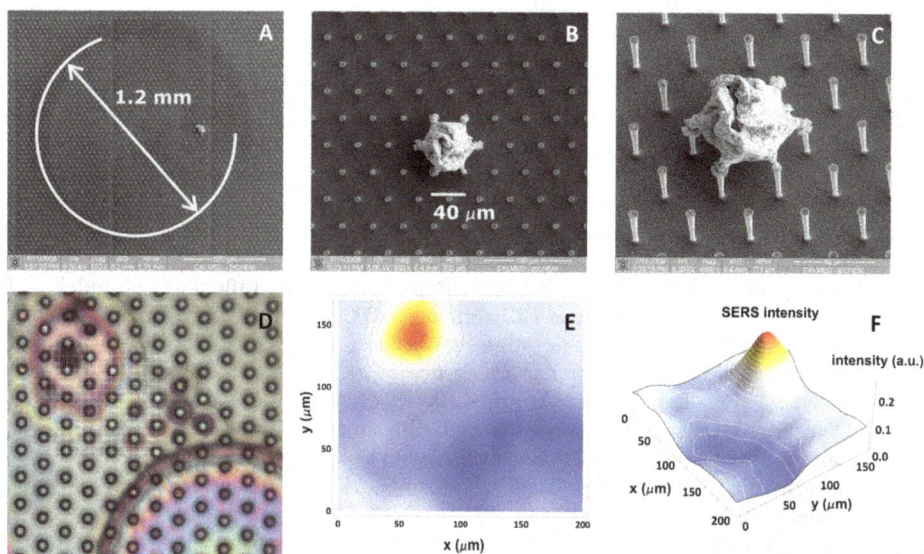

Fig. 4.5. SEM images of a metallic photonic crystal for SERS measurement of Rhodamine.

Fig. 4.5D is a microscope image reporting the deposit of a Rhodamine evaporation process as above. To prove that the identification of the residue as Rhodamine is correct and it is not instead constituted by debris or other refuses, SERS spectroscopy measurements were carried out on the sample. While conventional Raman intensity is directly proportional to the number of molecules probed, in the case of SERS, solely the molecules that are in close proximity of the nano-metallic substrate assure the enhancement of the electric field. In the contour and 3D plot as in Fig. 4.5E and F the SERS intensity signal is consistent with the matter distribution as in Fig. 4.5D, and this would prove the hypothesis above. The mapping analysis was performed by referring the band centred at 1650 cm^{-1}.

5. Acknowledgment

Authors would like to thanks Dr. Marco Salerno for providing APA template.

The authors also gratefully acknowledge support from European Projects SMD FP7-NMP-2008-SMALL-2 proposal No. CP-FP 229375-2 and Nanoantenna FP7-HEALTH-2009, Grant

No. 241818. FOCUS project proposal #270483- ICT-2009 8.7 - FET proactive 7: Molecular Scale Devices and Systems.

6. References

Accardo, A.; Gentile, F.; Mecarini, F.; De Angelis, F.; Burghammer, M.; Di Fabrizio, E. & Riekel, A. (2011). Ultrahydrophobic PMMA micro- and nano- textured surfaces fabricated by optical lithography and plasma etching for X-Ray diffraction studies. *Microelectronic Eng.*, Vol. 88, pp. 1660-1663.

Bermel, P.; Luo, C.; Zeng, L.; Kimerling, L. C. & Joannopoulos, J. D. (2007). Improving thin-film crystalline silicon solar cell efficiencies with photonic crystals. *Optics Express*, Vol. 15, No. 25, pp.16986-17000

Birks, T. A.; Knight, J. C. & Russell, P. St. J. (1997). Endlessly single-mode photonic crystal fiber. *Opt. Lett.*, Vol. 22, No. 13, pp. 961-963

Blossey, R. (2003). Self-cleaning surfaces - virtual realities. *Nature Materials*, Vol. 2, pp. 301-306

Cabrini, S., Carpentiero, A., Kumar, R., Businaro, L., Candeloro, P., Prasciolu, M., Gosparini, A., Andreani, A., De Vittorio, M., Stomeo, T. & Di Fabrizio, E. (2005). Focused ion beam lithography for two dimensional array structures for photonic applications. *Micr. Electr. Eng.*, Vol. 78, pp. 11-15

Cassie, A. B. D. & Baxter, S. (1944). Wettability of porous surfaces, *Trans. Faraday Soc.*, Vol. 40, pp. 546-551

Coluccio, M. L.; Das, G.; Mecarini, F.; Gentile, F.; Pujia, A.; Bava, L.; Tallerico, R.; Candeloro, P.; Liberale, C.; De Angelis, F. & Di Fabrizio, E. (2009a). Silver-based surface enhanced Raman scattering (SERS) substrate fabrication using nanolithography and site selective electroless deposition. *Microelectronic Eng.*, Vol. 86, No. 4, pp. 1085-1088

Constantino, C. J. L .; Lemma, T.; Antunes, P. A. & Aroca, R. (2001). Single-Molecule Detection Using Surface-Enhanced Resonance Raman Scattering and Langmuir–Blodgett Monolayers. *Anal. Chem.*, Vol. 73, No. 15, pp. 3674-3678

Das, G.; Mecarini, F.; Gentile, F.; De Angelis, F.; Kumar, M. H. G.; Candeloro, P.; Liberale, C.; Cuda, G. & Di Fabrizio, E. (2009). *Biosens. Bioelectron.*, Vol. 24, No. 6, pp. 1693-1699

De Angelis, F.; Patrini, M.; Das, G.; Maksymov, I.; Galli, M.; Businaro, L.; Andreani, L. C. & Di Fabrizio, E. (2008). A Hybrid Plasmonic–Photonic Nanodevice for Label-Free Detection of a Few Molecules. *Nano Letters*, Vol. 8, No. 8, pp. 2321-2327

De Angelis, F.; Das, G.; Candeloro, P.; Patrini, M.; Galli, M.; Bek. A.; Lazzarino, M.; Maksymov, I.; Liberale, C.; Andreani, L. C. & Di Fabrizio, E. (2010). Nanoscale chemical mapping using three-dimensional adiabatic compression of surface plasmon polaritons. *Nature Nanotech.*, Vol. 5, pp. 67-72

De Angelis, F.; Gentile, F.; Mecarini, F.; Das, G.; Moretti, M.; Candeloro, P.; Coluccio, M. L.; Cojoc, G.; Accardo, A.; Liberale, C.; Proietti Zaccaria, R.; Perozziello, G.; Tirinato, L.; Toma, A.; Cuda, G.; Cingolani R. & Di Fabrizio, E. (2011). Breaking the diffusion limit with super-hydrophobic delivery of molecules to plasmonic nanofocusing SERS structures. *Nature Photonics*, Vol. 5, pp. 682-687.

Deubel, M.; von Freymann, G.; Wegener, M.; Pereira, S.; Busch, K. & Soukoulis, C. M. (2004). Direct laser writing of three-dimensional photonic-crystal templates for telecommunications. *Nature Mater.*, Vol. 3, pp. 444-447

Di Fabrizio, E.; Cuda, G.; Mecarini, F.; De Angelis, F. & Gentile, F. (2008). Italian patent; Title: Dispositivo concentratore e localizzatore di un soluto e procedimento per concentrare e localizzare un soluto. CALMED s.r.l.; Italian Patent deposited: nr. TO2008A000646.

Dulkeith, E.; Morteani, A. C.; Niedereichholz, T.; Klar, T. A. & Feldmann, J. (2002). Fluorescence Quenching of Dye Molecules near Gold Nanoparticles: Radiative and Nonradiative Effects. *Phys. Rev. Lett.*, Vol. 89, No. 20, pp. 203002-1/4

Galli, M.; Agio, M.; Andreani, L. C.; Atzeni, L.; Bajoni, D.; Guizzetti, G.; Businaro, L.; Di Fabrizio, E.; Romanato, F. & Passaseo, A. (2002). Optical properties and photonic bands of GaAs photonic crystal waveguides with tilted square lattice. *Eur.. Phys. J. B*, Vol. 27, No. 1, pp. 79-87

Gentile, F.; Das, G.; Coluccio, M. L.; Mecarini, F.; Accardo, A.; Tirinato, L.; Tallerico, R.; Cojoc, G.; Liberale, C.; Candeloro, P.; Decuzzi, P.; De Angelis, F. & Di Fabrizio, E. (2010). Ultra low concentrated molecular detection using super hydrophobic surface based biophotonic devices. *Microelectronic Eng.*, Vol. 87, pp. 798-801

Gentile, F.; Accardo, A.; Coluccio, M. L.; Asande, M.; Cojoc, G.; Mecarini, F.; Das, G.; Liberale, C.; De Angelis, F.; Candeloro, P.; Decuzzi, P. & Di Fabrizio, E. (2011a). NanoPorous-MicroPatterned- SuperHydrophobic Surfaces as Concentrating/ Harvesting Agents For Low Molecolar Weight Proteins. *Microelectronic Eng.*, Vol. 88, pp. 1749-1752

Gentile, F.; Battista, E.; Accardo, A.; Coluccio, M. L.; Asande, M.; Perozziello, G.; Das, G.; Liberale, C.; De Angelis, F.; Candeloro, P.; Decuzzi, P. & Di Fabrizio E. (2011b). Fractal Structure Can Explain the Increased Hydrophobicity of NanoPorous Silicon Films, *Microelectronic Eng.*, Vol. 88, pp. 2537-2540

Haynes, C. L.; McFarland, A. D. & VanDuyne, R. P. (2005). Surface-Enhanced Raman Spectroscopy. *Anal. Chem.*, Vol. 77, No. 17, pp. 338A–346A

Ito, T. & and Sakoda, K. (2001). Photonic bands of metallic systems. II. Features of surface plasmon polaritons. *Phys. Rev. B*, Vol. 64, No. 4, pp. 045117-1/8

Joannopoulos, J. D.; Meade, R. D. & Winn, J. N. (1995). *Photonic Crystals: Molding the Flow of light* (first edition). Princeton University Press, ISBN 0-691-03744-2, UK

Kahl, M.; Voges, E.; Kostrewa, S.; Viets, C. & Hill, W. (1998). Periodically structured metallic substrates for SERS. *Sens. Actuators B: Chem.*, Vol. 51, No. 1, pp. 285-291

Kneipp, K.; Wang, Y.; Kneipp, H.; Perelman, L.T.; Itzkan, I.; Dasari, R. R. & Feld, M. S. (1997). Single Molecule Detection Using Surface-Enhanced Raman Scattering (SERS). *Phys. Rev. Lett.*, Vol. 78, No. 9, pp. 1667-1670

Kudelski, A. (2005). Raman studies of rhodamine 6G and crystal violet sub-monolayers on electrochemically roughened silver substrates: Do dye molecules adsorb preferentially on highly SERS-active sites? *Chem Phys. Lett.*, Vol. 414, No.4, pp. 271-275

Lafuma, A. & Quéré, D. (2003). Superhydrophobic states. *Nature Materials*, Vol. 2, pp. 457-460

Lezec, H. J. & Thio, T. (2004). Diffracted evanescent wave model for enhanced and suppressed optical transmission through subwavelength hole arrays. *Opt. Exp.*, Vol. 12, No. 16, pp. 3629-3651

Li, X. M.; Reinhoudt, D. & Crego-Calama, M. (2007). What do we need for a superhydrophobic surface? A review on the recent progress in the preparation of superhydrophobic surfaces. *Chem. Soc. Rev.*, Vol. 36, pp. 1350-1368

Liotta, L.A.; Ferrari M. & Petricoin, E. (2003). Clinical proteomics: Written in blood. *Nature*, Vol. 425, pp. 905 mm

Malvezzi, A. M.; Vecchi, G.; Patrini, M.; Guizzetti, G.; Andreani, L. C.; Romanato, F.; Businaro, L.; Di Fabrizio, E.; Passaseo, A. & De Vittorio, M. K. (2003). Resonant second-harmonic generation in a GaAs photonic crystal waveguide. Phys. Rev. B, Vol. 68, pp.161306-1/4

Malvezzi, A. M.; Cattaneo, F.; Vecchi, G.; Falasconi, M.; Guizzetti, G.; Andreani, L. C.; Romanato, F.; Businaro, L.; Di Fabrizio, E.; Passaseo, A. & De Vittorio, M. (2002).

Second-harmonic generation in reflection and diffraction by a GaAs photonic-crystal waveguide. *J. Opt. Soc. Am. B*, Vol. 19, No. 9, pp. 2122-2128

Nie, S. & Emory, S. R. (1997). Probing Single Molecules and Single Nanoparticles by Surface-Enhanced Raman Scattering. *Science*, Vol. 275, pp. 1102-1106

Posadas, E. M.; Simpkins, F.; Liotta, L. A.; MacDonald, C. & Kohn, E. C. (2005). *Annals of oncology*, Vol. 16, pp. 16-22

Proietti Zaccaria, R.; Shoji, S.; Sun, H. B. & Kawata, S. (2008a). Multi-shot interference approach for any kind of Bravais lattice. *Appl. Phys. B.*, Vol. 93, No. 1, pp.251-256.

Proietti Zaccaria, R.; Verma, P.; Kawaguchi, S.; Shoji, S. & Kawata, S. (2008b). Manipulating full photonic band gaps in two dimensional birefringent photonic crystals. *Optics Express*, Vol. 16, No. 19, pp. 14812-14820

Raether, H. (1988). *Surface Plasmons on smooth and rough surfaces and on gratings*. Springer-Verlag, ISBN 3-540-17363-3, Berlin, Germany

Rakic´, A. D.; Djuris˘ic´, A. B.; Elazar, J. M. & Majewski, M. L. (1998). Optical properties of metallic films for vertical-cavity optoelectronic devices. *Applied Optics*, Vol. 37, No. 22, pp. 5271-5283

Sajeev, J. (1987). Strong localization of photons in certain disordered dielectric superlattices. *Phys. Rew. Lett.*, Vol. 58, No. 23, pp. 2486-2489.

Sakoda, K.; Kawai, K.; Ito, T.; Chutinan, A.; Noda S.; Mitsuyu, T. & Hirao, K. (2001). Photonic bands of metallic systems. I. Principle of calculation and accuracy. *Phys. Rev. B*, Vol. 64, No.4, pp. 045116-1/8

Song, J. F.; Proietti Zaccaria, R.; Yu, M. B. & Sun, X. W. (2006). Tunable Fano resonance in photonic crystal slabs. *Opt. Exp.*, Vol. 14, No. 19, pp. 8812-8826

Song, J. F. & Proietti Zaccaria, R. (2007). Manipulation of light transmission through sub-wavelength hole array. *J. Opt. A.: Pure Appl. Opt.*, Vol. 9, No. 9, pp. s450-s457

Takahashi, S.; Okano, M.; Imada, M. & and Noda, S. (2006). Three-dimensional photonic crystals based on double-angled etching and wafer-fusion techniques. Sppl. Phys. Lett., Vol. 89, pp. 123106-1/3

Vogel, E.; Gbureck, A. & Kiefer, W. (2000). Vibrational spectroscopic studies on the dyes cresyl violet and coumarin 152. *J. Mol. Struct.*, Vol. 550, 177-190

Wang, D. D.; Wang, Y. S.; Zhang, X. Q.; He, Z. Q.; YI, L. X.; Deng, L. E.; Zhang, C. X. & Han, X. (2005). Enlargement of complete two-dimensional band gap by using photonic crystal heterostructure, *Applied Physics B: Lasers and Optics*, Vol. 81, No. 4, pp. 465-467

Yablonovitch, E. (1987). Inhibited Spontaneous Emission in Solid-State Physics and Electronics. *Phys. Rew. Lett.*, Vol. 58, No. 20, pp. 2059-2062.

Yokouchi, N.; Danner, A. J. & Choquette, K. D. (2003). Vertical-cavity surface-emitting laser operating with photonic crystal seven-point defect structure. *Appl. Phys. Lett.*, Vol. 82, No. 21 , pp.3608-3610

Young, T. (1805). *Phil. Trans.*, Vol. 84, pp.1

Zhang, X.; Whitney, A. V.; Zhao, J.; Hicks, E.M. & Van Duyne, R. (2006). *Journ. Nanosci. Nanotech.*, Vol. 6, No. 7, pp. 1920-1934

Zhao, H.; Proietti Zaccaria, R.; Song, J. F.; Kawata, S. & Sun, H. B. (2009). Photonic quasicrystals exhibit zero-transmission regions due to translational arrangement of constituent parts. *Phys. Rev. B*, Vol. 79, No. 11, pp.115118-1/7

Zhao, H.; Proietti Zaccaria, R.; Verma, P.; Song, J. F. & Sun H. B. (2010). Validity of the V parameter for photonic quasi-crystal fibers. *Opt. Lett.*, Vol. 35, No. 7, pp.1064-1066

16

Negative Index Photonic Crystals Superlattices and Zero Phase Delay Lines

C. W. Wong et al.,[*]
Columbia University, New York, NY
USA

1. Introduction

An intense interest in negative index metamaterials (NIMs) [1-2] has been witnessed over the last years. Metal based NIMs [3-11] have been demonstrated at both microwave and infrared frequencies with a motivation mainly coming from the unusual physical properties and potential use in many technological applications [12-21]; however, they usually have large optical losses in their metallic components. As an alternative, dielectric based photonic crystals (PhCs) have been shown to emulate the basic physical properties of NIMs [22-26] and, in addition, have relatively small absorption loss at optical frequencies. Equally important, PhCs can be nanofabricated using currently available silicon chip-scale foundry processing, allowing significant potential in the development of future electronic-photonic integrated circuits.

One particular type of PhC can be obtained by cascading alternating layers of NIMs and positive index materials (PIMs) [27 – 32]. This photonic structure (with an example shown in Figure 1) is postulated to show unusual and unique optical properties including new types of surface states and gap solitons [33], unusual transmission and emission properties [34 – 38], complete photonic bandgaps [39], and phase-invariant field that can be effectively used in cloaking applications [40]. Moreover, a remarkable property of these binary photonic structures is the existence of an omnidirectional bandgap that is insensitive to the wave polarization, angle of incidence, structure periodicity, and structural disorder [41 – 43]. The main reason for the occurrence of a bandgap with such unusual properties is the existence of a frequency band at which the path-averaged refractive index is equal to zero [27 – 32, 34]. Specifically, at this frequency the Bragg condition, $k\Lambda = (\bar{n}\,\omega/c)\Lambda = m\pi$, is satisfied for $m = 0$,

[*] S. Kocaman[1], M. S. Aras[1], P. Hsieh[1], J. F. McMillan[1], C. G. Biris[2], N. C. Panoiu[2], M. B. Yu[3], D. L. Kwong[3] and A. Stein[4]
[1]*Columbia University, New York, NY,*
[2]*University College of London, London,*
[3]*Institute of Microelectronics, Singapore,*
[4]*Brookhaven National Laboratory, Upton, NY,*
[1,4]*USA*
[2]*UK*
[3]*Singapore*

irrespective of the period Λ of the superlattice; here, k and ω are the wave vector and frequency, respectively, and $\bar{n}$ is the averaged refractive index. Because of this property this photonic bandgap is called zero-$\bar{n}$, or zero-order, bandgap [30, 34].

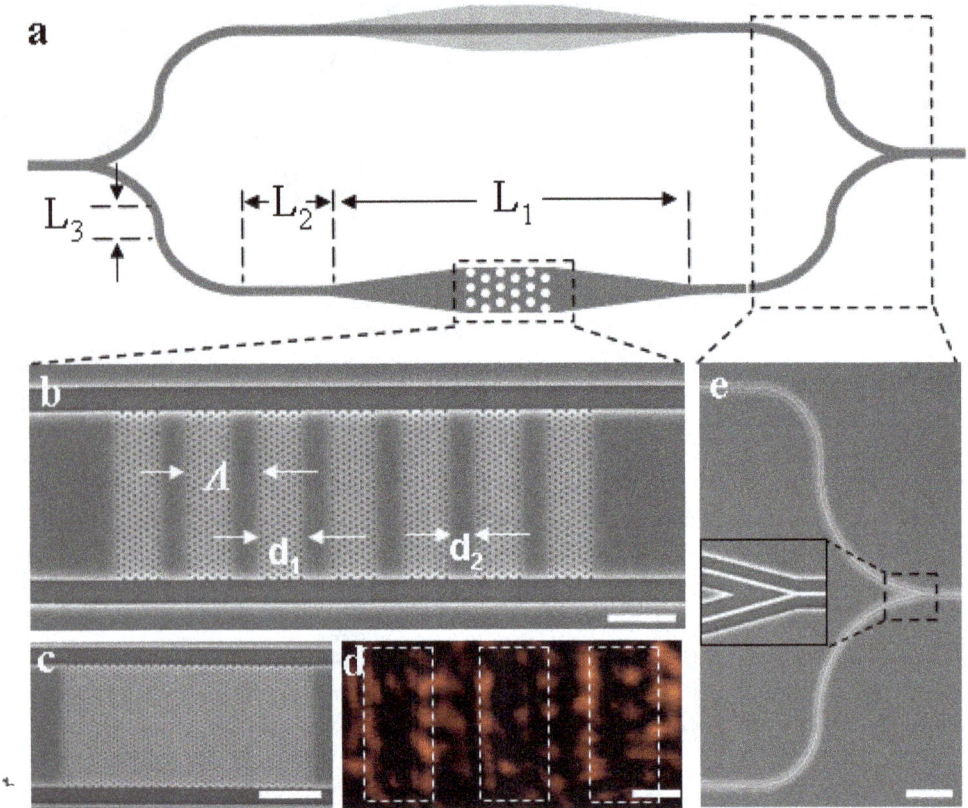

Fig. 1. **Schematic of a Mach-Zehnder interferometer (MZI) and SEM images of the fabricated device. a,** Schematic representation of the MZI. $L_1 \sim 850$ μm and $L_2 \sim 250$ μm. **b,** SEM of a fabricated superlattice with 7 super-periods. Each PhC layer contains 7 unit cells of PhC ($d_1 = 2.564$ μm, $\Lambda = 4.51$ μm) with $a = 423$ nm, $r/a = 0.276$, and $t/a = 0.756$ (the scale bar = 5 μm). **c,** SEM of a sample, showing only the PhC layer with same parameters as in **b** (the scale bar = 10 μm). **d,** Near-field image of a supperlattice with each PhC layer containing 9 unit cells ($d_1 = 3.297$ μm) (the scale bar = 2.5 μm). **e,** SEM of the Y-branch with a zoomed-in image in the inset (the scale bar = 25 μm).

Near-zero index materials have a series of exciting potential applications, such as diffraction-free beam propagation over thousands of wavelengths via beam self-collimation [34], extremely convergent lenses and control of spontaneous emission [35], strong field enhancement in thin-film layered structures [37], and cloaking devices [40]. Moreover, the vanishingly small value of the refractive index of near-zero index materials can be used to engineer the phase front of electromagnetic waves emitted by optical sources or antennas, namely, to reshape curved wave fronts into planar ones [36], or to transfer into the far-field the phase information contained in the near-field. In addition, at the frequencies at which the refractive index becomes vanishingly small the electromagnetic field has an unusual dual character, i.e., it is static in the spatial domain (the phase difference between arbitrary spatial locations is equal to zero) while remaining dynamic in the time domain, thus allowing energy transport. This remarkable property, which is also the main topic of our study, has exciting technological applications to delay lines with zero phase difference, information processing devices, and the development of new optical phase control and measurement techniques.

In this chapter, we show unequivocally that optical beams propagating in path-averaged zero-index photonic crystal superlattices can simultaneously have zero phase delay. The nanofabricated superlattices consist of alternating stacks of negative index photonic crystals and positive index homogeneous dielectric media, where the phase differences corresponding to consecutive primary unit cells are measured with integrated Mach-Zehnder interferometers. These measurements demonstrate that at path-averaged zero-index frequencies the phase accumulation remains constant despite increases in the physical path lengths. We further demonstrate experimentally for the first time that these superlattice zero-$\bar{n}$ bandgaps can either remain invariant to geometrical changes of the photonic structure or have a center frequency which is deterministically tunable. The properties of the zero-$\bar{n}$ gap frequencies, optical phase, and the effective refractive indices agree well between the series of measurements and the complete theoretical analysis and simulations.

2. Negative refraction photonic crystal superlattices

2.1 Theory

The photonic structures examined consist of dielectric PhC superlattices with alternating layers of negative index PhC and positive index homogeneous slabs, as shown in Figure 1 and Figure 2, that can give rise to the zero-$\bar{n}$ gaps [29]. The hexagonal PhC region (Figure 1c) is made of air holes etched into a dielectric Si slab (n_{Si}=3.48), with a lattice period a = 423 nm, a slab thickness t = 320 nm, placed on top of a silica substrate (n_{SiO2}=1.5). The band diagram of the PhC with a hole-to-lattice constant (r/a) ratio of 0.276 ($r \sim$ 117 nm) is shown in Figure 3a-b. Particularly the two-dimensional (2D) hexagonal PhC base unit has a negative index within the interested spectral band of 0.271 to 0.278 in normalized frequency of $\omega a/2\pi c$, or 1520 to 1560 nm wavelengths, such as reported earlier for near-field imaging [22]. The zero-$\bar{n}$ superlattices are then integrated with Mach–Zehnder interferometers (MZI) to facilitate the phase delay measurements. As illustrated in Figure 1a, the unbalanced interferometer is designed such that after splitting from the Y-branch (Figure 1e); a single mode input channel waveguide adiabatically tapers (over $\sim$ 400 µm) to match the width of the superlattice structures. On the reference arm, there is either a slab with exactly the same geometry to match the index variations and hence isolate the additional phase contribution of the PhC structures, or a channel waveguide leading to a large index difference and hence to distinctive Fabry-Perot fringes. For the one-dimensional (1D) binary superlattice of Figure 1b, a near-field scanning optical microscope image is taken (Figure 1d) to confirm transmission near the zero-$\bar{n}$ gap edge (1560 nm). The period of the superlattice is equal to

$\Lambda = d_1 + d_2$ where $d_{1(2)}$ is the thickness of the PhC (PIM) layer in the primary unit cell. Since a zero-$\overline{n}$ bandgap is formed when the spatially averaged index is zero, it is insensitive to the variation of the superlattice period, as long as the condition of zero-average index is satisfied [27 – 31, 34].

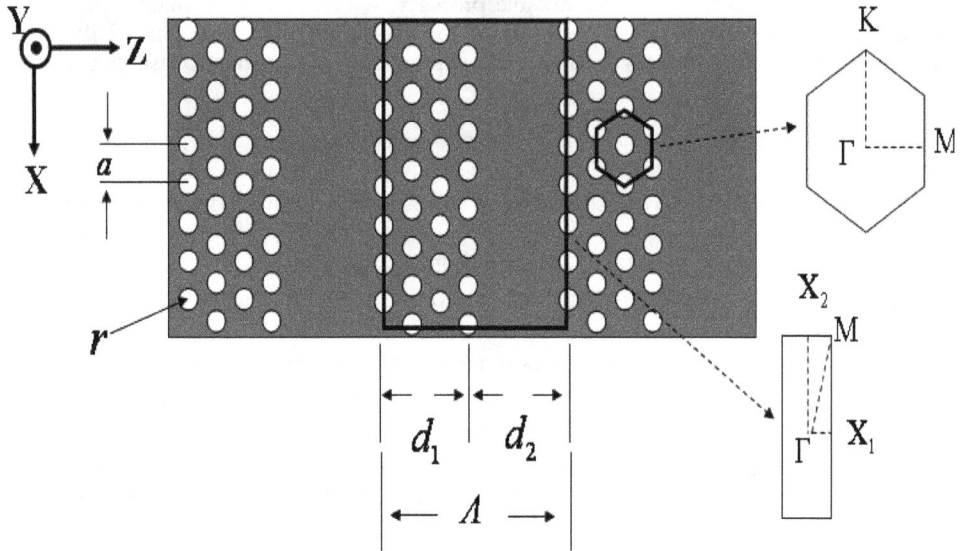

Fig. 2. **Schematic representation of the photonic superlattice.** There are two Brillouin zones defined as follows: one for the hexagonal photonic crystal lattice and one for the photonic superlattice. a is the lattice period and r is the radius of the holes forming the hexagonal lattice. d_1 is the length of the PhC layer and d_2 is the length of the PIM region. $d_1 + d_1 = \Lambda$ is equal to the superperiod (SP) of the photonic superlattice.

The existence of the zero-$\overline{n}$ bandgaps can be explained with the Bloch theorem, where for a 1D binary periodic lattice the trace of the transfer matrix, T, of a primary unit cell can be expressed as [27,29]

$$\text{Tr}\left[T(\omega)\right] = 2\cos(\kappa\Lambda) = 2\cos\left(\frac{\overline{n}\omega\Lambda}{c}\right) - \left(\frac{Z_1}{Z_2} + \frac{Z_2}{Z_1} - 2\right)\sin\left(\frac{n_1\omega d_1}{c}\right)\sin\left(\frac{n_2\omega d_2}{c}\right) \quad (1)$$

where $n_{1(2)}$ and $Z_{1(2)}$ are the refractive index and impedance of the first (second) layer, respectively, κ is the Bloch wave vector of the electromagnetic mode, $\overline{n}$ is the average refractive index, $\overline{n}(x) = \frac{1}{\Lambda}\int_0^\Lambda n(x)dx$. In the general case, when $Z_2 \neq Z_1$, if $\kappa_0\Lambda = \frac{\overline{n}\omega\Lambda}{c} = m\pi$, with m an integer, the relation

$$\left|\text{Tr}\left[T(\omega)\right]\right| = \left|2 + \left(\frac{Z_1}{Z_2} + \frac{Z_2}{Z_1} - 2\right)\sin^2\left(\frac{n_1\omega d_1}{c}\right)\right| \geq 2 \quad (2)$$

holds. This relation implies that the dispersion relation has no real solution for κ unless $\dfrac{n_1 \omega d_1}{c}$ is an integer multiple of π, which is the Bragg condition and thus photonic bandgaps are formed at the corresponding frequencies. However, if the lattice satisfies the special condition of a spatially averaged zero refractive index ($\bar{n} = 0$), again $\left| \mathrm{Tr}\left[T(\omega) \right] \right| \geq 2$, thereby leading to imaginary solutions for the wave vector κ and thus to a spectral gap [30 – 32]. We also note that the 1D binary superlattice and the hexagonal PhC have different symmetry properties and therefore different first Brillouin zones (see Figure 2a-insets). Schematic representation of a superlattice with 3 superperiods is shown in Figure 2. The superlattice consists of alternating layers of hexagonal PhCs and homogeneous slabs.

In our fabricated devices, the longitudinal direction of the superlattice (z-axis) coincides with the Γ-M axis of the hexagonal PhC. Moreover, within our operating wavelength range (Figure 3b) the PhC has two TM-like bands, one with positive index and the other one with negative index, and an almost complete TE-like bandgap (see Methods). The effective refractive indices corresponding TM-like bands (Figure3b) are determined from the relation $k = \omega |n|/c$ and plotted in Figure 3c (note that for the second band the effective index of refraction is negative since k decreases with ω [22]).

2.2 Mach-Zehnder interferometer with negative index photonic crystal

To examine effective index differences between different bands in the band diagram experimentally, we designed and fabricated 100 unit cells of PhC and a geometrically identical homogeneous slab on the two arms of the MZI. Example scanning electron micrographs (SEMs) are shown in Figure 1. Transmission is measured with amplified spontaneous emission source, in-line fiber polarizer with a polarization controller to couple the light in with a tapered lensed fiber, and an optical spectrum analyzer. In the transmission (Figure 3d; black), the MZ interference spectra has two steep variations, first at the end of the first band (negative index band) and second at the start of the second band (positive index band). This is a clear indication of an abrupt refractive index change (Figure 3c) that is only possible when there is an abrupt interband transition between two bands. The non-MZI transmission spectrum of a similar structure is also shown in Figure3d (red) for reference.

To characterize this steep index change further we placed on the two arms of the MZI PhC sections with different radius r. We kept a unchanged in order to have the same total physical length on both arms, for the same number of unit cells in the PhC sections. With this approach, the MZI sections that do not contain PhC regions are identical and hence one isolates the two PhC sections as the only source for the measured phase difference. For instance, we set r_2 to 5/6 of the original value of the radius r_1 (r_2/a= 0.283 × 5/6=0.236). Figure 4a illustrates the difference between the band structures of the two PhC designs, namely, a frequency shift of the photonic bands. Due to this shifted band structure, the accumulated phase difference between the two arms is almost independent of wavelength, except for a steep variation that again corresponds to a steep refractive index change (moving from band to band). When we place a section of 62 PhC unit cells in both arms of the MZI, the transmission spectra presents two spectral domains, 1525 nm to 1550 nm and

1580 nm to 1615 nm, where the interference transmission is rather constant (red curve in Figure 4b) with ~ 14dB transmission difference between the two domains. In the next section, we show high spatial resolution images for this experiment.

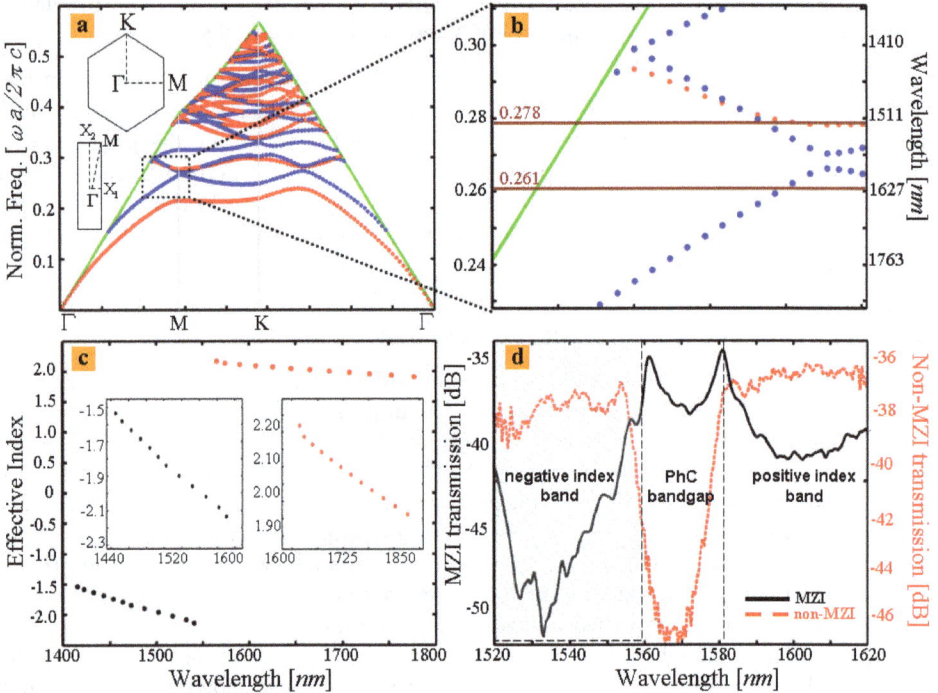

Fig. 3. **Band diagram of the PhC and the calculated effective index. a,** Band diagram of the PhC with the parameters given in Figure1. Insets: first Brillouin zones of the hexagonal PhC (top) and the 1D superlattice (bottom). The TM-like (TE-like) photonic bands are depicted in blue (darker) [red (lighter)]. The light cone is denoted by the green lines. **b,** A zoom-in of the spectral domain corresponding to experimental region of interest. Experiments were performed in the spectral region marked by the two horizontal lines. **c,** Calculated effective index of refraction of the PhC, corresponding to the two TM-like bands shown in Figure2b. Insets: zoom-in of the two bands. **d,** Black (solid) line: MZI transmission with 100 unit cells of PhC on one arm and an homogeneous slab waveguide on the other arm. Red (dashed) line: transmission spectrum for non-MZI PhC superlattice with 60 unit cells.

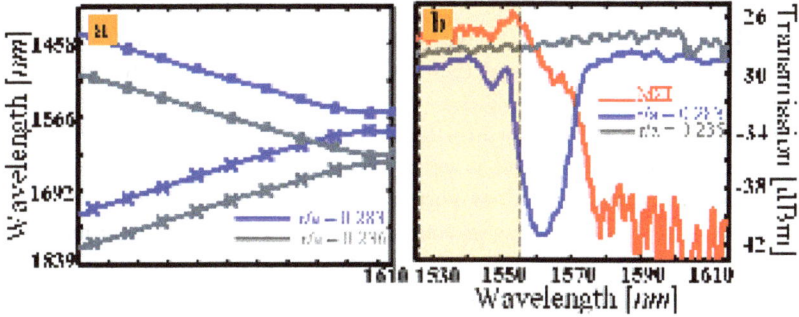

Fig. 4. **Mach-Zehnder interferences with negative refraction PhCs. a,** Band diagram shifts to lower frequency when the *r/a* ratio changes. Blue is the original design and gray is the design with *r/a*=0.236. **c.** Red line: MZI transmission with 62 unit cells of PhC on one arm with *r/a*=0.283 and 62 unit cells of PhC with *r/a*=0.236 on the other arm; the lattice period (*a*) is the same in both cases. Blue (grey) line: Transmission spectrum for PhC superlattice with *r/a*=0.283 (0.236) and 80 unit cells of PhC. Different index differences (*Δn*) from 1525 nm to 1550 nm and from 1580 nm to 1615 nm give different phase difference (*φ*) and different interference output.

2.3 Spatial field distribution

In addition, we performed high spatial resolution imaging of the radiated input-output ports for the devices that have been used for the experiment presented in Figure 4. Results are illustrated in the Figure 5 as follows: In the case of the reference arm (**i-iii**), we see light transmission for all three wavelengths, which corroborates the characteristics of the transmission spectrum in Figure 4b. For the device arm (**iv-vi**) there is transmission for 1600 nm and 1530 nm but not for 1570 nm. This agrees with the transmission spectra in Figure 4b. Note that although there is similar transmission for both arms at 1530 and 1600 nm, the interference output has 14dB difference.

3. Existence of zero-n gap

The band diagram in Figure 3a is calculated by using *RSoft's BandSOLVE* [45], a commercially available software that implements a numerical method based on the plane wave expansion of the electromagnetic field. 3D simulations have been performed to calculate 30 bands and for each band the corresponding values of the effective refractive index have also been determined. In all these numerical simulations a convergence tolerance of 10^{-8} has been used. The photonic bands have been divided into TM-like and TE-like, according to their parity symmetry. The path-averaged index of the superlattice has been calculated by using the negative effective index of the second TM-like band and the effective modal index of the homogeneous asymmetric slab waveguide.

3.1 Numerical simulations

In order to investigate numerically the spectral properties of the transmission characterizing a specific photonic superlattice, we have employed three-dimensional (3D) simulations based on the finite-difference time-domain (FDTD) algorithm; for this, we have used Rsoft's FullWAVE [45]. In our calculations we have studied three superlattices, each with a different period Λ: they have 7, 11 and 15 unit cells along the z-axis, in the PhC layer, so that the thickness of this layer is $d_1 = 3.5\sqrt{3}\,a$, $d_1 = 5.5\sqrt{3}\,a$, and $d_1 = 7.5\sqrt{3}\,a$, respectively. The corresponding thickness of the PIM layer has been calculated by requiring that the average of the index of refraction is zero that is $\bar{n} = (n_1 d_1 + n_2 d_2)/\Lambda = 0$ [28], while keeping the ratio d_2/d_1 unchanged. Importantly, n_1 and n_2 are the *effective indices* of the modes in the PhC and homogeneous layers, respectively, at the corresponding wavelength. We performed this numerical analysis for superlattices with ratio d_2/d_1=0.746 (design 1) and d_2/d_1=0.794 (design 2). Thus by changing the ratio d_2/d_1 we can investigate the dependence of the wavelength at which the zero-$\bar{n}$ gap is observed on the period of the superlattice. We expect to see the zero order gap in the design 2 at a different frequency, as compared to the frequency of the zero order gap in the design 1, a frequency at which the effective indices again satisfy the equation $\bar{n} = (n_1 d_1 + n_2 d_2)/\Lambda = 0$. The effective index of the PIM layer, n_2, can be calculated analytically, as it corresponds to the TM modes of an asymmetric slab waveguide; at λ=1550 nm it is n_2=2.648. Similarly, the effective index corresponding to the second TM-like band shown in Figure 1d is determined from the relation $k = \omega |n|/c$ (note

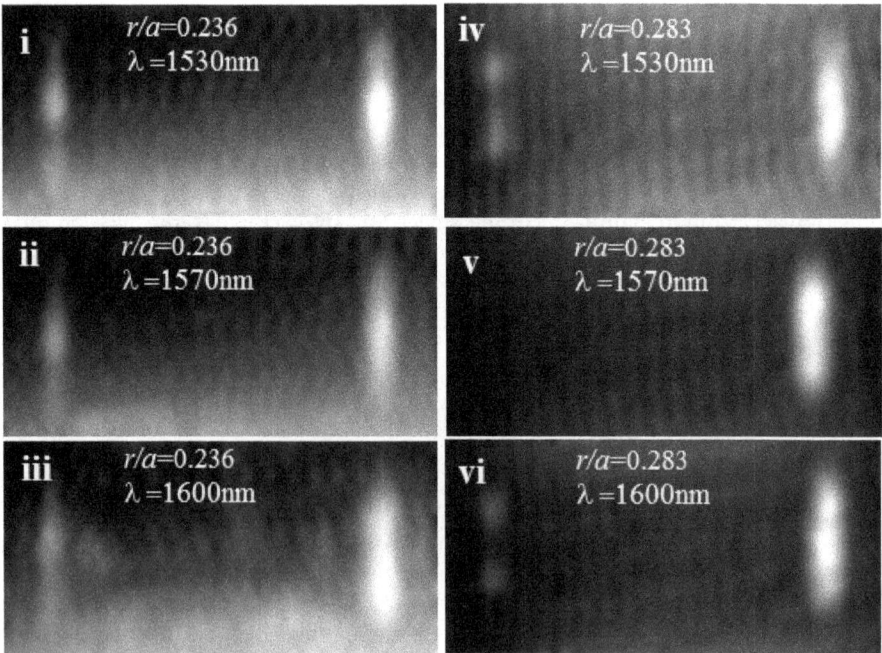

Fig. 5. **Infrared images pertaining to the experiments corresponding to Fig3c. i-iii** is taken from the reference arm containing a PhC structure with r/a=0.236 and **iv-vi** is taken from the device arm where the ratio r/a=0.283. In all the images the input beam is impinging onto the structure from the right, which means that light scattering at the left facet of the device indicates light transmission.

that for the second band the effective index is negative since ω decreases with k [31]). At normalized frequencies of $\omega a/2\pi c = 0.295$ and $\omega a/2\pi c = 0.276$, the effective index of refraction is approximately -1.604 and -1.988, respectively.

In Figure 6a, the results of the 3D FDTD simulations are summarized, for all three superlattices (7, 11, and 15 unit cells), each have a ratio $d_2/d_1=0.746$. In order to observe clear gaps, 5 stacks were used in the case of 7 unit cells in the PhC layer, whereas in the cases with 11 and 15 unit cells in the PhC layer 3 stacks sufficed. The results of similar calculations, for $d_2/d_1=0.794$ are shown in Figure 6b. The grid size resolution in all our numerical simulations is $0.0833a$ (35 nm). Furthermore, we highlight the region where the PhC has negative index of refraction, which is the region of interest in our study. As illustrated in Figure 6a and Figure 3.6b, the transmission spectra show several gaps; however, except for an invariant gap located at $\omega a/2\pi c = 0.276$ (Figure 6a; design 1) and $\omega a/2\pi c = 0.272$ (Figure 6b; design 2), the mid-gap locations of the gaps change with the period of the superlattice. The shift in the mid-gap frequency with respect to the period is typical for regular Bragg gaps; the presence of an invariant gap, however, demonstrates the existence of zero-$\bar{n}$ gaps.

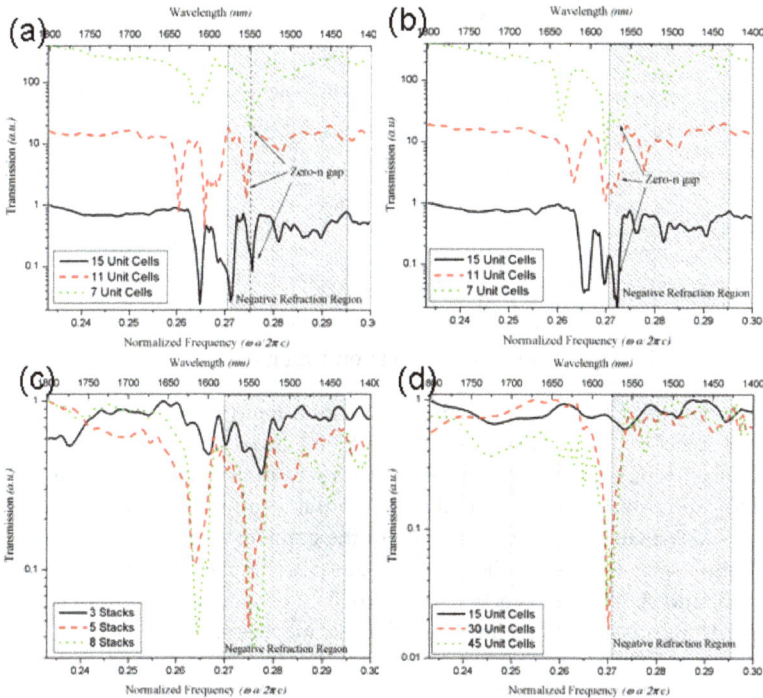

Fig. 6. (a) Transmission for a superlattice with $d_2/d_1= 0.746$, containing 7, 11 and 15 unit cells in each PhC slab. 3 stacks in the PhC superlattice are used in the case when the PhC layers contain 11 and 15 unit cells, and 5 stacks for the case of PhC layers with 7 unit cells. (b) The same as in a), but $d_2/d_1 = 0.794$. (c) Transmission for a superlattice with $d_2/d_1 = 0.794$, containing 3, 5 and 8 stacks. Each PhC layer contains 7 unit cells. (d) Transmission trough a PhC slab containing 15, 30 and 45 unit cells. All results are obtained through full 3D FDTD numerical simulations.

To further investigate the nature of these photonic gaps, we next calculate the order, m, for the Bragg condition. The average index $\bar{n}$ and the value of $k_0\Lambda/\pi$ for the two deigns are summarized in Table 1. The average indices are -0.007 and 0.001 whereas the corresponding $k_0\Lambda/\pi$ values are 0.044 and 0.007 for design 1 and design 2, respectively. It is thus clear that these normalized gap frequencies ($\omega a/2\pi c = 0.276$ for design 1 and $\omega a/2\pi c = 0.272$ for design 2) correspond to zero-order gaps. Moreover, for both superlattices, in the frequency range of negative index of refraction there are other spectral gaps. For example, the gap at $\omega a/2\pi c = 0.283$, corresponds to an average index of refraction of 0.091 (design 1) and 0.153 (design 2), with the corresponding $k_0\Lambda/\pi$ being 0.874 and 0.962, and thus it is a first-order gap. None of these gaps is due solely to the presence of the PhC layers, as the PhC band gap for the TM polarization is at $\omega a/2\pi c = 0.27$ (see Figure3b).

Figure	Unit Cells of PhC	Gap Frequency	d_2/d_1	Effective Index		Average index, $\bar{n}$	$k_0\Lambda/\pi$	Gap order
				PhC	Slab			
Fig. 3.3a (design 1)	7	0.276	0.746	-1.988	2.648	-0.007	-0.044	0
Fig. 3.3a (design 1)	11	0.283	0.746	-1.844	2.684	0.091	0.874	1
Fig. 3.3b (design 2)	7	0.272	0.794	-2.080	2.622	0.001	0.007	0
Fig. 3.3b (design 2)	7	0.283	0.794	-1.856	2.683	0.153	0.962	1

Table 1. **Average refractive index of the corresponding gaps and the gaps' order**

As an additional proof that the invariant gap is not a band gap of the PhC, we present in Figure 6d the calculated transmission spectra of a PhC layer with a number of 15, 30, and 45 unit cells (no layers of homogeneous material is present in this case). Thus, this figure shows that the PhC gap is shifted by almost 40 nm from the location of the zero-order gap in gap Figure 6a. We also examined the dependence of the gap locations on the number of stacks in the superlattice. The results of these calculations are presented in Figure 6c for stack numbers 3, 5, and 8. We observe that the zero-$\bar{n}$ gap location has not changed and, as expected, it becomes deeper as the number of stacks increases.

3.2 Electric field distribution

We have also calculated the electric field distribution in the zero-n superlattice for different wavelengths in our experimental region in continuous wave excitation type. In order to be able to get this distribution, we have run the simulation until it gets to steady state and then save the field. Next, we launched another simulation with an input field by using this saved field and ran for only one cycle by saving electric field in small steps. Then we have averaged the $|E|^2$ and plotted. Figure 7 shows the results.

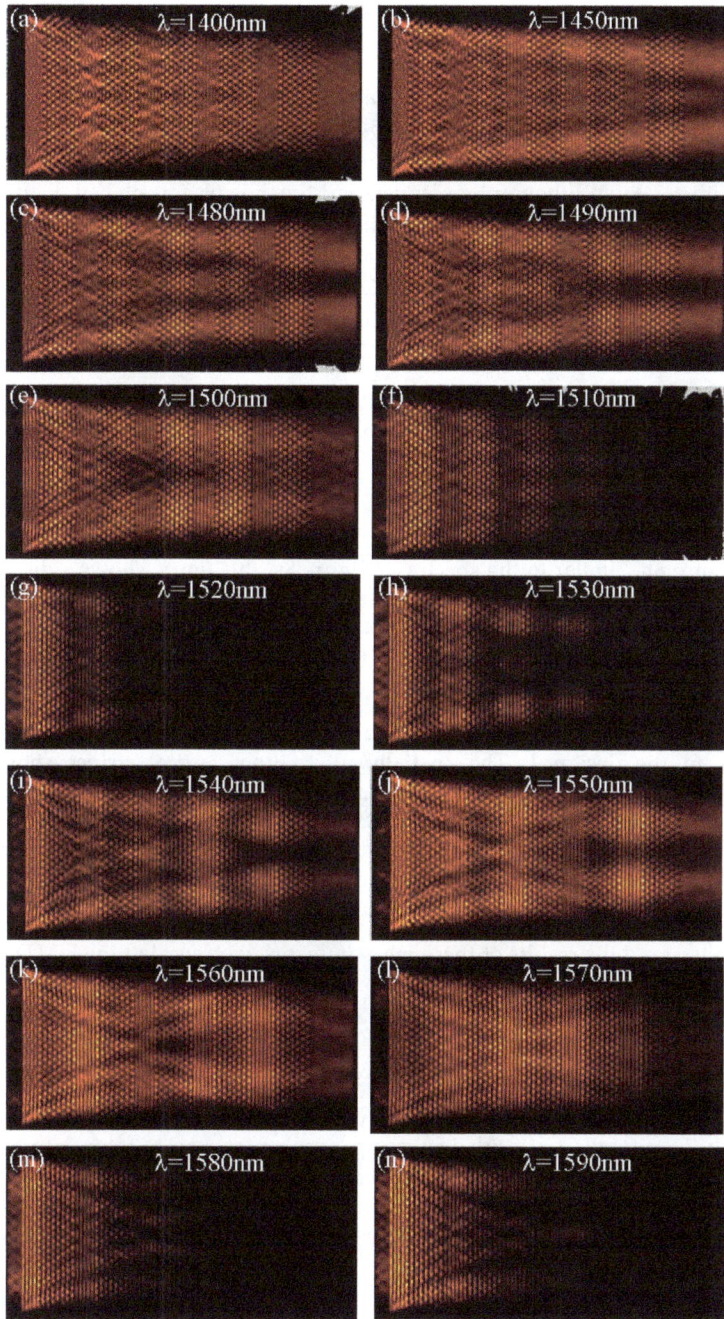

Fig. 7. A time-averaged steady-state distribution of the field intensity, $|E|^2$, corresponding to a propagating mode with different wavelengths within the experimental region.

3.3 Device nanofabrication and experiments

Our theoretical predictions are validated by a series of experiments. Thus, we have fabricated in a single-crystal silicon-on-insulator substrate samples with 3, 5 and 8 stacks whose PhC layers have thickness of $d_1 = 3.5\sqrt{3}$ a . The silicon device height is 320 nm and the silicon oxide cladding thickness is 1 μm. The PhC superlattice is lithographically patterned with a 248-nm lithography scanner, and the Si is plasma-etched. Figure 1b shows an example of a fabricated photonic crystal superlattice with 8 stacks. The fabrication disorder in the PhC slab was statistically parameterized [37], with resulting hole radius 122.207 ± 1.207 nm, lattice period 421.78 ± 1.26 nm (~ 0.003a), and ellipticity 1.21 nm ± 0.56 nm. These small variations are below ~0.05a disorder theoretical target [46].

Incident light from tunable lasers between 1480 nm to 1690 nm (0.248 to 0.284 in normalized frequency of $\omega a/2\pi c$) is coupled into the chip *via* tapered lensed fibers with manual fiber polarization control. The transmission for the TM polarization is measured, with each transmission measurement averaged over three scans. Figure 8a shows the transmission spectrum for design 1 (with $d_2/d_1 = 0.746$); it shows two distinct spectral dips, centered at 1520 nm ($\omega a/2\pi c \approx 0.276$) and 1585 nm ($\omega a/2\pi c \approx 0.265$). We then repeat these transmission measurements for a second design (design 2; $d_2/d_1 = 0.794$); the corresponding results are shown in Figure 8b. Similar to the spectra in Figure 8a, this figure shows a distinct spectral dip, located near the normalized frequency $\omega a/2\pi c \approx 0.272$, *i.e.* at λ=1543 nm. Furthermore, the frequency spectral dip at $\omega a/2\pi c \approx 0.262$ is weaker than in design 1, which is due to the fact that below a certain frequency, $\omega a/2\pi c \approx 0.265$, the detected power is not high enough for observing the spectral features.

Figure 8c shows the near-infrared image captured with incident laser at 1550 nm and corresponds to the design 1. The spatially alternating vertical stripes show radiation scattered at the interfaces between the PhC and the homogeneous layers and confirm the transmission of the light through the superlattice. The near-infrared images also confirm the existence of the dip in the transmission spectrum, with most of the light being reflected and only a small amount propagating out of the output facet of the third stack (in this figure , the incident laser is tuned to the zero-$\bar{n}$ gap frequency).

To better understand the results of these measurements, we repeated the FDTD simulations for the case of the fabricated devices. A good match between the results of the measurements and those of simulations, in terms of *absolute* values of the frequencies, has been observed for both values of the ratio d_2/d_1 (design 1 and 2) and varying stack numbers (design 1). The theoretical predictions are shown as the dotted lines in Figure 8a and 3.8b.

Figure 8d further shows the results of a series of experiments for PhC superlattices of design 1, each with 7 unit cells but with increasing number of stacks, from 3 to 5 and 8. As observed, both gaps become deeper with increasing number of stacks. It should be noted that by improving the impedance matching between the negative and positive index materials, transmission spectrum for even larger number of stacks in the PhC superlattice can be observed. The center-frequency of the gaps does not change when the number of stacks increases, the slight deviations in the gap frequency being attributable to small variations in the dimensions of the fabricated superlattice structures. This series of measurements further reinforce the observation of the zero-$\bar{n}$ gap in cascaded negative- and positive-refraction superlattices.

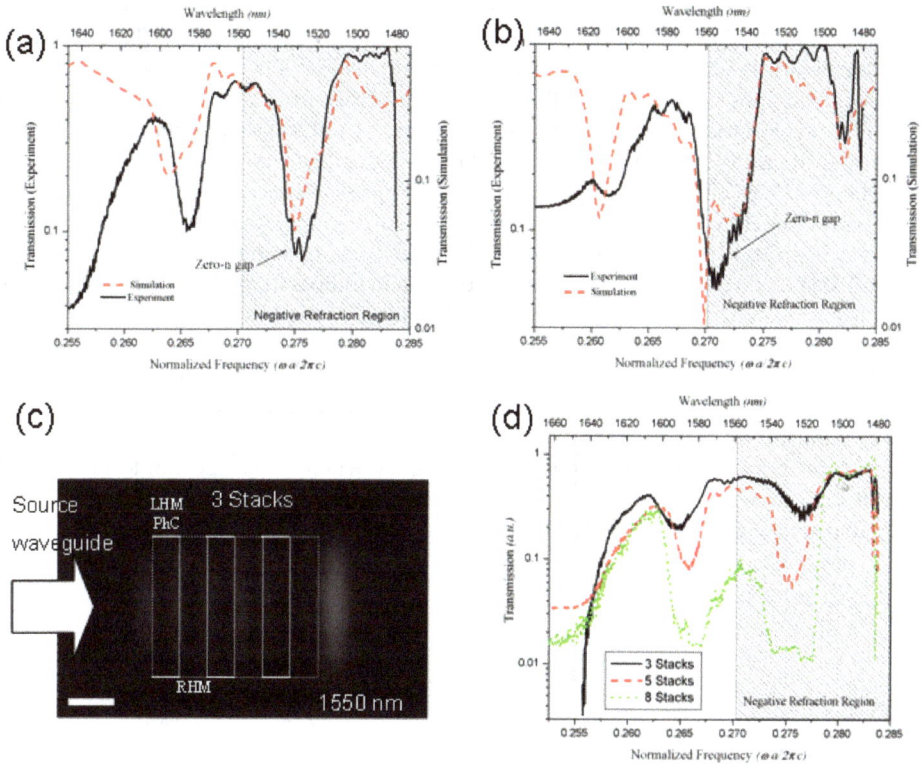

Fig. 8. (a) Measured transmission for a superlattice with d_2/d_1= 0.746, with 7 unit cells in the PhC layers and 5 stacks; for comparison, results of numerical simulations are also shown. (b) The same as in a), but for a superlattice with 0.794. (c) Example of near-infrared top-view image of a device with 3 stacks, from transmission measurement at 1550 nm. Superimposed are the locations of the negative refraction PhC and positive index material in the superlattice. Scale bar: 2 μm. (d) Measured transmission for a superlattice with d_2/d_1= 0.746, with 3, 5 and 8 stacks and 7 unit cells in the PhC layers. Both gaps become deeper as the number of stacks increases.

It has been pointed out that zero-$\bar{n}$ gaps can be omnidirectional [28, 41]; however, in our case, due to the anisotropy of the index of refraction of the PhC, the zero-$\bar{n}$ gap is not omnidirectional. Moreover, varying the lattice period, radius, and the thickness of the superlattice, and thus changing the frequency at which the average effective index of refraction is equal to zero, the frequency of the zero-$\bar{n}$ gap can be easily tuned as we show in the next section. Importantly, we note the demonstration of these zero-$\bar{n}$ gap structures can have potential applications as delay lines with zero phase differences which we also show later in this chapter.

4. Tunability of zero-n gap

Next, in order to demonstrate the tunable character of the zero-$\bar{n}$ bandgaps, we performed transmission experiments on four sets of binary superlattices, with each set having different superlattice ratios: d_2/d_1=0.74 (Figure 9a), d_2/d_1=0.76 (Figure 9b), d_2/d_1=0.78 (Figure 9d), and d_2/d_1=0.8 (Figure 9e). In all our experiments the negative index PhC has the same parameters as those given above. Each set has three devices of different periods Λ, with the negative index PhC layer in the superlattice spanning 7, 9, and 11 unit cells along the z-axis similar to the numerical study in 3.1, so that the thickness of this layer is $d_1 = 3.5\sqrt{3}$ a (2.564 μm), $d_1 = 4.5\sqrt{3}$ a (3.297 μm), and $d_1 = 5.5\sqrt{3}$ a (4.029 μm), respectively. The corresponding thickness of the PIM layer is determined by requiring that the average index is zero [$\bar{n} = (n_1d_1 + n_2d_2)/\Lambda = 0$], while keeping the ratio d_2/d_1 unchanged for all devices in each set (see Table 2). Here, n_1 and n_2 are the effective mode indices in the PhC and homogeneous layers respectively at the corresponding wavelengths. For the three devices in each set, we designed 7 super-periods (SPs) for the devices with 7 unit cells of PhC and 5 SPs for those with 9 and 11 unit cells of PhC (these designs ensure a sufficient signal-to-noise ratio for the transmission measurements). In these experiments we have tested both the existence of the zero-$\bar{n}$ bandgap as well as its tunability. For the three devices belonging to each set, we observed the zero-$\bar{n}$ bandgap at the same frequency whereas the spectral locations of the other bandgaps were observed to shift with the frequency – this confirms the zero-$\bar{n}$ bandgap does not depend on the total superperiod length Λ (gap existence dependent only

Figure 9a			
# of unit cells	d_1	d_2	Λ
7	2.56	1.90	4.46
9	3.30	2.44	5.74
11	4.03	2.98	7.01
Figure 9b			
# of unit cells	d_1	d_2	Λ
7	2.56	1.95	4.51
9	3.30	2.51	5.80
11	4.03	3.06	7.09
Figure 9d			
# of unit cells	d_1	d_2	Λ
7	2.56	2.00	4.56
9	3.30	2.57	5.87
11	4.03	3.14	7.17
Figure 9e			
# of unit cells	d_1	d_2	Λ
7	2.56	2.05	4.62
9	3.30	2.64	5.93
11	4.03	3.22	7.25

Table 2. **Calculated parameters of the devices in the Figure 9 (units in μm).**

on the condition of path-averaged zero index: $n_1d_1 + n_2d_2 = 0$) while the frequency of the regular 1D PhC Bragg bandgaps does depend on Λ. Our measurements show that the invariant, zero-$\overline{n}$, bandgap is located at 1525.5 nm, 1535.2 nm, 1546.3 nm, and 1556.5 nm, respectively (averaged over the three devices in each set). The slight red-shift with increasing number of unit cells in each set is due to effects of edge termination between the PhC and the homogeneous slab.

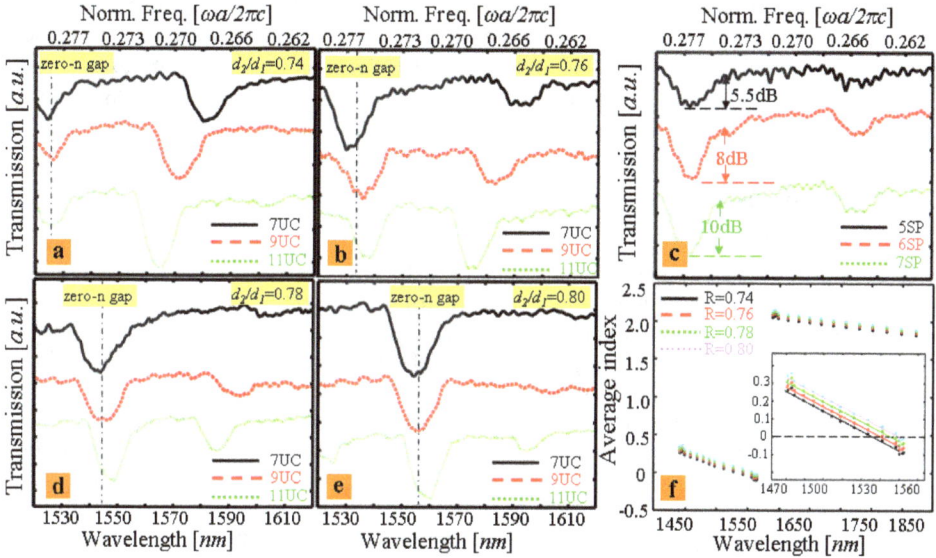

Fig. 9. **Experimental verification of period-invariance and tunability of zero-$\overline{n}$ bandgaps.** **a,** Experimental verification of the zero-$\overline{n}$ bandgap in superlattices with varying period (Λ). The ratio d_2/d_1 =0.74 and Λ=4.46 µm for black (solid), 5.74 µm for red (dashed), and 7.01 µm for green (dotted) curves (*a.u.* arbitrary units). The lightly shaded regions in all panels denote the negative index regions. **b,** Same as in **a**, but for d_2/d_1 =0.76. Λ=4.51 µm for black (solid), 5.80 µm for red (dashed), and 7.09 µm for green (dotted) curves. **c,** Transmission spectra for superlattices with d_2/d_1 =0.76, containing 5, 6, and 7 unit cells (UC). Each PhC layer contains 7 unit cells (d_1=2.564 µm , Λ=4.51µm). **d,** Same as in **a**, but for d_2/d_1 =0.78. Λ=4.56 µm for black (solid), 5.87µm for red (dashed), and 7.17 µm for green (dotted) curves. **e,** Same as in **a**, but for d_2/d_1 =0.80. Λ=4.61 µm for black (solid), 5.93 µm for red (dashed), and 7.25 µm for green (dotted) curves. **f,** Calculated effective index of refraction for the superlattices with the ratios in **a**, **b**, **d**, and **e**. The wavelengths at which the average index of refraction cancels agree very well with the measured values.

Furthermore, when we tuned the ratio d_2/d_1 and repeated these same experiments we observed a redshift of the zero-$\overline{n}$ mid-gap frequency as we increased the ratio d_2/d_1. This result is explained by the fact that for the negative index band the refractive index of the 2D hexagonal PhC decreases with respect to the wavelength (see Figure 3c) and therefore when the length of the PIM layer in the 1D binary superlattice increases (higher d_2/d_1), the wavelength at which the effective index cancels is red-shifted. The effective index of the PIM layer, n_2, is calculated numerically and for the asymmetric TM slab waveguide mode corresponds to, for example, 2.648 at 1550 nm. By using these n_1 (Figure 3c) and n_2 values, we determined the average refractive index for the different d_2/d_1 ratios as summarized in Figure 9f. A distinctive red-shift in the zero-$\bar{n}$ gap location is observed with increasing d_2/d_1 ratios from the numerically modeling, demonstrating good agreement with the experimental measurements (Figure 9a, 9b, 9d, and 9e) without any parameter fitting in the analysis. Furthermore, Figure 9c shows how the spectral features of the zero-$\bar{n}$ bandgap changes with increasing the number of superperiods and the results are similar to those in Figure8d as expected. We note that this is the first rigorous and complete experimental confirmation of invariant and tunable character of zero-$\bar{n}$ bandgaps in photonic superlattices containing negative index PhCs.

5. Zero phase delay lines

Next, we prove that the total phase accumulation in the superlattice is zero. For this, we performed phase measurements for the designs with three different sets of measurements: (a) d_2/d_1=0.78 and 7 unit cells in the PhC layer; (b) d_2/d_1= 0.8 and 7 unit cells in the PhC layer; and (c) d_2/d_1= 0.8 and 9 unit cells in the PhC layer. In these series of measurements we used a single mode channel waveguide for the reference arm of the MZI – this enables a series of interference fringes at the output, which can be used to determine the phase change by analyzing the spectral location of the fringes and their free spectral range (FSR). In most free-space interferometric applications, the phase difference leading to interference originates from the physical length difference between the two arms, but in integrated photonic circuits this delay can easily be modulated by the imbalance in the refractive indices of two arms [47]. For (a) and (c), we examined three devices, namely, superlattices with 5, 6, and 7 SPs and for (b) we tested superlattices with 5 and 7 SPs. When we designed these devices, we modified the MZI such that when we added a SP to the superlattice the length of the adiabatic transition arms was carefully increased by $\Lambda/2$, making the horizontal single mode channel waveguides shorter (from L_2 to L_2-$\Lambda/2$, at both sides in Figure 1a). This change is compensated by adding the same length to the vertical part (from L_3 to L_3+$\Lambda/2$ on both sides in Figure 1a). As a result, the only phase difference between devices is due to the additional SPs. This procedure is explained in Section 5.1 in detail.

5.1 Device modification for phase measurements

Figure 10 shows a schematic representation of a device with 2 superperiods and the integrated Mach Zehnder Interferometer is modified after introducing the third superperiod. The adiabatic region remains unchanged if L_1 is increased to L_1+Λ and L_2 is shortened by $\Lambda/2$, in both the input and output sides of the device. To keep the total length of the waveguide unchanged, the length L_3 is increased to L_3+ $\Lambda/2$. This procedure is used each time a superperiod is added to the structure. In addition, to be able to compare devices with different number of unit cells in the PhC layer, a common reference point is used for all devices that have the same d_2/d_1 ratio.

In our implementation, the interferometer output intensity is given as:

$$I = I_1 + I_2 + 2\sqrt{I_1 I_2} \cos\phi \tag{3}$$

where ϕ is the phase difference (or imbalance) between the modes propagating in the two arms (denoted by subscript 1 and 2). Considering our implementation, the phase ϕ can be decomposed as:

$$\phi = \phi_1 - \phi_2 = \frac{2\pi}{\lambda}\left[\left(n_{wg}L_{wg\_1} + n_{slab\_1}L_{slab\_1} + n_{sl}L_{sl}\right) - n_{wg}L_{wg\_2}\right]$$

$$= \frac{2\pi}{\lambda}\left[n_{wg}\left(L_{wg\_1} - L_{wg\_2}\right) + n_{slab\_1}L_{slab\_1}\right] + \frac{2\pi}{\lambda}n_{sl}L_{sl} \tag{4}$$

where n_{wg}, n_{slab_1}, n_{sl} are the effective mode refractive indices of the channel waveguide, the adiabatic slab in arm 1, and the zero-index superlattice, respectively. L_i denotes the corresponding lengths.

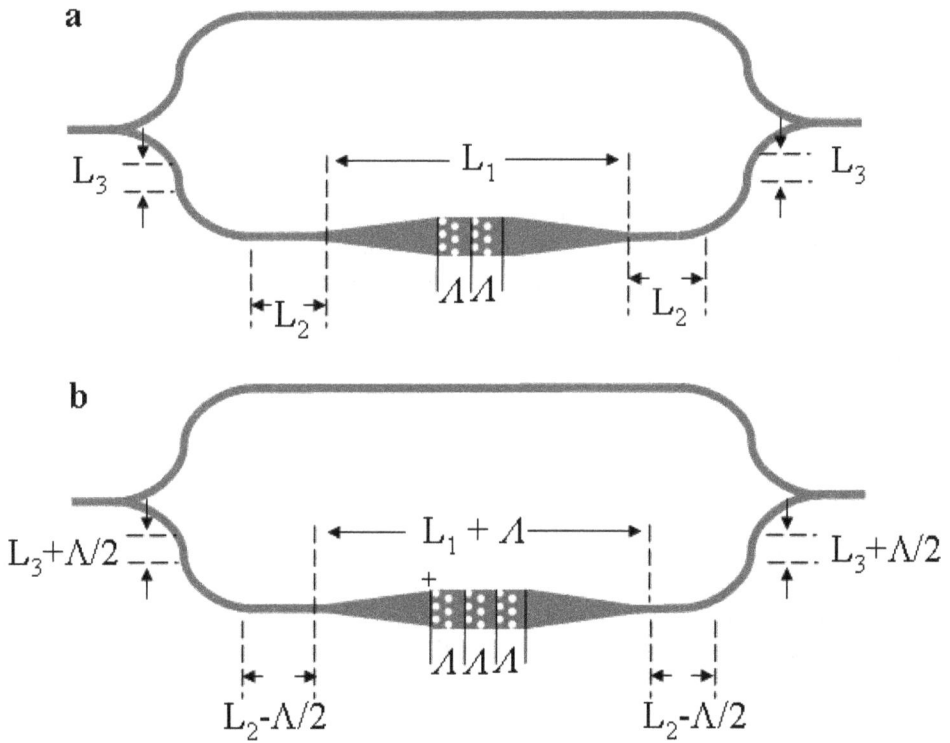

Fig. 10. **Schematic representation of the device modification induced by adding a superperiod. a,** Integrated MZI of a device with 2 superperiods and a channel waveguide. **b,** Device modifications after the third superperiod is added. The length of the channel waveguide remains the same so as the effect of the additional superperiod is isolated.

We note that the difference between the physical path length of the channel waveguides on both arms is designed to be equal to the physical path length of the tapering slab. Thus we have:

$$\cos[\phi] = \cos\left[\frac{2\pi}{\lambda}\left(\left(-n_{wg}L_{Slab\_1} + n_{Slab\_1}L_{Slab\_1}\right) + n_{sl}L_{sl}\right)\right]$$

$$= \cos\left[\frac{2\pi}{\lambda}\left(\left(n_{Slab\_1} - n_{wg}\right)L_{Slab\_1} + n_{sl}L_{sl}\right)\right] \tag{5}$$

$$= \cos\left[\phi_{slab-wg} + \phi_{sl}\right]$$

The mode indices n_{wg}, n_{slab_1}, n_{sl} have different frequency dispersion. We kept the phase (ϕ_{slab_wg}), arising from $\frac{2\pi}{\lambda}\left(n_{slab\_1} - n_{wg}\right)L_{slab\_1}$, constant between different devices in each set of measurements by simply ensuring that the physical lengths and widths of the slabs are the same for each nanofabricated device. The remaining phase variation therefore is generated only by the photonic crystal superlattice ($\phi_{sl} = \frac{2\pi}{\lambda}n_{sl}L_{sl}$). If n_{sl} is equal to zero, ϕ_{sl} is zero too, hence the total phase difference ϕ in the interferometer arises only from the ϕ_{slab_wg} component and is the same for all the devices in each set. Therefore, the sinusoidal oscillations in the transmission and the free spectral range are determined only by ϕ_{slab_wg}.

5.2 Experimental results for zero phase

Figure 11a shows the interference pattern for d_2/d_1=0.78 with 7 unit cells in the PhC layer. As can be seen in this figure, outside the zero-$\bar{n}$ spectral region the fringes differ from each other both in wavelength and the FSR, but overlap almost perfectly within the zero-$\bar{n}$ spectral domain.

To illustrate the phase evolution, we show in Figure 12a the FSR values for each of the devices examined – specifically we calculate the spectral spacing between the transmission minima and plot its dependence on the center wavelength between the two neighboring minima. As these measurements illustrate, in the zero-$\bar{n}$ spectral domain the FSR corresponding to each of the devices approaches the same value, indicating that the corresponding phase difference is zero or, alternatively, that the optical path remains unchanged. This is a surprising conclusion since the physical path is certainly not the same in all the cases. This apparent paradox has a simple explanation: although the physical path varies among the three cases the optical path is the same (and equal to zero) as the spatially averaged refractive index of the three superlattices vanishes. In other words, within the zero-$\bar{n}$ spectral region the photonic superlattice emulates the properties of a zero phase delay line. The output corresponding to the structures with d_2/d_1=0.8 and 7 unit cells in the PhC layer is shown in Figure 11b whereas the FSR values are plotted in Figure 12b. Finally, Figure 11c and Figure 12c show the interference patterns for the case of d_2/d_1=0.8 and 9 unit cells in the PhC layer. Again, both the FSR (Figure 12b-c) and the absolute wavelength values (Figure 11b-c) overlap, proving the zero phase variation across the superlattice.

It should be noted that in all our plots of experimental data we have used the raw data and as such there is no data post-processing, except for the intensity rescaling. Measurements are taken 3 times with 500 pm resolution for Figure 3d and Figure 11a-e; 100 pm and 500 pm resolutions for Figure11a and 200 pm and 500 pm resolutions for Figure 11b-c. There is ~ 1% deviation between Figure 11e and Figure 11b-c in terms of the center frequency of the zero-$\bar{n}$ region. This is so because of the fabrication differences between the samples. For Figure 11b-c, the r/a ratio was ~5% smaller (~0.264) resulting in the shift of the band structure to lower frequencies, and, consequently to a shift of the zero-$\bar{n}$ bandgap. We verified the location of the zero-$\bar{n}$ bandgap (~1565 nm) by performing the transmission measurements described before. Thus, the spectral location of the zero-$\bar{n}$ bandgap can be tracked from the phase measurements, as the spectral region of small amplitude oscillations in the transmission spectra correspond to the zero-$\bar{n}$ bandgaps.

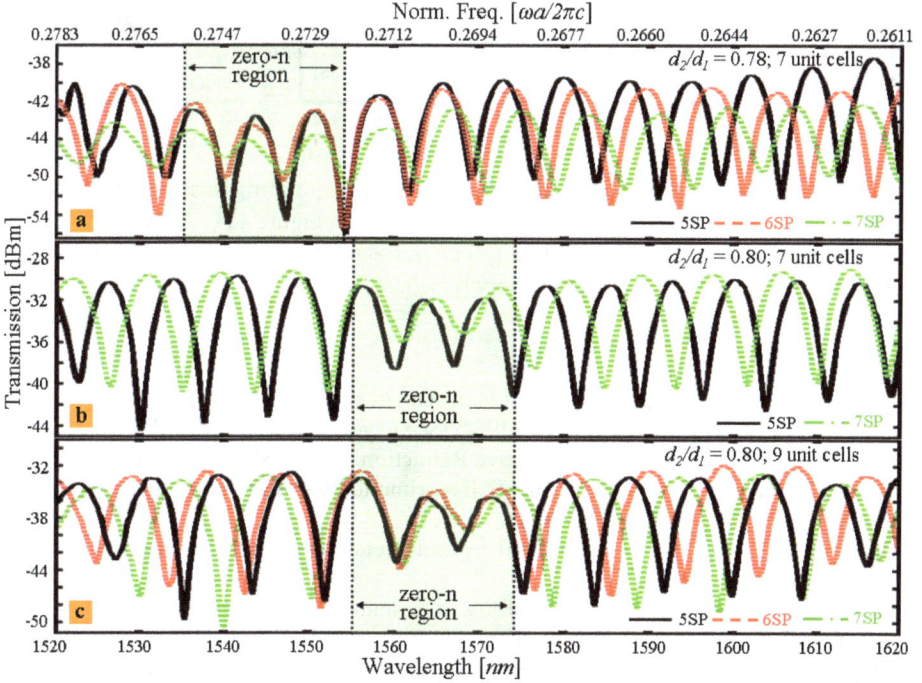

Fig. 11. **Phase measurements. a,** Output of the MZI with increasing number of superperiods (5, 6, and 7 SP) on one arm and with length-adjusted single mode channel waveguides on the other arm. Each PhC layer contains 7 unit cells (d_1=2.564 μm) and d_2/d_1=0.78. **b,** Same as in **a,** but for d_2/d_1=0.8. **c,** Same as in b, with each PhC layer containing 9 unit cells (d_1=3.297 μm).

In summary, we have demonstrated for the first time zero-phase delay in negative-positive index superlattices, in addition to the simultaneous observations of deterministic zero-$\bar{n}$ gaps that can remain invariant to geometric changes and band-to-band transitions in negative-positive index photonic crystal superlattices. Through the interferometric measurements, the transmissive binary superlattices with varying lengths are shown unequivocally to enable the absolute control of the optical phase. The engineered control of the phase delay in these near-zero superlattices can be implemented in chip-scale transmission lines with deterministic phase array control, even with technological potential in phase-insensitive image processing, phase-invariant field for electromagnetic cloaking, and the arbitrary radiation wavefront reshaping of antennae from first principles.

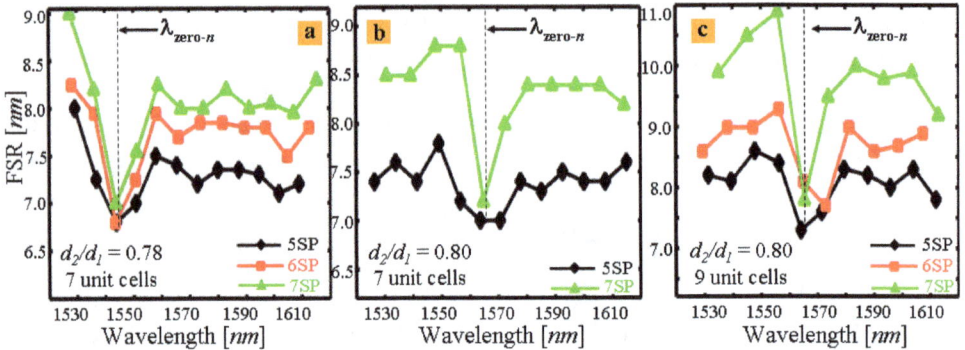

Fig. 12. **Free spectral range wavelength dependence corresponding to superlattices in Figure 4. a-c,** Free spectral range extracted from the data in Figure 4a-c. At the zero-$\bar{n}$ bandgap wavelength, the free spectral range converges to the same value, which proves the zero phase contribution from the added superperiods.

6. References

[1] Veselago, V. G. The electrodynamics of substances with simultaneously negative values of ε and μ. *Usp.Fiz.Nauk* 92, 517 (1964) [*Sov.Phys.Usp.* 10, 509 (1968)].

[2] Pendry, J. B. Time Reversal and Negative Refraction. *Science* 322, 71 (2008).

[3] Shelby, R. A., Smith, D. R. & Schultz, S. Experimental verification of a negative index of refraction. *Science* 292, 77 (2001).

[4] Valentine, J. et al. Three-dimensional optical metamaterial with a negative refractive index. *Nature* 455, 376 (2008).

[5] Li, N. et al. Three-dimensional photonic metamaterials at optical frequencies. *Nature Mat.* 7, 31 (2008).

[6] Gramotnev, D. K. & Bozhevolnyi, S. I. Plasmonics beyond the diffraction limit. *Nature Phot.* 4, 83 (2010).

[7] Shalaev, V. M. Optical negative-index metamaterials. *Nature Photonics* 1, 41 (2007).

[8] Kundtz, N. & Smith, D. R. Extreme-angle broadband metamaterial lens. *Nature Mat.* 9, 129 (2010).

[9] Panoiu, N. C. & Osgood, R. M. Influence of the dispersive properties of metals on the transmission characteristics of left-handed materials. *Phys. Rev. E* 68, 016611 (2003).

[10] Zhang, S. et al. Demonstration of near-infrared negative-index metamaterials. *Phys. Rev. Lett.* 95, 137404 (2005).

[11] Liu, R. et al. Experimental Demonstration of Electromagnetic Tunneling Through an Epsilon-Near-Zero Metamaterial at Microwave Frequencies. *Phys. Rev. Lett.* 100, 023903 (2008).

[12] Tsakmakidis, K. L., Boardman, A. D. & Hess, O. 'Trapped rainbow' storage of light in metamaterials. *Nature* 450, 397 (2007).

[13] Yao, J. et al. Optical Negative Refraction in Bulk Metamaterials of Nanowires. *Science* 321, 930 (2008).

[14] Xi, S. et al. Experimental Verification of Reversed Cherenkov Radiation in Left-Handed Metamaterial. *Phys. Rev. Lett.* 103, 194801 (2009)

[15] Hoffman, A. J. et al. Negative refraction in semiconductor metamaterials. *Nature Mat.* 6, 946 (2007).

[16] Grigorenko, A. N. et al. Nanofabricated media with negative permeability at visible frequencies. *Nature* 438, 335 (2005).

[17] Dolling, G., Enkrich, C., Wegener, M., Soukoulis, C. M. & Linden, S. Simultaneous Negative Phase and Group Velocity of Light in a Metamaterials. *Science* 312, 892 (2006).

[18] Pollard, R. J. et al. Optical Nonlocalities and Additional Waves in Epsilon-Near-Zero Metamaterials. *Phys. Rev. Lett.* 102, 127405 (2009).

[19] Jin, Y., Zhang, P. & He, S. Squeezing electromagnetic energy with a dielectric split ring inside a permeability-near-zero metamaterial. *Phys. Rev. B* 81, 085117 (2010)

[20] Hsueh, W. J., Chen, C. T. & Chen, C. H. Omnidirectional band gap in Fibonacci photonic crystals with metamaterials using a band-edge formalism. *Phys. Rev. A* 78, 013836 (2008).

[21] Xu, J.-P., Yang, Y.-P., Chen, H. & Zhu, S.-Y. Spontaneous decay process of a two-level atom embedded in a one-dimensional structure containing left-handed material. *Phys. Rev. A* 76, 063813 (2007).

[22] Chatterjee, R. et al. Achieving subdiffraction imaging through bound surface states in negative-refraction photonic crystals in the near-infrared range. *Phys. Rev. Lett.* 100, 187401 (2008).

[23] Decoopman, T., Tayeb, G., Enoch, S., Maystre, D. & Gralak, B. Photonic Crystal Lens: From Negative Refraction and Negative Index to Negative Permittivity and Permeability. *Phys. Rev. Lett.* 97, 073905 (2006).

[24] Lu, Z. et al. Three-Dimensional Subwavelength Imaging by a Photonic-Crystal Flat Lens Using Negative Refraction at Microwave Frequencies. *Phys. Rev. Lett.* 95, 153901 (2005).

[25] Parimi, P. V., Lu, W. T., Vodo, P. & Sridhar, S. Photonic crystals: Imaging by flat lens using negative refraction. *Nature* 426, 404 (2003).

[26] Notomi, M. Theory of light propagation in strongly modulated photonic crystals: Refractionlike behavior in the vicinity of the photonic band gap. *Phys. Rev. B* 62, 10696 (2000).

[27] Li, J., Zhou, L., Chan, C. T. & Sheng, P. Photonic Band Gap from a Stack of Positive and Negative Index Materials. *Phys. Rev. Lett.* 90, 083901 (2003).

[28] Panoiu, N. C., Osgood, R. M., Zhang, S. & Brueck, S. R. J. Zero-n bandgap in photonic crystal superlattices. *J. Opt. Soc. Am. B* 23, 506 (2006).

[29] Kocaman, S. et al. Observations of zero-order bandgaps in negative-index photonic crystal superlattices at the near-infrared. *Phys. Rev. Lett.* 102, 203905 (2009).

[30] Kocaman, S et al. Zero phase delay in negative-index photonic crystal superlattices, *Nature Photonics* 5, 499 (2011).

[31] Yuan, Y. et al. Experimental verification of zero order bandgap in a layered stack of left-handed and right-handed materials. *Opt. Express* 14, 2220 (2006).

[32] Zhang, L., Zhang, Y., He, L., Li, H. & Chen, H. Experimental investigation on zero-n_{eff} gaps of photonic crystals containing single-negative materials. *Eur. Phys. J. B.* 62, 1 (2008).

[33] Hegde, R. S. & Winful, H. G. Zero-n gap soliton. *Opt. Lett.* 30, 1852 (2005).

[34] Mocella, V. et al. Self-Collimation of Light over Millimeter-Scale Distance in a Quasi-Zero-Average-Index Metamaterial. *Phys. Rev. Lett.* 102, 133902 (2009).

[35] Enoch, S., Tayeb, G., Sabouroux, P., Guérin, N. & Vincent, P. A Metamaterial for Directive Emission. *Phys. Rev. Lett.* 89, 213902 (2002).

[36] Ziolkowski, R. W. Propagation in and scattering from a matched metamaterial having a zero index of refraction. *Phys. Rev. E* 70, 046608 (2004).

[37] Litchinitser, N. M., Maimistov, A. I., Gabitov, I. R., Sagdeev, R. Z. & Shalaev, V. M. Metamaterials: electromagnetic enhancement at zero-index transition. *Opt. Lett.* 33, 2350 (2008).

[38] Wang, L.-G., Li, G.-X. & Zhu, S.-Y. Thermal emission from layered structures containing a negative-zero-positive index metamaterial. *Phys. Rev. B* 81, 073105 (2010).

[39] Shadrivov, I. V., Sukhorukov, A. A. & Kivshar, Y. S. Complete band gaps in one-dimensional left-handed periodic structures. *Phys. Rev. Lett* 95, 193903 (2005).

[40] Hao, J., Yan, W. & Qiu, M. Super-reflection and cloaking based on zero index metamaterial. *Appl. Phys. Lett.* 96, 101109 (2010).

[41] Jiang, H., Chen, H., Li, H., Zhang, Y. & Zhu, S. Omnidirectional gap and defect mode of one-dimensional photonic crystals containing negative-index materials. *Appl. Phys. Lett.* 83, 5386 (2003).

[42] Bria, D. et al. Band structure and omnidirectional photonic band gap in lamellar structures with left-handed materials. *Phys. Rev. E* 69, 066613 (2004).

[43] A.R. Davoyan, I. V. Shadrivov, A.A. Sukhorukov, and Y.S. Kivshar, Bloch oscillations in chirped layered structures with metamaterials. *Opt. Express* 16, 3299 (2008).

[44] M.D. Henry, C. Welch, and A. Scherer. Techniques of cryogenic reactive ion etching in silicon for fabrication of sensors. *J. Vac. Sci. Technol. A* 27, 1211 (2009).

[45] Rsoft Design Group Inc, FullWAVE™, Ossining, NY.

[46] X. Wang *et. al.*, "Effects of disorder on subwavelength lensing in two-dimensional photonic crystal slabs", Phys. Rev. B 71, 085101 (2005).

[47] Vlasov, Y.A., O'Boyle, M., Hamann, H.F., and McNab, S.J. Active control of slow light on a chip with photonic crystal waveguides, *Nature* 438, 65 (2005).

Permissions

The contributors of this book come from diverse backgrounds, making this book a truly international effort. This book will bring forth new frontiers with its revolutionizing research information and detailed analysis of the nascent developments around the world.

We would like to thank Dr. Eng. Alessandro Massaro, for lending his expertise to make the book truly unique. He has played a crucial role in the development of this book. Without his invaluable contribution this book wouldn't have been possible. He has made vital efforts to compile up to date information on the varied aspects of this subject to make this book a valuable addition to the collection of many professionals and students.

This book was conceptualized with the vision of imparting up-to-date information and advanced data in this field. To ensure the same, a matchless editorial board was set up. Every individual on the board went through rigorous rounds of assessment to prove their worth. After which they invested a large part of their time researching and compiling the most relevant data for our readers. Conferences and sessions were held from time to time between the editorial board and the contributing authors to present the data in the most comprehensible form. The editorial team has worked tirelessly to provide valuable and valid information to help people across the globe.

Every chapter published in this book has been scrutinized by our experts. Their significance has been extensively debated. The topics covered herein carry significant findings which will fuel the growth of the discipline. They may even be implemented as practical applications or may be referred to as a beginning point for another development. Chapters in this book were first published by InTech; hereby published with permission under the Creative Commons Attribution License or equivalent.

The editorial board has been involved in producing this book since its inception. They have spent rigorous hours researching and exploring the diverse topics which have resulted in the successful publishing of this book. They have passed on their knowledge of decades through this book. To expedite this challenging task, the publisher supported the team at every step. A small team of assistant editors was also appointed to further simplify the editing procedure and attain best results for the readers.

Our editorial team has been hand-picked from every corner of the world. Their multi-ethnicity adds dynamic inputs to the discussions which result in innovative outcomes. These outcomes are then further discussed with the researchers and contributors who give their valuable feedback and opinion regarding the same. The feedback is then collaborated with the researches and they are edited in a comprehensive manner to aid the understanding of the subject.

Apart from the editorial board, the designing team has also invested a significant amount of their time in understanding the subject and creating the most relevant covers. They scrutinized every image to scout for the most suitable representation of the subject and create an appropriate cover for the book.

The publishing team has been involved in this book since its early stages. They were actively engaged in every process, be it collecting the data, connecting with the contributors or procuring relevant information. The team has been an ardent support to the editorial, designing and production team. Their endless efforts to recruit the best for this project, has resulted in the accomplishment of this book. They are a veteran in the field of academics and their pool of knowledge is as vast as their experience in printing. Their expertise and guidance has proved useful at every step. Their uncompromising quality standards have made this book an exceptional effort. Their encouragement from time to time has been an inspiration for everyone.

The publisher and the editorial board hope that this book will prove to be a valuable piece of knowledge for researchers, students, practitioners and scholars across the globe.

List of Contributors

Petcu Andreea Cristina
National Research and Development Institute for Gas Turbines Bucharest, Romania

Tien-Chang Lu, Ting-Chun Liu, Peng-Hsiang Weng, Hao-Chung Kuo and Shing-Chung Wang
Department of Photonic and Institute of Electro-Optical Engineering, National Chiao Tung University, Hsinchu, Taiwan, R.O.C.

Shih-Wei Chen
Green Energy & Environment Research Labs, Industrial Technology Research Institute, Hsinchu, Taiwan, R.O.C
Department of Photonic and Institute of Electro-Optical Engineering, National Chiao Tung University, Hsinchu, Taiwan, R.O.C.

Yongqiang Ning and Guangyu Liu
Changchun Institute of Optics, Fine Mechanics and Physics Chinese Academy of Sciences Changchun, China

Alireza Bananej and S. Morteza Zahedi
Laser and Optics Research School, NSTRI, I. R. Iran

S. M. Hamidi
Laser and Plasma research Institute, Shahid Beheshti University, Evin, Tehran, I.R. Iran

Amir Hassanpour
Department of physics, K. N. Toosi University of Technology, Tehran, I.R. Iran

S. Amiri
Institute for Research in Fundamental Sciences, Tehran, I. R. Iran

Binbin Weng and Zhisheng Shi
School of Electrical and Computer Engineering, University of Oklahoma, Norman Oklahoma, USA

Mohammad Danaie and Hassan Kaatuzian
Photonics Research Lab., Amirkabir University of Technology, Iran

Bartłomiej Salski and Kamila Leśniewska-Matys
Warsaw University of Technology, Poland

Paweł Szczepański
National Institute of Telecommunications, Poland
Warsaw University of Technology, Poland

Emmanuel K. Akowuah, Terry Gorman, Huseyin Ademgil, Shyqyri Haxha, Gary Robinson and Jenny Oliver
University of Kent, Canterbury, United Kingdom

Rossen Todorov, Jordanka Tasseva and Tsvetanka Babeva
Institute of Optical Materials and Technologies "Acad. J. Malinowski", Bulgarian Academy of Sciences, Bulgaria

Hanben Niu and Jun Yin
College of Optoelectronic Engineering, Shenzhen University, Shenzhen, China
Key Laboratory of Optoelectronic Devices and Systems of Ministry of Education and Guangdong Province, Shenzhen University, Shenzhen, China

Jaime García-Rupérez, Veronica Toccafondo and Javier García Castelló
Nanophotonics Technology Center, Universidad Politécnica de Valencia, Spain

Rajneesh Kumar
Plasmonics and Metamaterials Lab, Department of Physics Indian Institute of Technology, Kanpur, India

Bayat and Baroughi
South Dakota State University, USA

Luca Marseglia
Centre for Quantum Photonics, H. H. Wills Physics Laboratory & Department of Electrical and Electronic Engineering, University of Bristol, BS8 1UB, United Kingdom

Remo Proietti Zaccaria
Nanobiotech Facility, Italian Institute of Technology, Genova, Italy

Anisha Gopalakrishnan, Gobind Das, Andrea Toma, Francesco De Angelis, Carlo Liberale, Federico Mecarini, Luca Razzari, Andrea Giugni and Roman Krahne
Nanobiotech Facility, Italian Institute of Technology, Genova, Italy

Francesco Gentile and Enzo Di Fabrizio
BIONEM lab., Departement of Clinical and Experimental Medicine, Magna Graecia University, viale Europa, Catanzaro, Italy
Nanobiotech Facility, Italian Institute of Technology, Genova, Italy

Ali Haddadpour
Department of Electrical and Computer Engineering, University of Tabriz, Iran

C. W. Wong
Columbia University, New York, NY, USA

S. Kocaman, M. S. Aras, P. Hsieh and J. F. McMillan
Columbia University, New York, NY, USA

C. G. Biris and N. C. Panoiu
University College of London, London, UK

M. B. Yu and D. L. Kwong
Institute of Microelectronics, Singapore,

A. Stein
Brookhaven National Laboratory, Upton, NY, USA

www.ingramcontent.com/pod-product-compliance
Lightning Source LLC
Chambersburg PA
CBHW072251210326
41458CB00073B/1010

9 7 8 1 6 3 2 4 0 4 0 8 4